T0215450

Tensor Analysis
with Applications
in Mechanics

Tensor Analysis with Applications in Mechanics

Leonid P. Lebedev
National University of Colombia
Southern Federal University, Russia

Michael J. Cloud
Lawrence Technological University, USA

Victor A. Eremeyev
Martin-Luther-University Halle-Wittenberg, Germany
Southern Scientific Center of Russian Academy of Science
Southern Federal University, Russia

World Scientific

NEW JERSEY · LONDON · SINGAPORE · BEIJING · SHANGHAI · HONG KONG · TAIPEI · CHENNAI

Published by

World Scientific Publishing Co. Pte. Ltd.

5 Toh Tuck Link, Singapore 596224

USA office: 27 Warren Street, Suite 401-402, Hackensack, NJ 07601

UK office: 57 Shelton Street, Covent Garden, London WC2H 9HE

British Library Cataloguing-in-Publication Data
A catalogue record for this book is available from the British Library.

First published 2010 (Hardcover)
Reprinted 2016 (in paperback edition)
ISBN 978-981-3203-64-8

TENSOR ANALYSIS WITH APPLICATIONS IN MECHANICS

ISBN-13 978-981-4313-12-4
ISBN-10 981-4313-12-2

Printed in Singapore

Foreword

Every science elaborates tools for the description of its objects of study. In classical mechanics we make extensive use of vectorial quantities: forces, moments, positions, velocities, momenta. Confining ourselves to a single coordinate frame, we can regard a vector as a fixed column matrix. The definitive trait of a vector quantity, however, is its objectivity; a vector does not depend on our choice of coordinate frame. This means that as soon as the components of a force are specified in one frame, the components of that force relative to any other frame can be found through the use of appropriate transformation rules.

But vector quantities alone do not suffice for the description of continuum media. The stress and strain at a point inside a body are also objective quantities; however, the specification of each of these relative to a given frame requires a square matrix of elements. Under changes of frame, these elements transform according to rules different from the transformation rules for vectors. Stress and strain tensors are examples of tensors of the second order. We could go on to cite other objective quantities that occur in the mechanics of continua. The set of elastic moduli associated with Hooke's law comprise a tensor of the fourth order; as such, these moduli obey yet another set of transformation rules. Despite the differences that exist between the transformation laws for the various types of objective quantities, they all fit into a unified scheme: the theory of tensors.

Tensor theory not only relieves our memory from a huge burden, but enables us to carry out differential operations with ease. This is the case even in curvilinear coordinate systems. Through the unmatched simplicity and brevity it affords, tensor analysis has attained the status of a general language that can be spoken across the various areas of continuum physics. A full comprehension of this language has become necessary for those working in electromagnetism, the theory of relativity, or virtually any other field-theoretic discipline. More modern books on physical subjects invariably contain appendices in which various vector and tensor identities are listed. These may suffice when one wishes to verify the steps in a development, but can leave one in doubt as to how the facts were established or, *a fortiori*, how they could be adapted to other circumstances. On the other hand, a comprehensive treatment of tensors (e.g., involving a full excursion into multilinear algebra) is necessarily so large as to be flatly inappropriate for the student or practicing engineer.

Hence the need for a treatment of tensor theory that does justice to the subject and is friendly to the practitioner. The authors of the present book have met these objectives with a presentation that is simple, clear, and sufficiently detailed. The concise text explains practically all those formulas needed to describe objects in three-dimensional space. Occurrences in physics are mentioned when helpful, but the discussion is kept largely independent of application area in order to appeal to the widest possible audience. A chapter on the properties of curves and surfaces has been included; a brief introduction to the study of these properties can be considered as an informative and natural extension of tensor theory.

I.I. Vorovich
Late Professor of Mechanics and Mathematics
Rostov State University, Russia
Fellow of Russian Academy of Sciences
(1920–2001)

Preface

The first edition of this book was written for students, engineers, and physicists who must employ tensor techniques. We did not present the material in complete generality for the case of n-dimensional space, but rather presented a three-dimensional version (which is easy to extend to n dimensions); hence we could assume a background consisting only of standard calculus and linear algebra.

We have decided to extend the book in a natural direction, adding two chapters on applications for which tensor analysis is the principal tool. One chapter is on linear elasticity and the other is on the theory of shells and plates. We present complete derivations of the equations in these theories, formulate boundary value problems, and discuss the problem of uniqueness of solutions, Lagrange's variational principle, and some problems on vibration. Space restrictions prohibited us from presenting an entire course on mechanics; we had to select those questions in elasticity where the role of tensor analysis is most crucial.

We should mention the essential nature of tensors in elasticity and shell theory. Of course, to solve a certain engineering problem, one should write things out in component form; sometimes this takes a few pages. The corresponding formulas in tensor notation are quite simple, allowing us to grasp the underlying ideas and perform manipulations with relative ease. Because tensor representation leads quickly and painlessly to component-wise representation, this technique is ideal for presenting continuum theories to students.

The first five chapters are largely unmodified, aside from some new problem sets and material on tensorial functions needed for the chapters on elasticity. The end-of-chapter problems are supplementary, whereas the integrated exercises are required for a proper understanding of the text.

In the first edition we used the term *rank* instead of *order*. This was common in the older literature. In the newer literature, the term "rank" is often assigned a different meaning.

Because the book is largely self-contained, we make no attempt at a comprehensive reference list. We merely list certain books that cover similar material, that extend the treatment slightly, or that may be otherwise useful to the reader.

We are deeply grateful to our World Scientific editor, Mr. Tjan Kwang Wei, for his encouragement and support.

L.P. Lebedev
Department of Mathematics
National University of Colombia, Colombia

M.J. Cloud
Department of Electrical and Computer Engineering
Lawrence Technological University, USA

V.A. Eremeyev
South Scientific Center of RASci
&

Department of Mathematics, Mechanics
and Computer Sciences
South Federal University, Russia

Preface to the First Edition

Originally a vector was regarded as an arrow of a certain length that could represent a force acting on a material point. Over a period of many years, this naive viewpoint evolved into the modern interpretation of the notion of vector and its extension to tensors. It was found that the use of vectors and tensors led to a proper description of certain properties and behaviors of real natural objects: those aspects that do not depend on the coordinate systems we introduce in space. This independence means that if we define such properties using one coordinate system, then in another system we can recalculate these characteristics using valid transformation rules. The ease with which a given problem can be solved often depends on the coordinate system employed. So in applications we must apply various coordinate systems, derive corresponding equations, and understand how to recalculate results in other systems. This book provides the tools necessary for such calculation.

Many physical laws are cumbersome when written in coordinate form but become compact and attractive looking when written in tensorial form. Such compact forms are easy to remember, and can be used to state complex physical boundary value problems. It is conceivable that soon an ability to merely formulate statements of boundary value problems will be regarded as a fundamental skill for the practitioner. Indeed, computer software is slowly advancing toward the point where the only necessary input data will be a coordinate-free statement of a boundary value problem; presumably the user will be able to initiate a solution process in a certain frame and by a certain method (analytical, numerical, or mixed), or simply ask the computer algorithm to choose the best frame and method. In this way, vectors and tensors will become important elements of the macro-language for the next generation of software in engineering and applied mathematics.

We would like to thank the editorial staff at World Scientific — especially Mr. Tjan Kwang Wei and Ms. Sook-Cheng Lim — for their assistance in the production of this book. Professor Byron C. Drachman of Michigan State University commented on the manuscript in its initial stages. Lastly, Natasha Lebedeva and Beth Lannon-Cloud deserve thanks for their patience and support.

L.P. Lebedev
Department of Mechanics and Mathematics
Rostov State University, Russia
&

Department of Mathematics
National University of Colombia, Colombia

M.J. Cloud
Department of Electrical and Computer Engineering
Lawrence Technological University, USA

Contents

PART 1
Tensor Analysis

Chapter 1

Preliminaries

1.1   The Vector Concept Revisited

The concept of a vector has been one of the most fruitful ideas in all of
mathematics, and it is not surprising that we receive repeated exposure to
the idea throughout our education. Students in elementary mathematics
deal with vectors in component form — with quantities such as

$$\mathbf{x} = (2, 1, 3)$$

for example. But let us examine this situation more closely. Do the components $2, 1, 3$ determine the vector $\mathbf{x}$? They surely do if we specify the basis
vectors of the coordinate frame. In elementary mathematics these are supposed to be mutually orthogonal and of unit length; even then they are not
fully characterized, however, because such a frame can be rotated. In the
description of many common phenomena we deal with vectorial quantities
like forces that have definite directions and magnitudes. An example is the
force your body exerts on a chair as you sit in front of the television set.
This force does not depend on the coordinate frame employed by someone
writing a textbook on vectors somewhere in Russia or China. Because the
vector $\mathbf{f}$ representing a particular force is something objective, we should
be able to write it in such a form that it ceases to depend on the details of
the coordinate frame. The simplest way is to incorporate the frame vectors
$\mathbf{e}_i$ $(i = 1, 2, 3)$ explicitly into the notation: if $\mathbf{x}$ is a vector we may write

$$\mathbf{x} = \sum_{i=1}^{3} x_i \mathbf{e}_i. \tag{1.1}$$

Then if we wish to change the frame, we should do so in such a way that
$\mathbf{x}$ remains the same. This of course means that we cannot change only the

frame vectors $\mathbf{e}_i$: we must change the components x_i correspondingly. So the components of a vector $\mathbf{x}$ in a new frame are not independent of those in the old frame.

1.2   A First Look at Tensors

In what follows we shall discuss how to work with vectors using different coordinate frames. Let us note that in mechanics there are objects of another nature. For example, there is a so-called tensor of inertia. This is an objective characteristic of a solid body, determining how the body rotates when torques act upon it. If the body is considered in a Cartesian frame, the tensor of inertia is described by a 3×3 matrix. If we change the frame, the matrix elements change according to certain rules. In textbooks on mechanics the reader can find lengthy discussions on how to change the matrix elements to maintain the same objective characteristic of the body when the new frame is also Cartesian. Although the tensor of inertia is objective (i.e., frame-independent), it is not a vector: it belongs to another class of mathematical objects. Many such *tensors of the second order* arise in continuum mechanics: tensors of stress, strain, etc. They characterize certain properties of a body at each point; again, their "components" should transform in such a way that the tensors themselves do not depend on the frame. The precise meaning of the term *order* will be explained later.

For both vectors and tensors we can introduce various operations. Of course, the introduction of any new operation should be done in such a way that the results agree with known special cases when such familiar cases are met. If we introduce, say, dot multiplication of a tensor by a vector, then in a Cartesian frame the operation should resemble the multiplication of a matrix by a column vector. Similarly, the multiplication of two tensors should be defined so that in a Cartesian frame the operation involves matrix multiplication. To this end we consider *dyads* of vectors. These are quantities of the form

$$\mathbf{e}_i\mathbf{e}_j.$$

A tensor may then be represented as

$$\sum_{i,j} a_{ij}\mathbf{e}_i\mathbf{e}_j$$

where the a_{ij} are the components of the tensor. We compare with equation (1.1) and notice the similarity in notation.

The quantity $\mathbf{e}_i\mathbf{e}_j$ is also called the *tensor product* of the vectors $\mathbf{e}_i$ and $\mathbf{e}_j$, and is sometimes denoted $\mathbf{e}_i \otimes \mathbf{e}_j$. Our notation (without the symbol $\otimes$) emphasizes that, for example, $\mathbf{e}_1\mathbf{e}_2$ is an elemental object belonging to the set of second-order tensors, in the same way that $\mathbf{e}_1$ is an elemental object belonging to the set of vectors. Note that $\mathbf{e}_2\mathbf{e}_1$ and $\mathbf{e}_1\mathbf{e}_2$ are different objects, however. The term "tensor product" indicates that the operation shares certain properties with the product we know from elementary algebra. We will discuss this further in Chapter 3.

Natural objects can possess characteristics described by tensors of higher order. For example, the elastic properties of a body are described by a tensor of the fourth order (i.e., a tensor whose elemental parts are of the form $\mathbf{abcd}$, where $\mathbf{a}, \mathbf{b}, \mathbf{c}, \mathbf{d}$ are vectors). This means that in general the properties of a body are given by a "table" consisting of $3 \times 3 \times 3 \times 3 = 81$ elements. The elements change according to certain rules if we change the frame.

Tensors also occur in electrodynamics, the general theory of relativity, and other sciences that deal with objects situated or distributed in space.

1.3 Assumed Background

In what follows we suppose a familiarity with the dot and cross products and their expression in Cartesian frames. Recall that if $\mathbf{a}$ and $\mathbf{b}$ are vectors, then by definition

$$\mathbf{a} \cdot \mathbf{b} = |\mathbf{a}||\mathbf{b}| \cos \theta,$$

where $|\mathbf{a}|$ and $|\mathbf{b}|$ are the magnitudes of $\mathbf{a}$ and $\mathbf{b}$ and θ is the (smaller) angle between $\mathbf{a}$ and $\mathbf{b}$. From now on we reserve the symbol $\mathbf{i}$ for the basis vectors of a Cartesian system. In a Cartesian frame with basis vectors $\mathbf{i}_1, \mathbf{i}_2, \mathbf{i}_3$ where $\mathbf{a}$ and $\mathbf{b}$ are expressed as

$$\mathbf{a} = a_1\mathbf{i}_1 + a_2\mathbf{i}_2 + a_3\mathbf{i}_3, \qquad \mathbf{b} = b_1\mathbf{i}_1 + b_2\mathbf{i}_2 + b_3\mathbf{i}_3,$$

we have

$$\mathbf{a} \cdot \mathbf{b} = a_1b_1 + a_2b_2 + a_3b_3.$$

Also recall that

$$\mathbf{a} \times \mathbf{b} = \begin{vmatrix} \mathbf{i}_1 & \mathbf{i}_2 & \mathbf{i}_3 \\ a_1 & a_2 & a_3 \\ b_1 & b_2 & b_3 \end{vmatrix}.$$

The dot product will play a role in our discussion from the very beginning. The cross product will be used as needed, and a fuller discussion will appear in § 2.6.

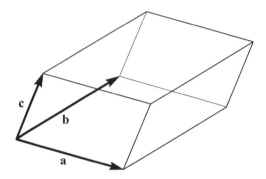

Fig. 1.1   Geometrical meaning of the scalar triple product.

Given three vectors $\mathbf{a}, \mathbf{b}, \mathbf{c}$ we can form the *scalar triple product*

$$\mathbf{a} \cdot (\mathbf{b} \times \mathbf{c}).$$

This may be interpreted as the volume of the parallelepiped having $\mathbf{a}, \mathbf{b}, \mathbf{c}$ as three of its co-terminal edges (Fig. 1.1). In rectangular components we have, according to the expressions above,

$$\mathbf{a} \cdot (\mathbf{b} \times \mathbf{c}) = \begin{vmatrix} a_1 & a_2 & a_3 \\ b_1 & b_2 & b_3 \\ c_1 & c_2 & c_3 \end{vmatrix}.$$

Permissible manipulations with the scalar triple product include cyclic interchange of the vectors:

$$\mathbf{a} \cdot (\mathbf{b} \times \mathbf{c}) = \mathbf{c} \cdot (\mathbf{a} \times \mathbf{b}) = \mathbf{b} \cdot (\mathbf{c} \times \mathbf{a}).$$

Exercise 1.1. What does the condition $\mathbf{a} \cdot (\mathbf{b} \times \mathbf{c}) \neq 0$ say about $\mathbf{a}$, $\mathbf{b}$, and $\mathbf{c}$? (Hints for this and many other exercises appear in Appendix B beginning on page 315.)

So far we have made reference to vectors in a three-dimensional space, and shall continue this practice throughout most of the book. It is also possible (and sometimes useful) to introduce vectors in a more general space of $n > 3$ dimensions, e.g.,

$$\mathbf{a} = \mathbf{i}_1 a_1 + \mathbf{i}_2 a_2 + \cdots + \mathbf{i}_n a_n.$$

It turns out that many (but not all) of the principles and techniques we shall learn have direct extensions to such higher-dimensional spaces. It is also true that many three-dimensional notions can be reconsidered in *two* dimensions. The reader should take the time to reduce each three-dimensional formula to its two-dimensional analogue to understand what happens with the corresponding assertion.

1.4 More on the Notion of a Vector

Before closing out this chapter we should mention the notions of a vector as a "directed line segment" or "quantity having magnitude and direction." We find these in elementary math and physics textbooks. But it is easy to point to a situation in which a quantity of interest has magnitude and direction but is not a vector. The total electric current flowing in a thin wire is one example: to describe the current we must specify the rate of flow of electrons, the orientation of the wire, and the sense of electron movement along the wire. However, if two wires lie in a plane and carry equal currents running perpendicular to each other, then we cannot duplicate their physical effects by replacing them with a third wire directed at a 45° angle with respect to the first two. Total electric currents cannot be considered as vector quantities since they do not combine according to the rule for vector addition.

Another problem concerns the notion of an n-dimensional Euclidean space. We sometimes hear it defined as the set each point of which is uniquely determined by n parameters. However, it is not reasonable to regard every list of n things as a possible vector. A tailor may take measurements for her client and compose an ordered list of lengths, widths, etc., and by the above "definition" the set of all such lists is an n-dimensional space. But in $\mathbb{R}^n$ any two points are joined by a vector whose components equal the differences of the corresponding coordinates of the points. In the tailor's case this notion is completely senseless. To make matters worse, in $\mathbb{R}^n$ one can multiply any vector by any real number and obtain another vector in the space. A tailor's list showing that someone is six meters tall would be pretty hard to find.

Such simplistic definitions can be dangerous. The notion of a vector was eventually elaborated in physics, more precisely in mechanics where the behavior of forces was used as the model for a general vector. However, forces have some rather strange features: for example, if we shift the point of application of a force acting on a solid body, then we must introduce a moment acting on the body or the motion of the body will change. So the mechanical forces that gave birth to the notion of a vector possess features not covered by the mathematical definition of a vector. In the classical mechanics of a rigid body we are allowed to move a force along its line of action but cannot simply shift it off that line. With a deformable body we cannot move a force anywhere because in doing so we immediately change the state of the body. We can vectorially add two forces acting on a rigid body (not forgetting about the moments arising during shift of the forces). On the other hand, if two forces act on two different material points then we can add the vectors that represent the forces, but will not necessarily obtain a new vector that is relevant to the physics of the situation. So we should understand that the idea of a vector in mathematics reflects only certain features of the real objects it describes.

We would like to mention something else about vectors. In elementary mathematics students use vectors and points quite interchangeably. However, these are objects of different natures: there are vectors in space and there are the points of a space. We can, for instance, associate with a vector in an n-dimensional Euclidean vector space a point in an n-dimensional Euclidean point space. We can then consider a vector $\mathbf{x}$ as a shift of all points of the point space by an amount specified by $\mathbf{x}$. The result of this map is the same space of points; each pair of points that correspond under the mapping define a vector $\mathbf{x}$ under which a point shifts into its image. This vector is what we find when we subtract the Cartesian coordinates of the initial point from those of the final point. If we add the fact that the composition of two such maps obeys the rules of vector addition, then we get a strict definition of the space introduced intuitively in elementary mathematics. Engineers might object to the imposition of such formality on an apparently simple situation, but mathematical rigor has proved its worth from many practical points of view. For example, a computer that processes information in the complete absence of intuition can deal properly with objects for which rigorous definitions and manipulation rules have been formulated.

This brings us to our final word, regarding the expression of vectors in component-free notation. The simple and compact notation $\mathbf{x}$ for a vector leads to powerful ways of exhibiting relationships between the many vector (and tensor) quantities that occur in mathematical physics. It permits us to accomplish manipulations that would take many pages and become quite confusing if done in component form. This is typical for problems of nonlinear physics, and for those where change of coordinate frame becomes necessary. The resulting formal nature of the manipulations means that computers can be expected to take on more and more of this sort of work, even at the level of original research.

1.5 Problems

1.1 Find $2\mathbf{a}_1 + 3\mathbf{a}_2$ for the vectors $\mathbf{a}_1$ and $\mathbf{a}_2$ given by

(a) $\mathbf{a}_1 = (1, 2, -1)$, $\mathbf{a}_2 = (1, 1, 2)$;
(b) $\mathbf{a}_1 = (-1, 2, 0)$, $\mathbf{a}_2 = (3, 1, -2)$;
(c) $\mathbf{a}_1 = (1, 3, 4)$, $\mathbf{a}_2 = (5, 1, -4)$.

1.2 Find $\mathbf{a}_1 + 3\mathbf{a}_2 - 2\mathbf{a}_3$ for the vectors $\mathbf{a}_1$, $\mathbf{a}_2$, $\mathbf{a}_3$ given by

(a) $\mathbf{a}_1 = (-1, 2, -2)$, $\mathbf{a}_2 = (1, -1, 2)$, $\mathbf{a}_3 = (1, -1, 2)$;
(b) $\mathbf{a}_1 = (1, 3, 2)$, $\mathbf{a}_2 = (2, -3, 2)$, $\mathbf{a}_3 = (3, 2, 3)$;
(c) $\mathbf{a}_1 = (2, 1, 2)$, $\mathbf{a}_2 = (4, 3, 0)$, $\mathbf{a}_3 = (1, 1, -2)$.

1.3 Find $\mathbf{x}$ satisfying the equation

(a) $\mathbf{a} + 2\mathbf{x} - 4\mathbf{b} = \mathbf{0}$;
(b) $2(\mathbf{a} + 2\mathbf{x}) + \mathbf{b} = 3(\mathbf{b} + \mathbf{x})$;
(c) $\mathbf{x} + 2\mathbf{a} + 16(\mathbf{x} - \mathbf{b}) + \mathbf{c} = 2(\mathbf{c} + 3\mathbf{x}) + \mathbf{a} - \mathbf{x}$.

1.4 Find the values of $\mathbf{a} \cdot \mathbf{b}$ and $\mathbf{a} \times \mathbf{b}$ if

(a) $\mathbf{a} = (0, 1, 1)$, $\mathbf{b} = (1, 1, 1)$;
(b) $\mathbf{a} = (1, 2, 3)$, $\mathbf{b} = (2, 3, 1)$;
(c) $\mathbf{a} = (1, 1, 1)$, $\mathbf{b} = (1, 1, 2)$;
(d) $\mathbf{a} = (-1, 1, -1)$, $\mathbf{b} = (2, -1, -1)$;
(e) $\mathbf{a} = (0, -1, 1)$, $\mathbf{b} = (-1, 1, 0)$;
(f) $\mathbf{a} = (-1, 1, 0)$, $\mathbf{b} = (2, 3, 0)$;
(g) $\mathbf{a} = (1, 2, -1)$, $\mathbf{b} = (1, 1, 2)$;
(h) $\mathbf{a} = (-1, 2, 0)$, $\mathbf{b} = (3, 1, -2)$;
(i) $\mathbf{a} = (1, 3, 4)$, $\mathbf{b} = (5, 1, -4)$.

1.5 Show that (a) $\mathbf{i}_1 \times \mathbf{i}_2 = \mathbf{i}_3$; (b) $\mathbf{i}_2 \times \mathbf{i}_3 = \mathbf{i}_1$; (c) $\mathbf{i}_3 \times \mathbf{i}_1 = \mathbf{i}_2$.

1.6 Show that the equation $\mathbf{a} \times \mathbf{i}_1 = -a_2\mathbf{i}_3 + a_3\mathbf{i}_2$ holds for an arbitrary vector $\mathbf{a} = a_1\mathbf{i}_1 + a_2\mathbf{i}_2 + a_3\mathbf{i}_3$.

1.7 Let $\mathbf{a} = a_1\mathbf{i}_1 + a_2\mathbf{i}_2$, $\mathbf{b} = b_1\mathbf{i}_1 + b_2\mathbf{i}_2$, and let $\mathbf{i}_1, \mathbf{i}_2, \mathbf{i}_3$ be a Cartesian basis. Show that $\mathbf{a} \times \mathbf{b} = (a_1b_2 - a_2b_1)\mathbf{i}_3$.

1.8 Suppose $\mathbf{a} \times \mathbf{x} = \mathbf{0}$, $\mathbf{a} \cdot \mathbf{x} = 0$, and $\mathbf{a} \neq \mathbf{0}$. Demonstrate that $\mathbf{x} = \mathbf{0}$.

1.9 Find: (a) $\mathbf{i}_1 \cdot (\mathbf{i}_2 \times \mathbf{i}_3)$; (b) $\mathbf{i}_1 \cdot (\mathbf{i}_3 \times \mathbf{i}_2)$; (c) $\mathbf{i}_1 \cdot (\mathbf{i}_3 \times \mathbf{i}_1)$.

1.10 Find $\mathbf{a}_1 \cdot (\mathbf{a}_2 \times \mathbf{a}_3)$ when

  (a) $\mathbf{a}_1 = (-1, 2, -2)$, $\mathbf{a}_2 = (1, -1, 2)$, $\mathbf{a}_3 = (1, -1, 3)$;
  (b) $\mathbf{a}_1 = (-1, 1, 0)$, $\mathbf{a}_2 = (1, 1, 1)$, $\mathbf{a}_3 = (1, -1, 2)$;
  (c) $\mathbf{a}_1 = (1, 1, 1)$, $\mathbf{a}_2 = (1, 2, 2)$, $\mathbf{a}_3 = (1, -3, 2)$;
  (d) $\mathbf{a}_1 = (-1, 2, -1)$, $\mathbf{a}_2 = (1, -2, 2)$, $\mathbf{a}_3 = (3, -1, 3)$;
  (e) $\mathbf{a}_1 = (9, 8, 4)$, $\mathbf{a}_2 = (7, 1, 3)$, $\mathbf{a}_3 = (5, 3, 6)$;
  (f) $\mathbf{a}_1 = (1, 2, 3)$, $\mathbf{a}_2 = (7, 2, 3)$, $\mathbf{a}_3 = (1, 4, 6)$;
  (g) $\mathbf{a}_1 = (-1, 2, -2)$, $\mathbf{a}_2 = (1, -1, 2)$, $\mathbf{a}_3 = (1, -1, 2)$;
  (h) $\mathbf{a}_1 = (1, 3, 2)$, $\mathbf{a}_2 = (2, -3, 2)$, $\mathbf{a}_3 = (3, 2, 3)$;
  (i) $\mathbf{a}_1 = (2, 1, 2)$, $\mathbf{a}_2 = (4, 3, 0)$, $\mathbf{a}_3 = (1, 1, -2)$.

1.11 Find $(\mathbf{a}_1 \times \mathbf{a}_2) \cdot (\mathbf{a}_3 \times \mathbf{a}_4)$, where

  (a) $\mathbf{a}_1 = (-1, 2, -2)$, $\mathbf{a}_2 = (1, -1, 2)$, $\mathbf{a}_3 = (1, -1, 3)$, $\mathbf{a}_4 = (1, 0, 0)$;
  (b) $\mathbf{a}_1 = (-1, 1, 0)$, $\mathbf{a}_2 = (1, 1, 1)$, $\mathbf{a}_3 = (1, -1, 2)$, $\mathbf{a}_4 = (0, -1, 0)$;
  (c) $\mathbf{a}_1 = (1, 1, 1)$, $\mathbf{a}_2 = (1, 2, 2)$, $\mathbf{a}_3 = (1, -3, 2)$, $\mathbf{a}_4 = (1, 1, 1)$;
  (d) $\mathbf{a}_1 = (-1, 2, -1)$, $\mathbf{a}_2 = (1, -2, 2)$, $\mathbf{a}_3 = (3, -1, 3)$, $\mathbf{a}_4 = (1, 0, 2)$;
  (e) $\mathbf{a}_1 = (9, 8, 4)$, $\mathbf{a}_2 = (7, 1, 3)$, $\mathbf{a}_3 = (5, 3, 6)$, $\mathbf{a}_4 = (2, 3, -1)$;
  (f) $\mathbf{a}_1 = (1, 2, 3)$, $\mathbf{a}_2 = (7, 2, 3)$, $\mathbf{a}_3 = (1, 4, 6)$, $\mathbf{a}_4 = (0, 0, 1)$;
  (g) $\mathbf{a}_1 = (-1, 2, -2)$, $\mathbf{a}_2 = (1, -1, 2)$, $\mathbf{a}_3 = (1, -1, 2)$, $\mathbf{a}_4 = (2, -2, 4)$;
  (h) $\mathbf{a}_1 = (1, 3, 2)$, $\mathbf{a}_2 = (2, -3, 2)$, $\mathbf{a}_3 = (3, 2, 3)$, $\mathbf{a}_4 = (1, -1, 1)$;
  (i) $\mathbf{a}_1 = (2, 1, 2)$, $\mathbf{a}_2 = (-6, -3, -6)$, $\mathbf{a}_3 = (1, 1, -2)$, $\mathbf{a}_4 = (1, 12, 3)$.

Chapter 2

Transformations and Vectors

2.1 Change of Basis

Let us reconsider the vector

$$\mathbf{x} = (2, 1, 3).$$

Fully written out in a given Cartesian frame $\mathbf{e}_i$ ($i = 1, 2, 3$), it is

$$\mathbf{x} = 2\mathbf{e}_1 + \mathbf{e}_2 + 3\mathbf{e}_3.$$

(This is one of the few times we do not use $\mathbf{i}$ as the symbol for a Cartesian frame vector.) Suppose we appoint a new frame $\tilde{\mathbf{e}}_i$ ($i = 1, 2, 3$) such that

$$\mathbf{e}_1 = \tilde{\mathbf{e}}_1 + 2\tilde{\mathbf{e}}_2 + 3\tilde{\mathbf{e}}_3,$$
$$\mathbf{e}_2 = 4\tilde{\mathbf{e}}_1 + 5\tilde{\mathbf{e}}_2 + 6\tilde{\mathbf{e}}_3,$$
$$\mathbf{e}_3 = 7\tilde{\mathbf{e}}_1 + 8\tilde{\mathbf{e}}_2 + 9\tilde{\mathbf{e}}_3.$$

From these expansions we could calculate the $\tilde{\mathbf{e}}_i$ and verify that they are non-coplanar. As $\mathbf{x}$ is an objective, frame-independent entity, we can write

$$\mathbf{x} = 2(\tilde{\mathbf{e}}_1 + 2\tilde{\mathbf{e}}_2 + 3\tilde{\mathbf{e}}_3) + (4\tilde{\mathbf{e}}_1 + 5\tilde{\mathbf{e}}_2 + 6\tilde{\mathbf{e}}_3) + 3(7\tilde{\mathbf{e}}_1 + 8\tilde{\mathbf{e}}_2 + 9\tilde{\mathbf{e}}_3)$$
$$= (2 + 4 + 21)\tilde{\mathbf{e}}_1 + (4 + 5 + 24)\tilde{\mathbf{e}}_2 + (6 + 6 + 27)\tilde{\mathbf{e}}_3$$
$$= 27\tilde{\mathbf{e}}_1 + 33\tilde{\mathbf{e}}_2 + 39\tilde{\mathbf{e}}_3.$$

In these calculations it is unimportant whether the frames are Cartesian; it is important only that we have the table of transformation

$$\begin{pmatrix} 1 & 2 & 3 \\ 4 & 5 & 6 \\ 7 & 8 & 9 \end{pmatrix}.$$

It is clear that we can repeat the same operation in general form. Let $\mathbf{x}$ be of the form

$$\mathbf{x} = \sum_{i=1}^{3} x^i \mathbf{e}_i \qquad (2.1)$$

with the table of transformation of the frame given as

$$\mathbf{e}_i = \sum_{j=1}^{3} A_i^j \tilde{\mathbf{e}}_j.$$

Then

$$\mathbf{x} = \sum_{i=1}^{3} x^i \sum_{j=1}^{3} A_i^j \tilde{\mathbf{e}}_j = \sum_{j=1}^{3} \tilde{\mathbf{e}}_j \sum_{i=1}^{3} A_i^j x^i.$$

So in the new basis we have

$$\mathbf{x} = \sum_{j=1}^{3} \tilde{x}^j \tilde{\mathbf{e}}_j \quad \text{where} \quad \tilde{x}^j = \sum_{i=1}^{3} A_i^j x^i.$$

Here we have introduced a new notation, placing some indices as subscripts and some as superscripts. Although this practice may seem artificial, there are fairly deep reasons for following it.

2.2 Dual Bases

To perform operations with a vector $\mathbf{x}$, we must have a straightforward method of calculating its components — ultimately, no matter how advanced we are, we must be able to obtain the x^i using simple arithmetic. We prefer formulas that permit us to find the components of vectors using dot multiplication only; we shall need these when doing frame transformations, etc. In a Cartesian frame the necessary operation is simple dot multiplication by the corresponding basis vector of the frame: we have

$$x^k = \mathbf{x} \cdot \mathbf{i}_k \qquad (k = 1, 2, 3).$$

This procedure fails in a more general non-Cartesian frame where we do not necessarily have $\mathbf{e}_i \cdot \mathbf{e}_j = 0$ for all $j \neq i$. However, it may still be possible to find a vector $\mathbf{e}^i$ such that

$$x^i = \mathbf{x} \cdot \mathbf{e}^i \qquad (i = 1, 2, 3)$$

in this more general situation. If we set

$$x^i = \mathbf{x} \cdot \mathbf{e}^i = \left(\sum_{j=1}^{3} x^j \mathbf{e}_j \right) \cdot \mathbf{e}^i = \sum_{j=1}^{3} x^j (\mathbf{e}_j \cdot \mathbf{e}^i)$$

and compare the left- and right-hand sides, we see that equality holds when

$$\mathbf{e}_j \cdot \mathbf{e}^i = \delta^i_j \tag{2.2}$$

where

$$\delta^i_j = \begin{cases} 1, & j = i, \\ 0, & j \neq i, \end{cases}$$

is the *Kronecker delta* symbol. In a Cartesian frame we have

$$\mathbf{e}^k = \mathbf{e}_k = \mathbf{i}_k$$

for each k.

Exercise 2.1. Show that $\mathbf{e}^i$ is determined uniquely by the requirement that $x^i = \mathbf{x} \cdot \mathbf{e}^i$ for every $\mathbf{x}$.

Now let us discuss the geometrical nature of the vectors $\mathbf{e}^i$. Consider, for example, the equations for $\mathbf{e}^1$:

$$\mathbf{e}_1 \cdot \mathbf{e}^1 = 1, \qquad \mathbf{e}_2 \cdot \mathbf{e}^1 = 0, \qquad \mathbf{e}_3 \cdot \mathbf{e}^1 = 0.$$

We see that $\mathbf{e}^1$ is orthogonal to both $\mathbf{e}_2$ and $\mathbf{e}_3$, and its magnitude is such that $\mathbf{e}_1 \cdot \mathbf{e}^1 = 1$. Similar properties hold for $\mathbf{e}^2$ and $\mathbf{e}^3$.

Exercise 2.2. Show that the vectors $\mathbf{e}^i$ are linearly independent.

By Exercise 2.2, the $\mathbf{e}^i$ constitute a frame or basis. This basis is said to be *reciprocal* or *dual* to the basis $\mathbf{e}_i$. We can therefore expand an arbitrary vector $\mathbf{x}$ as

$$\mathbf{x} = \sum_{i=1}^{3} x_i \mathbf{e}^i. \tag{2.3}$$

Note that superscripts and subscripts continue to appear in our notation, but in a way complementary to that used in equation (2.1). If we dot-multiply the representation (2.3) of $\mathbf{x}$ by $\mathbf{e}_j$ and use (2.2) we get x_j. This explains why the frames $\mathbf{e}^i$ and $\mathbf{e}_i$ are dual: the formulas

$$\mathbf{x} \cdot \mathbf{e}^i = x^i, \qquad \mathbf{x} \cdot \mathbf{e}_i = x_i,$$

look quite similar. So the introduction of a reciprocal basis gives many potential advantages.

Let us discuss the reciprocal basis in more detail. The first problem is to find suitable formulas to define it. We derive these formulas next, but first let us note the following. The use of reciprocal vectors may not be practical in those situations where we are working with only two or three vectors. The real advantages come when we are working intensively with many vectors. This is reminiscent of the solution of a set of linear simultaneous equations: it is inefficient to find the inverse matrix of the system if we have only one forcing vector. But when we must solve such a problem repeatedly for many forcing vectors, the calculation and use of the inverse matrix is reasonable.

Writing out $\mathbf{x}$ in the $\mathbf{e}_i$ and $\mathbf{e}^i$ bases, we used a combination of indices (i.e., subscripts and superscripts) and summation symbols. From now on we shall omit the symbol of summation when we meet matching subscripts and superscripts: we shall write, say,

$$x_i a^i \quad \text{for the sum} \quad \sum_i x_i a^i.$$

That is, whenever we see i as a subscript and a superscript, we shall understand that a summation is to be carried out over i. This rule shall apply to situations involving vectors as well: we shall understand, for example,

$$x^i \mathbf{e}_i \quad \text{to mean the summation} \quad \sum_i x^i \mathbf{e}_i.$$

This rule is called the *rule of summation over repeated indices*.[1] Note that a repeated index is a *dummy index* in the sense that it may be replaced by any other index not already in use: we have

$$x_i a^i = x_1 a^1 + x_2 a^2 + x_3 a^3 = x_k a^k$$

for instance. An index that occurs just once in an expression, for example the index i in

$$A_i^k x_k,$$

is called a *free index*. In tensor discussions each free index is understood to range independently over a set of values — presently this set is $\{1, 2, 3\}$.

---

[1] The rule of summation was first introduced not by mathematicians but by Einstein, and is sometimes referred to as the *Einstein summation convention*. In a paper where he introduced this rule, Einstein used Cartesian frames and therefore did not distinguish superscripts from subscripts. However, we shall continue to make the distinction so that we can deal with non-Cartesian frames.

Let us return to the task of deriving formulas for the reciprocal basis vectors $\mathbf{e}^i$ in terms of the original basis vectors $\mathbf{e}_i$. We construct $\mathbf{e}^1$ first. Since the cross product of two vectors is perpendicular to both, we can satisfy the conditions

$$\mathbf{e}_2 \cdot \mathbf{e}^1 = 0, \qquad \mathbf{e}_3 \cdot \mathbf{e}^1 = 0,$$

by setting

$$\mathbf{e}^1 = c_1(\mathbf{e}_2 \times \mathbf{e}_3)$$

where c_1 is a constant. To determine c_1 we require

$$\mathbf{e}_1 \cdot \mathbf{e}^1 = 1.$$

We obtain

$$c_1[\mathbf{e}_1 \cdot (\mathbf{e}_2 \times \mathbf{e}_3)] = 1.$$

The quantity $\mathbf{e}_1 \cdot (\mathbf{e}_2 \times \mathbf{e}_3)$ is a scalar whose absolute value is the volume of the parallelepiped described by the vectors $\mathbf{e}_i$. Denoting it by V, we have

$$\mathbf{e}^1 = \frac{1}{V}(\mathbf{e}_2 \times \mathbf{e}_3).$$

Similarly,

$$\mathbf{e}^2 = \frac{1}{V}(\mathbf{e}_3 \times \mathbf{e}_1), \qquad \mathbf{e}^3 = \frac{1}{V}(\mathbf{e}_1 \times \mathbf{e}_2).$$

The reader may verify that these expressions satisfy (2.2). Let us mention that if we construct the reciprocal basis to the basis $\mathbf{e}^i$ we obtain the initial basis $\mathbf{e}_i$. Hence we immediately get the dual formulas

$$\mathbf{e}_1 = \frac{1}{V'}(\mathbf{e}^2 \times \mathbf{e}^3), \qquad \mathbf{e}_2 = \frac{1}{V'}(\mathbf{e}^3 \times \mathbf{e}^1), \qquad \mathbf{e}_3 = \frac{1}{V'}(\mathbf{e}^1 \times \mathbf{e}^2),$$

where

$$V' = \mathbf{e}^1 \cdot (\mathbf{e}^2 \times \mathbf{e}^3).$$

Within an algebraic sign, V' is the volume of the parallelepiped described by the vectors $\mathbf{e}^i$.

Exercise 2.3. Show that $V' = 1/V$.

Let us now consider the forms of the dot product between two vectors

$$\mathbf{a} = a^i\mathbf{e}_i = a_j\mathbf{e}^j, \qquad \mathbf{b} = b^p\mathbf{e}_p = b_q\mathbf{e}^q.$$

We have

$$\mathbf{a} \cdot \mathbf{b} = a^i \mathbf{e}_i \cdot b^p \mathbf{e}_p = a^i b^p \mathbf{e}_i \cdot \mathbf{e}_p.$$

Introducing the notation

$$g_{ip} = \mathbf{e}_i \cdot \mathbf{e}_p, \tag{2.4}$$

we have

$$\mathbf{a} \cdot \mathbf{b} = a^i b^p g_{ip}.$$

(As a short exercise the reader should write out this expression in full.) Using the reciprocal component representations we get

$$\mathbf{a} \cdot \mathbf{b} = a_j \mathbf{e}^j \cdot b_q \mathbf{e}^q = a_j b_q g^{jq}$$

where

$$g^{jq} = \mathbf{e}^j \cdot \mathbf{e}^q. \tag{2.5}$$

Finally, using a mixed representation we get

$$\mathbf{a} \cdot \mathbf{b} = a^i \mathbf{e}_i \cdot b_q \mathbf{e}^q = a^i b_q \delta_i^q = a^i b_i$$

and, similarly,

$$\mathbf{a} \cdot \mathbf{b} = a_j b^j.$$

Hence

$$\mathbf{a} \cdot \mathbf{b} = a^i b^j g_{ij} = a_i b_j g^{ij} = a^i b_i = a_i b^i.$$

We see that when we use mixed bases to represent $\mathbf{a}$ and $\mathbf{b}$ we get formulas that resemble the equation

$$\mathbf{a} \cdot \mathbf{b} = a_1 b_1 + a_2 b_2 + a_3 b_3$$

from § 1.3; otherwise we get more terms and additional multipliers. We will encounter g_{ij} and g^{ij} often. They are the components of a unique tensor known as the *metric tensor*. In Cartesian frames we obviously have

$$g_{ij} = \delta_i^j, \qquad g^{ij} = \delta_j^i.$$

2.3 Transformation to the Reciprocal Frame

How do the components of a vector $\mathbf{x}$ transform when we change to the reciprocal frame? We simply set

$$x_i \mathbf{e}^i = x^i \mathbf{e}_i$$

and dot both sides with $\mathbf{e}_j$ to get

$$x_i \mathbf{e}^i \cdot \mathbf{e}_j = x^i \mathbf{e}_i \cdot \mathbf{e}_j$$

or

$$x_j = x^i g_{ij}. \tag{2.6}$$

In the system of equations

$$\begin{pmatrix} x_1 \\ x_2 \\ x_3 \end{pmatrix} = \begin{pmatrix} g_{11} & g_{21} & g_{31} \\ g_{12} & g_{22} & g_{32} \\ g_{13} & g_{23} & g_{33} \end{pmatrix} \begin{pmatrix} x^1 \\ x^2 \\ x^3 \end{pmatrix}$$

the matrix of the components of the metric tensor g_{ij} is also called the *Gram matrix*. A theorem in linear algebra states that its determinant is not zero if and only if the vectors $\mathbf{e}_i$ are linearly independent.

Exercise 2.4. (a) Show that if the Gram determinant vanishes, then the $\mathbf{e}_i$ are linearly dependent. (b) Prove that the Gram determinant equals V^2.

We called the basis $\mathbf{e}_i$ dual to the basis $\mathbf{e}^i$. In $\mathbf{e}^i$ the metric components are given by g^{ij}, so we can immediately write an expression dual to (2.6):

$$x^i = x_j g^{ij}. \tag{2.7}$$

We see from (2.6) and (2.7) that, using the components of the metric tensor, we can always change subscripts to superscripts and vice versa. These actions are known as the *raising* and *lowering of indices*. Finally, (2.6) and (2.7) together imply

$$x_i = g_{ij} g^{jk} x_k,$$

hence

$$g_{ij} g^{jk} = \delta_i^k.$$

Of course, this means that the matrices of g_{ij} and g^{ij} are mutually inverse.

Quick summary

Given a basis $\mathbf{e}_i$, the vectors $\mathbf{e}^i$ given by the requirement that

$$\mathbf{e}_j \cdot \mathbf{e}^i = \delta^i_j$$

are linearly independent and form a basis called the reciprocal or dual basis. The definition of dual basis is motivated by the equation $x^i = \mathbf{x} \cdot \mathbf{e}^i$. The $\mathbf{e}^i$ can be written as

$$\mathbf{e}^i = \frac{1}{V}(\mathbf{e}_j \times \mathbf{e}_k)$$

where the ordered triple (i, j, k) equals $(1, 2, 3)$ or one of the cyclic permutations $(2, 3, 1)$ or $(3, 1, 2)$, and where

$$V = \mathbf{e}_1 \cdot (\mathbf{e}_2 \times \mathbf{e}_3).$$

The dual of the basis $\mathbf{e}^k$ (i.e., the dual of the dual) is the original basis $\mathbf{e}_k$. A given vector $\mathbf{x}$ can be expressed as

$$\mathbf{x} = x^i \mathbf{e}_i = x_i \mathbf{e}^i$$

where the x_i are the components of $\mathbf{x}$ with respect to the dual basis.

Exercise 2.5. (a) Let $\mathbf{x} = x^k \mathbf{e}_k = x_k \mathbf{e}^k$. Write out the modulus of $\mathbf{x}$ in all possible forms using the metric tensor. (b) Write out all forms of the dot product $\mathbf{x} \cdot \mathbf{y}$.

2.4   Transformation Between General Frames

Having transformed the components x^i of a vector $\mathbf{x}$ to the corresponding components x_i relative to the reciprocal basis, we are now ready to take on the more general task of transforming the x^i to the corresponding components $\tilde{x}^i$ relative to *any* other basis $\tilde{\mathbf{e}}_i$. Let the new basis $\tilde{\mathbf{e}}_i$ be related to the original basis $\mathbf{e}_i$ by

$$\mathbf{e}_i = A^j_i \tilde{\mathbf{e}}_j. \tag{2.8}$$

This is, of course, compact notation for the system of equations

$$\begin{pmatrix} \mathbf{e}_1 \\ \mathbf{e}_2 \\ \mathbf{e}_3 \end{pmatrix} = \underbrace{\begin{pmatrix} A^1_1 & A^2_1 & A^3_1 \\ A^1_2 & A^2_2 & A^3_2 \\ A^1_3 & A^2_3 & A^3_3 \end{pmatrix}}_{\equiv A, \text{ say}} \begin{pmatrix} \tilde{\mathbf{e}}_1 \\ \tilde{\mathbf{e}}_2 \\ \tilde{\mathbf{e}}_3 \end{pmatrix}.$$

Before proceeding, we note that in the symbol A_i^j the subscript indexes the row number in the matrix A, while the superscript indexes the column number. Throughout our development we shall often take the time to write various equations of interest in matrix notation. It follows from (2.8) that

$$A_i^j = \mathbf{e}_i \cdot \tilde{\mathbf{e}}^j.$$

Exercise 2.6. A Cartesian frame is rotated about its third axis to give a new Cartesian frame. Find the matrix of transformation.

A vector $\mathbf{x}$ can be expressed in the two forms

$$\mathbf{x} = x^k \mathbf{e}_k, \qquad \mathbf{x} = \tilde{x}^i \tilde{\mathbf{e}}_i.$$

Equating these two expressions for the same vector $\mathbf{x}$, we have

$$\tilde{x}^i \tilde{\mathbf{e}}_i = x^k \mathbf{e}_k,$$

hence

$$\tilde{x}^i \tilde{\mathbf{e}}_i = x^k A_k^j \tilde{\mathbf{e}}_j. \tag{2.9}$$

To find $\tilde{x}^i$ in terms of x^i, we may expand the notation and write (2.9) as

$$\tilde{x}^1 \tilde{\mathbf{e}}_1 + \tilde{x}^2 \tilde{\mathbf{e}}_2 + \tilde{x}^3 \tilde{\mathbf{e}}_3 = x^1 A_1^j \tilde{\mathbf{e}}_j + x^2 A_2^j \tilde{\mathbf{e}}_j + x^3 A_3^j \tilde{\mathbf{e}}_j$$

where, of course,

$$A_1^j \tilde{\mathbf{e}}_j = A_1^1 \tilde{\mathbf{e}}_1 + A_1^2 \tilde{\mathbf{e}}_2 + A_1^3 \tilde{\mathbf{e}}_3,$$
$$A_2^j \tilde{\mathbf{e}}_j = A_2^1 \tilde{\mathbf{e}}_1 + A_2^2 \tilde{\mathbf{e}}_2 + A_2^3 \tilde{\mathbf{e}}_3,$$
$$A_3^j \tilde{\mathbf{e}}_j = A_3^1 \tilde{\mathbf{e}}_1 + A_3^2 \tilde{\mathbf{e}}_2 + A_3^3 \tilde{\mathbf{e}}_3.$$

Matching coefficients of the $\tilde{\mathbf{e}}_i$ we find

$$\tilde{x}^1 = x^1 A_1^1 + x^2 A_2^1 + x^3 A_3^1 = x^j A_j^1,$$
$$\tilde{x}^2 = x^1 A_1^2 + x^2 A_2^2 + x^3 A_3^2 = x^j A_j^2,$$
$$\tilde{x}^3 = x^1 A_1^3 + x^2 A_2^3 + x^3 A_3^3 = x^j A_j^3,$$

hence

$$\tilde{x}^i = x^j A_j^i. \tag{2.10}$$

It is possible to obtain (2.10) from (2.9) in a succinct manner. On the right-hand side of (2.9) the index j is a dummy index which we can replace with

i and thereby obtain (2.10) immediately. The matrix notation equivalent of (2.10) is

$$\begin{pmatrix} \tilde{x}^1 \\ \tilde{x}^2 \\ \tilde{x}^3 \end{pmatrix} = \begin{pmatrix} A_1^1 & A_2^1 & A_3^1 \\ A_1^2 & A_2^2 & A_3^2 \\ A_1^3 & A_2^3 & A_3^3 \end{pmatrix} \begin{pmatrix} x^1 \\ x^2 \\ x^3 \end{pmatrix}$$

and thus involves multiplication by A^T, the transpose of A.

We shall also need the equations of transformation from the frame $\tilde{\mathbf{e}}_i$ back to the frame $\mathbf{e}_i$. Since the direct transformation is linear the inverse must be linear as well, so we can write

$$\tilde{\mathbf{e}}_i = \tilde{A}_i^j \mathbf{e}_j \tag{2.11}$$

where

$$\tilde{A}_i^j = \tilde{\mathbf{e}}_i \cdot \mathbf{e}^j.$$

Let us find the relation between the matrices of transformation A and $\tilde{A}$. By (2.11) and (2.8) we have

$$\tilde{\mathbf{e}}_i = \tilde{A}_i^j \mathbf{e}_j = \tilde{A}_i^j A_j^k \tilde{\mathbf{e}}_k,$$

and since the $\tilde{\mathbf{e}}_i$ form a basis we must have

$$\tilde{A}_i^j A_j^k = \delta_i^k.$$

The relationship

$$A_i^j \tilde{A}_j^k = \delta_i^k$$

follows similarly. The product of the matrices $(\tilde{A}_i^j)$ and (A_j^k) is the unit matrix and thus these matrices are *mutually inverse*.

Exercise 2.7. Show that $x^i = \tilde{x}^k \tilde{A}_k^i$.

Formulas for the relations between reciprocal bases can be obtained as follows. We begin with the obvious identities

$$\mathbf{e}^j (\mathbf{e}_j \cdot \mathbf{x}) = \mathbf{x}, \qquad \tilde{\mathbf{e}}^j (\tilde{\mathbf{e}}_j \cdot \mathbf{x}) = \mathbf{x}.$$

Putting $\mathbf{x} = \tilde{\mathbf{e}}^i$ in the first of these gives

$$\tilde{\mathbf{e}}^i = A_j^i \mathbf{e}^j,$$

while the second identity with $\mathbf{x} = \mathbf{e}^i$ yields

$$\mathbf{e}^i = \tilde{A}_j^i \tilde{\mathbf{e}}^j.$$

From these follow the transformation formulas

$$\tilde{x}_i = x_k \tilde{A}_i^k, \qquad x_i = \tilde{x}_k A_i^k.$$

2.5 Covariant and Contravariant Components

We have seen that if the basis vectors transform according to the relation

$$\mathbf{e}_i = A_i^j \tilde{\mathbf{e}}_j,$$

then the components x_i of a vector $\mathbf{x}$ must transform according to

$$x_i = A_i^j \tilde{x}_j.$$

The similarity in form between these two relations results in the x_i being termed the *covariant components* of the vector $\mathbf{x}$. On the other hand, the transformation law

$$x^i = \tilde{A}_j^i \tilde{x}^j$$

shows that the x^i transform like the $\mathbf{e}^i$. For this reason the x^i are termed the *contravariant components* of $\mathbf{x}$. We shall find a further use for this nomenclature in Chapter 3.

Quick summary

If frame transformations

$$\mathbf{e}_i = A_i^j \tilde{\mathbf{e}}_j, \qquad \mathbf{e}^i = \tilde{A}_j^i \tilde{\mathbf{e}}^j,$$
$$\tilde{\mathbf{e}}_i = \tilde{A}_i^j \mathbf{e}_j, \qquad \tilde{\mathbf{e}}^i = A_j^i \mathbf{e}^j,$$

are considered, then $\mathbf{x}$ has the various expressions

$$\mathbf{x} = x^i \mathbf{e}_i = x_i \mathbf{e}^i = \tilde{x}^i \tilde{\mathbf{e}}_i = \tilde{x}_i \tilde{\mathbf{e}}^i$$

and the transformation laws

$$x_i = A_i^j \tilde{x}_j, \qquad x^i = \tilde{A}_j^i \tilde{x}^j,$$
$$\tilde{x}_i = \tilde{A}_i^j x_j, \qquad \tilde{x}^i = A_j^i x^j,$$

apply. The x^i are termed contravariant components of $\mathbf{x}$, while the x_i are termed covariant components. The transformation laws are particularly simple when the frame is changed to the dual frame. Then

$$x_i = g_{ji} x^j, \qquad x^i = g^{ij} x_j,$$

where

$$g_{ij} = \mathbf{e}_i \cdot \mathbf{e}_j, \qquad g^{ij} = \mathbf{e}^i \cdot \mathbf{e}^j,$$

are components of the metric tensor.

2.6   The Cross Product in Index Notation

In mechanics a major role is played by the quantity called torque. This quantity is introduced in elementary physics as the product of a force magnitude and a length ("force times moment arm"), along with some rules for algebraic sign to account for the sense of rotation that the force would encourage when applied to a physical body. In more advanced discussions in which three-dimensional problems are considered, torque is regarded as a vectorial quantity. If a force $\mathbf{f}$ acts at a point which is located relative to an origin O by position vector $\mathbf{r}$, then the associated torque $\mathbf{t}$ about O is normal to the plane of the vectors $\mathbf{r}$ and $\mathbf{f}$. Of the two possible unit normals, $\mathbf{t}$ is conventionally (but arbitrarily) associated with the vector $\hat{\mathbf{n}}$ given by the familiar *right-hand rule*: if the forefinger of the right hand is directed along $\mathbf{r}$ and the middle finger is directed along $\mathbf{f}$, then the thumb indicates the direction of $\hat{\mathbf{n}}$ and hence the direction of $\mathbf{t}$. The magnitude of $\mathbf{t}$ equals $|\mathbf{f}||\mathbf{r}|\sin\theta$, where θ is the smaller angle between $\mathbf{f}$ and $\mathbf{r}$. These rules are all encapsulated in the brief symbolism

$$\mathbf{t} = \mathbf{r} \times \mathbf{f}.$$

The definition of torque can be taken as a model for a more general operation between vectors: the *cross product*. If $\mathbf{a}$ and $\mathbf{b}$ are any two vectors, we define

$$\mathbf{a} \times \mathbf{b} = \hat{\mathbf{n}}|\mathbf{a}||\mathbf{b}|\sin\theta$$

where $\hat{\mathbf{n}}$ and θ are defined as in the case of torque above. Like any other vector, $\mathbf{c} = \mathbf{a} \times \mathbf{b}$ can be expanded in terms of a basis; we choose the reciprocal basis $\mathbf{e}^i$ and write

$$\mathbf{c} = c_i\mathbf{e}^i.$$

Because the magnitudes of $\mathbf{a}$ and $\mathbf{b}$ enter into $\mathbf{a} \times \mathbf{b}$ in multiplicative fashion, we are prompted to seek c_i in the form

$$c_i = \epsilon_{ijk}a^j b^k. \tag{2.12}$$

Here the ϵ's are formal coefficients. Let us find them. We write

$$\mathbf{a} = a^j\mathbf{e}_j, \qquad \mathbf{b} = b^k\mathbf{e}_k,$$

and employ the well-known distributive property

$$(\mathbf{u} + \mathbf{v}) \times \mathbf{w} \equiv \mathbf{u} \times \mathbf{w} + \mathbf{v} \times \mathbf{w}$$

to obtain

$$\mathbf{c} = a^j \mathbf{e}_j \times b^k \mathbf{e}_k = a^j b^k (\mathbf{e}_j \times \mathbf{e}_k).$$

Then

$$\mathbf{c} \cdot \mathbf{e}_i = c_m \mathbf{e}^m \cdot \mathbf{e}_i = c_i = a^j b^k [(\mathbf{e}_j \times \mathbf{e}_k) \cdot \mathbf{e}_i]$$

and comparison with (2.12) shows that

$$\epsilon_{ijk} = (\mathbf{e}_j \times \mathbf{e}_k) \cdot \mathbf{e}_i.$$

Now the value of $(\mathbf{e}_j \times \mathbf{e}_k) \cdot \mathbf{e}_i$ depends on the values of the indices i, j, k. Here it is convenient to introduce the idea of a permutation of the ordered triple $(1, 2, 3)$. A permutation of $(1, 2, 3)$ is called *even* if it can be brought about by performing any even number of interchanges of pairs of these numbers; a permutation is *odd* if it results from performing any odd number of interchanges. We saw before that $(\mathbf{e}_j \times \mathbf{e}_k) \cdot \mathbf{e}_i$ equals the volume of the frame parallelepiped if i, j, k are distinct and the ordered triple (i, j, k) is an even permutation of $(1, 2, 3)$. If i, j, k are distinct and the ordered triple (i, j, k) is an odd permutation of $(1, 2, 3)$, we obtain minus the volume of the frame parallelepiped. If any two of the numbers i, j, k are equal we obtain zero. Hence

$$\epsilon_{ijk} = \begin{cases} +V, & (i, j, k) \text{ an even permutation of } (1, 2, 3), \\ -V, & (i, j, k) \text{ an odd permutation of } (1, 2, 3), \\ 0, & \text{two or more indices equal.} \end{cases}$$

Moreover, it can be shown (Exercise 2.4) that

$$V^2 = g$$

where g is the determinant of the matrix formed from the elements $g_{ij} = \mathbf{e}_i \cdot \mathbf{e}_j$ of the metric tensor. Note that $|V| = 1$ for a Cartesian frame.

The *permutation symbol* ϵ_{ijk} is useful in writing formulas. For example, the determinant of a matrix $A = (a_{ij})$ can be expressed succinctly as

$$\det A = \epsilon_{ijk} a_{1i} a_{2j} a_{3k}.$$

Much more than a notational device however, ϵ_{ijk} represents a tensor (the so-called *Levi–Civita tensor*). We discuss this further in Chapter 3.

Exercise 2.8. The contravariant components of a vector $\mathbf{c} = \mathbf{a} \times \mathbf{b}$ can be expressed as

$$c^i = \epsilon^{ijk} a_j b_k$$

for suitable coefficients ϵ^{ijk}. Use the technique of this section to find the coefficients. Then establish the identity

$$\epsilon_{ijk}\epsilon^{pqr} = \begin{vmatrix} \delta_i^p & \delta_i^q & \delta_i^r \\ \delta_j^p & \delta_j^q & \delta_j^r \\ \delta_k^p & \delta_k^q & \delta_k^r \end{vmatrix}$$

and use it to show that

$$\epsilon_{ijk}\epsilon^{pqk} = \delta_i^p\delta_j^q - \delta_i^q\delta_j^p.$$

Use this in turn to prove that

$$\mathbf{a} \times (\mathbf{b} \times \mathbf{c}) = \mathbf{b}(\mathbf{a} \cdot \mathbf{c}) - \mathbf{c}(\mathbf{a} \cdot \mathbf{b}) \qquad (2.13)$$

for any vectors $\mathbf{a}, \mathbf{b}, \mathbf{c}$.

Exercise 2.9. Establish Lagrange's identity

$$(\mathbf{a} \times \mathbf{b}) \cdot (\mathbf{c} \times \mathbf{d}) = (\mathbf{a} \cdot \mathbf{c})(\mathbf{b} \cdot \mathbf{d}) - (\mathbf{a} \cdot \mathbf{d})(\mathbf{b} \cdot \mathbf{c}).$$

2.7   Norms on the Space of Vectors

We often need to characterize the intensity of some vector field locally or globally. For this, the notion of a norm is appropriate. The well-known Euclidean norm of a vector $\mathbf{a} = a_k\mathbf{i}_k$ written in a Cartesian frame is

$$\|\mathbf{a}\| = \left(\sum_{k=1}^{3} a_k^2\right)^{1/2}.$$

This norm is related to the inner product of two vectors $\mathbf{a} = a_k\mathbf{i}_k$ and $\mathbf{b} = b_k\mathbf{i}_k$: we have $\mathbf{a} \cdot \mathbf{b} = a_k b_k$ so that

$$\|\mathbf{a}\| = (\mathbf{a} \cdot \mathbf{a})^{1/2}.$$

In a non-Cartesian frame, the components of a vector depend on the lengths of the frame vectors and the angles between them. Since the sum of squared components of a vector depends on the frame, we cannot use it to characterize the vector. But the formulas connected with the dot product are invariant under change of frame, so we can use them to characterize the intensity of the vector — its length. Thus for two vectors $\mathbf{x} = x^i\mathbf{e}_i$ and $\mathbf{y} = y^j\mathbf{e}_j$ written in the arbitrary frame, we can introduce a scalar product (i.e., a simple dot product)

$$\mathbf{x} \cdot \mathbf{y} = x^i\mathbf{e}_i \cdot y^j\mathbf{e}_j = x^iy^jg_{ij} = x_iy_jg^{ij} = x^iy_i.$$

Note that only in mixed coordinates does this resemble the scalar product in a Cartesian frame. Similarly, the norm of a vector $\mathbf{x}$ is

$$\|\mathbf{x}\| = (\mathbf{x} \cdot \mathbf{x})^{1/2} = \left(x^i x^j g_{ij}\right)^{1/2} = \left(x_i x_j g^{ij}\right)^{1/2} = \left(x^i x_i\right)^{1/2}.$$

This dot product and associated norm have all the properties required from objects of this nature in algebra or functional analysis. Indeed, it is necessary only to check whether all the axioms of the inner product are satisfied.

(i) $\mathbf{x} \cdot \mathbf{x} \geq 0$, and $\mathbf{x} \cdot \mathbf{x} = 0$ if and only if $\mathbf{x} = \mathbf{0}$. This property holds because all the quantities involved can be written in a Cartesian frame where it holds trivially. By the same reasoning, we confirm satisfaction of the property

(ii) $\mathbf{x} \cdot \mathbf{y} = \mathbf{y} \cdot \mathbf{x}$. The reader should check that this holds for any representation of the vectors. Finally,

(iii) $(\alpha\mathbf{x} + \beta\mathbf{y}) \cdot \mathbf{z} = \alpha(\mathbf{x} \cdot \mathbf{z}) + \beta(\mathbf{y} \cdot \mathbf{z})$ where α and β are arbitrary real numbers and $\mathbf{z}$ is a vector.

By the general theory then, the expression

$$\|\mathbf{x}\| = (\mathbf{x} \cdot \mathbf{x})^{1/2} \tag{2.14}$$

satisfies all the axioms of a norm:

(i) $\|\mathbf{x}\| \geq 0$, with $\|\mathbf{x}\| = 0$ if and only if $\mathbf{x} = \mathbf{0}$.
(ii) $\|\alpha\mathbf{x}\| = |\alpha| \, \|\mathbf{x}\|$ for any real α.
(iii) $\|\mathbf{x} + \mathbf{y}\| \leq \|\mathbf{x}\| + \|\mathbf{y}\|$.

In addition we have the *Schwarz inequality*

$$\|\mathbf{x} \cdot \mathbf{y}\| \leq \|\mathbf{x}\| \, \|\mathbf{y}\|, \tag{2.15}$$

where in the case of nonzero vectors the equality holds if and only if $\mathbf{x} = \lambda\mathbf{y}$ for some real λ.

The set of all three-dimensional vectors constitutes a three-dimensional linear space. A linear space equipped with the norm (2.14) becomes a normed space. In this book, the principal space is $\mathbb{R}^3$. Note that we can introduce more than one norm in any normed space, and in practice a variety of norms turn out to be necessary. For example, $2\|\mathbf{x}\|$ is also a norm in $\mathbb{R}^3$. We can introduce other norms, quite different from the above. One norm can be introduced as follows. Let $\mathbf{e}_k$ be a basis of $\mathbb{R}^3$ and let

$\mathbf{x} = x^k \mathbf{e}_k$. For $p \geq 1$, we introduce

$$\|\mathbf{x}\|_p = \left(\sum_{k=1}^{3} |x^k|^p \right)^{1/p}.$$

Norm axioms (i) and (ii) obviously hold. Axiom (iii) is a consequence of the classical Minkowski inequality for finite sums. The reader should be aware that this norm is given in a certain basis. If we use it in another basis, the value of the norm of a vector will change in general. An advantage of the norm (2.14) is that it is independent of the basis of the space.

Later, when investigating the eigenvalues of a tensor, we will need a space of vectors with complex components. It can be introduced similarly to the space of complex numbers. We start with the space $\mathbb{R}^3$ having basis $\mathbf{e}_k$, and introduce multiplication of vectors in $\mathbb{R}^3$ by complex numbers. This also yields a linear space, but it is complex and denoted by $\mathbb{C}^3$. An arbitrary vector $\mathbf{x}$ in $\mathbb{C}^3$ takes the form

$$\mathbf{x} = (a^k + ib^k)\mathbf{e}_k,$$

where i is the imaginary unit ($i^2 = -1$). Analogous to the conjugate number is the conjugate vector to $\mathbf{x}$, defined by

$$\overline{\mathbf{x}} = (a^k - ib^k)\mathbf{e}_k.$$

The real and imaginary parts of $\mathbf{x}$ are $a^k \mathbf{e}_k$ and $b^k \mathbf{e}_k$, respectively. Clearly, a basis in $\mathbb{C}^3$ may contain vectors that are not in $\mathbb{R}^3$. As an exercise, the reader should write out the form of the real and imaginary parts of $\mathbf{x}$ in such a basis.

In $\mathbb{C}^3$, the dot product loses the property that $\mathbf{x} \cdot \mathbf{x} \geq 0$. However, we can introduce the inner product of two vectors $\mathbf{x}$ and $\mathbf{y}$ as

$$\langle \mathbf{x}, \mathbf{y} \rangle = \mathbf{x} \cdot \overline{\mathbf{y}}.$$

It is easy to see that this inner product has the following properties. Let $\mathbf{x}, \mathbf{y}, \mathbf{z}$ be arbitrary vectors of $\mathbb{C}^3$. Then

(i) $\mathbf{x} \cdot \mathbf{x} \geq 0$, and $\mathbf{x} \cdot \mathbf{x} = 0$ if and only if $\mathbf{x} = \mathbf{0}$.

(ii) $\mathbf{x} \cdot \mathbf{y} = \overline{\mathbf{y} \cdot \mathbf{x}}$.

(iii) $(\alpha\mathbf{x} + \beta\mathbf{y}) \cdot \mathbf{z} = \alpha(\mathbf{x} \cdot \mathbf{z}) + \beta(\mathbf{y} \cdot \mathbf{z})$ where α and β are arbitrary complex numbers.

The reader should verify these properties. Now we can introduce the norm related to the inner product,

$$\|\mathbf{x}\| = \langle \mathbf{x}, \mathbf{x} \rangle^{1/2},$$

and verify that it satisfies all the axioms of a norm in a complex linear space. As a consequence of the general properties of the inner product, Schwarz's inequality (2.15) also holds in $\mathbb{C}^3$.

2.8 Closing Remarks

We close by repeating something we said in Chapter 1:

A vector is an objective entity.

In elementary mathematics we learn to think of a vector as an ordered triple of components. There is, of course, no harm in this if we keep in mind a certain Cartesian frame. But if we fix those components then in any other frame the vector is determined uniquely. Absolutely uniquely! So a vector is something objective, but as soon as we specify its components in one frame we can find them in any other frame by the use of certain rules.

We emphasize this because the situation is exactly the same with tensors. A tensor is an objective entity, and fixing its components relative to one frame, we determine the tensor uniquely — even though its components relative to other frames will in general be different.

2.9 Problems

2.1 Find the dual basis to $\mathbf{e}_i$.

(a) $\mathbf{e}_1 = 2\mathbf{i}_1 + \mathbf{i}_2 - \mathbf{i}_3$, $\mathbf{e}_2 = 2\mathbf{i}_2 + 3\mathbf{i}_3$, $\mathbf{e}_3 = \mathbf{i}_1 + \mathbf{i}_3$;

(b) $\mathbf{e}_1 = \mathbf{i}_1 + 3\mathbf{i}_2 + 2\mathbf{i}_3$, $\mathbf{e}_2 = 2\mathbf{i}_1 - 3\mathbf{i}_2 + 2\mathbf{i}_3$, $\mathbf{e}_3 = 3\mathbf{i}_1 + 2\mathbf{i}_2 + 3\mathbf{i}_3$;

(c) $\mathbf{e}_1 = \mathbf{i}_1 + \mathbf{i}_2$, $\mathbf{e}_2 = \mathbf{i}_1 - \mathbf{i}_2$, $\mathbf{e}_3 = 3\mathbf{i}_3$;

(d) $\mathbf{e}_1 = \cos\phi\,\mathbf{i}_1 + \sin\phi\,\mathbf{i}_2$, $\mathbf{e}_2 = -\sin\phi\,\mathbf{i}_1 + \cos\phi\,\mathbf{i}_2$, $\mathbf{e}_3 = \mathbf{i}_3$.

2.2 Let

$$\tilde{\mathbf{e}}_1 = -2\mathbf{i}_1 + 3\mathbf{i}_2 + 2\mathbf{i}_3, \qquad \mathbf{e}_1 = 2\mathbf{i}_1 + \mathbf{i}_2 - \mathbf{i}_3,$$
$$\tilde{\mathbf{e}}_2 = -2\mathbf{i}_1 + 2\mathbf{i}_2 + \mathbf{i}_3, \qquad \mathbf{e}_2 = 2\mathbf{i}_2 + 3\mathbf{i}_3,$$
$$\tilde{\mathbf{e}}_3 = -\mathbf{i}_1 + \mathbf{i}_2 + \mathbf{i}_3, \qquad \mathbf{e}_3 = \mathbf{i}_1 + \mathbf{i}_3.$$

Find the matrix A_i^j of transformation from the basis $\tilde{\mathbf{e}}_i$ to the basis $\mathbf{e}_j$.

2.3 Let

$$\tilde{\mathbf{e}}_1 = \mathbf{i}_1 + 2\mathbf{i}_2, \qquad\qquad \mathbf{e}_1 = \mathbf{i}_1 - 6\mathbf{i}_3,$$
$$\tilde{\mathbf{e}}_2 = -\mathbf{i}_2 - \mathbf{i}_3, \qquad\qquad \mathbf{e}_2 = -3\mathbf{i}_1 - 4\mathbf{i}_2 + 4\mathbf{i}_3,$$
$$\tilde{\mathbf{e}}_3 = -\mathbf{i}_1 + 2\mathbf{i}_2 - 2\mathbf{i}_3, \qquad \mathbf{e}_3 = \mathbf{i}_1 + \mathbf{i}_2 + \mathbf{i}_3.$$

Find the matrix of transformation of the basis $\tilde{\mathbf{e}}_i$ to $\mathbf{e}_j$.

2.4 Find

(a) $a_j \delta^{jk}$,

(b) $a_i a^j \delta^i_j$,

(c) δ^i_i,

(d) $\delta_{ij} \delta^{jk}$,

(e) $\delta_{ij} \delta^{ji}$,

(f) $\delta^j_i \delta^k_j \delta^i_k$.

2.5 Show that $\epsilon_{ijk} \epsilon^{ijl} = 2\delta^l_k$.

2.6 Show that $\epsilon_{ijk} \epsilon^{ijk} = 6$.

2.7 Find

(a) $\epsilon_{ijk} \delta^{jk}$,

(b) $\epsilon_{ijk} \epsilon^{mkj} \delta^i_m$,

(c) $\epsilon_{ijk} \delta^k_m \delta^j_n$,

(d) $\epsilon_{ijk} a^i a^j$,

(e) $\epsilon_{ijk} |\epsilon^{ijk}|$,

(f) $\epsilon_{ijk} \epsilon^{imn} \delta^j_m$.

2.8 Find $(\mathbf{a} \times \mathbf{b}) \times \mathbf{c}$.

2.9 Show that $(\mathbf{a} \times \mathbf{b}) \cdot \mathbf{a} = 0$.

2.10 Show that $\mathbf{a} \cdot (\mathbf{b} \times \mathbf{c})\mathbf{d} = (\mathbf{a} \cdot \mathbf{d})\mathbf{b} \times \mathbf{c} + (\mathbf{b} \cdot \mathbf{d})\mathbf{c} \times \mathbf{a} + (\mathbf{c} \cdot \mathbf{d})\mathbf{a} \times \mathbf{b}$.

2.11 Show that $(\mathbf{e} \times \mathbf{a}) \times \mathbf{e} = \mathbf{a}$ if $|\mathbf{e}| = 1$ and $\mathbf{e} \cdot \mathbf{a} = 0$.

2.12 Let $\mathbf{e}_k$ be a basis of $\mathbb{R}^3$, let $\mathbf{x} = x^k \mathbf{e}_k$, and suppose h_1, h_2, h_3 are fixed positive numbers. Show that $h_k |x^k|$ is a norm in $\mathbb{R}^3$.

Chapter 3

Tensors

3.1 Dyadic Quantities and Tensors

We have met sets of quantities like g^{ij} or g_{ij}. Such a table of $3 \times 3 = 9$ coefficients could be considered as a vector in a nine-dimensional space, but we must reject this idea for an important reason: if we change the frame vectors and calculate the relations between the new and old components, the results differ in form from those that apply to vector components. The components of the metric tensor transform according to certain rules, however, and it is found that these transformation rules also apply to various quantities encountered in physical science. We indicated in Chapter 1 that these quantities, represented by 3×3 matrices, form a class of objects known as second-order tensors. Our plan is to present the relevant theory in a way that parallels the vector presentation of Chapter 2.

We begin to realize this program with the introduction of the *dyad* (or *tensor product*) of two vectors $\mathbf{a}$ and $\mathbf{b}$, denoted $\mathbf{a} \otimes \mathbf{b}$. We assume that the tensor product satisfies many usual properties of a product:

$$(\lambda \mathbf{a}) \otimes \mathbf{b} = \mathbf{a} \otimes (\lambda \mathbf{b}) = \lambda(\mathbf{a} \otimes \mathbf{b}),$$
$$(\mathbf{a} + \mathbf{b}) \otimes \mathbf{c} = \mathbf{a} \otimes \mathbf{c} + \mathbf{b} \otimes \mathbf{c},$$
$$\mathbf{a} \otimes (\mathbf{b} + \mathbf{c}) = \mathbf{a} \otimes \mathbf{b} + \mathbf{a} \otimes \mathbf{c}, \tag{3.1}$$

where λ is an arbitrary real number. However, the tensor product is not symmetric: if $\mathbf{a}$ is not proportional to $\mathbf{b}$ then $\mathbf{a} \otimes \mathbf{b} \neq \mathbf{b} \otimes \mathbf{a}$. From now on, we shall write out the dyad without the $\otimes$ symbol: $\mathbf{ab} = \mathbf{a} \otimes \mathbf{b}$.

Let us once again consider the space of three-dimensional vectors with the frame $\mathbf{e}_i$. Using the expansion of the vectors in the basis vectors and the properties (3.1), we represent the dyad $\mathbf{ab}$ as

$$\mathbf{ab} = a^i \mathbf{e}_i b^j \mathbf{e}_j = a^i b^j \mathbf{e}_i \mathbf{e}_j.$$

This introduces exactly nine different dyads $\mathbf{e}_i\mathbf{e}_j$. We now consider a linear space whose basis is this set of nine dyads and call it the space of second-order tensors (or tensors of order two). The numerical coefficients of the dyads are called the components of the tensor. Thus an element of this space, a tensor $\mathbf{A}$, has the representation

$$\mathbf{A} = a^{ij}\mathbf{e}_i\mathbf{e}_j.$$

To maintain the property of objectivity of the elements of this space, we require that upon transformation of the frame the components of $\mathbf{A}$ transform correspondingly. Note that we have introduced superscript indices for the components of $\mathbf{A}$. This was done in keeping with the development of Chapter 2.

In preparation for the next section let us introduce the dot product of a dyad $\mathbf{ab}$ by a vector $\mathbf{c}$:

$$\mathbf{ab} \cdot \mathbf{c} = (\mathbf{b} \cdot \mathbf{c})\mathbf{a}. \tag{3.2}$$

So the result is a vector co-oriented with $\mathbf{a}$. Analogously we can introduce the dot product from the left:

$$\mathbf{c} \cdot \mathbf{ab} = (\mathbf{c} \cdot \mathbf{a})\mathbf{b}. \tag{3.3}$$

Exercise 3.1. (a) A dyad of the form $\mathbf{ee}$, where $\mathbf{e}$ is a unit vector, is sometimes called a *projection dyad*. Explain. (b) Write down matrices for the dyads $\mathbf{i}_1\mathbf{i}_1$, $\mathbf{i}_2\mathbf{i}_2$, and $\mathbf{i}_3\mathbf{i}_1$.

3.2　Tensors From an Operator Viewpoint

An alternative to viewing a second-order tensor as a weighted sum of dyads is to view the tensor as an *operator*. From this standpoint a tensor $\mathbf{A}$ is considered to map a vector $\mathbf{x}$ into a vector $\mathbf{y}$ according to the equation

$$\mathbf{y} = \mathbf{A} \cdot \mathbf{x}.$$

Conversely, a given linear relation between $\mathbf{x}$ and $\mathbf{y}$ will define the operator $\mathbf{A}$ uniquely. Thus if we have $\mathbf{A} \cdot \mathbf{x} = \mathbf{B} \cdot \mathbf{x}$ for all $\mathbf{x}$, then we have $\mathbf{A} = \mathbf{B}$. Let us show that the components are really uniquely defined in any basis by the equality $\mathbf{y} = \mathbf{A} \cdot \mathbf{x}$. The tensor $\mathbf{A}$ is represented by the expression $a^{ij}\mathbf{e}_i\mathbf{e}_j$ in some basis $\mathbf{e}_i$. It is clear that the operation $\mathbf{A} \cdot \mathbf{x}$ is linear in $\mathbf{x}$, so we define $\mathbf{A}$ uniquely if we specify its action on all three vectors of a basis. Taking $\mathbf{x} = \mathbf{e}^k$, the corresponding $\mathbf{y}$ is

$$\mathbf{A} \cdot \mathbf{x} = a^{ij}\mathbf{e}_i\mathbf{e}_j \cdot \mathbf{e}^k = a^{ik}\mathbf{e}_i.$$

Dot multiplying this by $\mathbf{e}^l$ we get

$$a^{lk} = \mathbf{e}^l \cdot \mathbf{A} \cdot \mathbf{e}^k.$$

In this way we can find the components of a tensor $\mathbf{A}$ in any basis:

$$a^{ij} = \mathbf{e}^i \cdot \mathbf{A} \cdot \mathbf{e}^j, \qquad a^i_{\cdot j} = \mathbf{e}^i \cdot \mathbf{A} \cdot \mathbf{e}_j,$$

$$a_{ij} = \mathbf{e}_i \cdot \mathbf{A} \cdot \mathbf{e}_j, \qquad a_i^{\cdot j} = \mathbf{e}_i \cdot \mathbf{A} \cdot \mathbf{e}^j.$$

Note that in "mixed components" we position the indices in such a way that their association with the various dyads remains clear.

Analyzing the above reasoning, we can find that we have proved the *quotient law* for tensors of order two. If $\mathbf{y}$ is a given vector and there is a linear transformation from $\mathbf{x}$ to $\mathbf{y}$ for an *arbitrary* vector $\mathbf{x}$, then the linear transformation is a tensor and we can write $\mathbf{y} = \mathbf{A} \cdot \mathbf{x}$. This statement is sometimes useful in establishing the tensorial character of a set of scalar quantities (i.e., the components of $\mathbf{A}$).

We may also define common algebraic operations from the operator viewpoint. Given tensors $\mathbf{A}$ and $\mathbf{B}$, the *sum* is the tensor $\mathbf{A} + \mathbf{B}$ uniquely defined by the requirement that

$$(\mathbf{A} + \mathbf{B}) \cdot \mathbf{x} = \mathbf{A} \cdot \mathbf{x} + \mathbf{B} \cdot \mathbf{x}$$

for all $\mathbf{x}$. If c is a scalar, $c\mathbf{A}$ is defined by the requirement that

$$(c\mathbf{A}) \cdot \mathbf{x} = c(\mathbf{A} \cdot \mathbf{x})$$

for all $\mathbf{x}$. In particular, any product of the form $0\mathbf{A}$ gives a *zero tensor* denoted $\mathbf{0}$. The dot product $\mathbf{A} \cdot \mathbf{B}$ is regarded as the composition of the operators $\mathbf{B}$ and $\mathbf{A}$:

$$(\mathbf{A} \cdot \mathbf{B}) \cdot \mathbf{x} \equiv \mathbf{A} \cdot (\mathbf{B} \cdot \mathbf{x}).$$

The dot product $\mathbf{y} \cdot \mathbf{A}$, called *pre-multiplication* of $\mathbf{A}$ by a vector $\mathbf{y}$, is defined by the requirement that

$$(\mathbf{y} \cdot \mathbf{A}) \cdot \mathbf{x} = \mathbf{y} \cdot (\mathbf{A} \cdot \mathbf{x})$$

for all vectors $\mathbf{x}$.

A simple but important tensor is the *unit tensor* denoted by $\mathbf{E}$ and defined by the requirement that for any $\mathbf{x}$

$$\mathbf{E} \cdot \mathbf{x} = \mathbf{x} \cdot \mathbf{E} = \mathbf{x}. \tag{3.4}$$

It is evident that in any Cartesian frame $\mathbf{i}_i$ we must have

$$\mathbf{E} = \sum_{i=1}^{3} \mathbf{i}_i \mathbf{i}_i. \qquad (3.5)$$

In any frame we have

$$\mathbf{E} = \mathbf{e}^i \mathbf{e}_i = \mathbf{e}_j \mathbf{e}^j \qquad (3.6)$$

for the mixed components. Consequently, the raising and lowering of indices gives

$$\mathbf{E} = g_{ij} \mathbf{e}^i \mathbf{e}^j = g^{ij} \mathbf{e}_i \mathbf{e}_j \qquad (3.7)$$

in non-mixed components. We see that the role of the unit tensor belongs to the metric tensor! Throughout our discussion of second-order tensors we shall emphasize the close analogy between tensor theory and matrix theory. Equations (3.5) and (3.6) show that the matrix representation of $\mathbf{E}$ in either Cartesian or mixed components is the 3×3 identity matrix

$$\begin{pmatrix} 1 & 0 & 0 \\ 0 & 1 & 0 \\ 0 & 0 & 1 \end{pmatrix}.$$

This does not hold for the non-mixed components of (3.7).

Exercise 3.2. Use (3.4) along with (2.4) and (2.5) to show that the various components of $\mathbf{E}$ are given by

$$e^{ij} = g^{ij}, \qquad\qquad e^i_{\cdot j} = \delta^i_j,$$

$$e_{ij} = g_{ij}, \qquad\qquad e^{\cdot i}_j = \delta^i_j.$$

Hence establish (3.6).

Our consideration of $\mathbf{A}$ as an operator leads us to introduce the notion of an *inverse tensor*: if

$$\mathbf{A} \cdot \mathbf{A}^{-1} = \mathbf{E},$$

then $\mathbf{A}^{-1}$ is called the inverse of $\mathbf{A}$. The inverse of a tensor is also a tensor. An important special case occurs when the matrix of the tensor is a *diagonal matrix*. If in a Cartesian frame $\mathbf{i}_i$ we have

$$\mathbf{A} = \sum_{i=1}^{3} \lambda_i \mathbf{i}_i \mathbf{i}_i$$

then the corresponding matrix representation is

$$\begin{pmatrix} \lambda_1 & 0 & 0 \\ 0 & \lambda_2 & 0 \\ 0 & 0 & \lambda_3 \end{pmatrix}.$$

If we take

$$\mathbf{B} = \sum_{j=1}^{3} \lambda_j^{-1} \mathbf{i}_j \mathbf{i}_j$$

to which there corresponds the matrix

$$\begin{pmatrix} \lambda_1^{-1} & 0 & 0 \\ 0 & \lambda_2^{-1} & 0 \\ 0 & 0 & \lambda_3^{-1} \end{pmatrix}$$

and form the dot product $\mathbf{A} \cdot \mathbf{B}$, we get

$$\mathbf{A} \cdot \mathbf{B} = \sum_{i=1}^{3} \lambda_i \mathbf{i}_i \mathbf{i}_i \cdot \sum_{j=1}^{3} \lambda_j^{-1} \mathbf{i}_j \mathbf{i}_j$$

$$= \sum_{i=1}^{3} \lambda_i \mathbf{i}_i \sum_{j=1}^{3} \lambda_j^{-1} (\mathbf{i}_i \cdot \mathbf{i}_j) \mathbf{i}_j$$

$$= \sum_{i=1}^{3} \lambda_i \mathbf{i}_i \lambda_i^{-1} \mathbf{i}_i$$

$$= \sum_{i=1}^{3} \mathbf{i}_i \mathbf{i}_i$$

$$= \mathbf{E}.$$

This means that $\mathbf{B} = \mathbf{A}^{-1}$. Correspondingly,

$$\begin{pmatrix} \lambda_1 & 0 & 0 \\ 0 & \lambda_2 & 0 \\ 0 & 0 & \lambda_3 \end{pmatrix} \begin{pmatrix} \lambda_1^{-1} & 0 & 0 \\ 0 & \lambda_2^{-1} & 0 \\ 0 & 0 & \lambda_3^{-1} \end{pmatrix} = \begin{pmatrix} 1 & 0 & 0 \\ 0 & 1 & 0 \\ 0 & 0 & 1 \end{pmatrix}.$$

Exercise 3.3. Establish the formula

$$(\mathbf{A} \cdot \mathbf{B})^{-1} = \mathbf{B}^{-1} \cdot \mathbf{A}^{-1}$$

for invertible tensors $\mathbf{A}$, $\mathbf{B}$.

A second-order tensor $\mathbf{A}$ is *singular* if $\mathbf{A} \cdot \mathbf{x} = \mathbf{0}$ for some $\mathbf{x} \neq \mathbf{0}$. Hence $\mathbf{A}$ is *nonsingular* if $\mathbf{A} \cdot \mathbf{x} = \mathbf{0}$ only when $\mathbf{x} = \mathbf{0}$. Recall that a matrix A is said to be nonsingular if and only if $\det A \neq 0$. The connection between the uses of this terminology in the two areas is as follows. If we take a mixed representation of the tensor $\mathbf{A}$, the equation $\mathbf{A} \cdot \mathbf{x} = \mathbf{0}$ yields a set of simultaneous equations in the components of $\mathbf{x}$; these equations have a nontrivial solution if and only if the determinant of the coefficient matrix (i.e., the matrix representing $\mathbf{A}$) is zero. Moreover, taking any other representation of $\mathbf{A}$ and a dual representation of the vector, we again arrive the same conclusion regarding the determinant. This brings the use of the term "singular" to the tensor $\mathbf{A}$. By definition the determinant of a second-order tensor $\mathbf{A}$, denoted $\det \mathbf{A}$, is the determinant of the matrix of its mixed components:

$$\det \mathbf{A} = |a_i^{\cdot j}| = |a^k_{\cdot m}| = \frac{1}{g}|a_{st}| = g|a^{pq}|.$$

The first equality is the definition as stated above; the rest are left for the reader to establish. Various other formulas such as

$$\det \mathbf{A} = \frac{1}{6}\epsilon_{ijk}\epsilon^{mnp}a_m^{\cdot i}a_n^{\cdot j}a_p^{\cdot k}$$

can be established for the determinant.

We close this section with an important remark. We can derive all desired properties of a tensor, and perform actions with the tensor, in any coordinate frame. Convenience will often dictate the use of Cartesian frames. But if we obtain an equation or expression through the use of a Cartesian frame and can subsequently represent this result in non-coordinate form, then we have provided rigorous justification of the latter. As we have said before, tensors are objective entities and ultimately all results pertaining to them must be frame independent.

3.3   Dyadic Components Under Transformation

The standpoint for deriving the transformation rules is that in any basis a tensor is the same element of some space, and only (3.1) and the rules we derived for vectors can govern the rules for transforming the components of a tensor. Let us begin with the transformation of the components when we go to the reciprocal basis. We set

$$a_{ij}\mathbf{e}^i\mathbf{e}^j = a^{ij}\mathbf{e}_i\mathbf{e}_j$$

and take dot products as in (3.2) and (3.3):

$$\mathbf{e}_k \cdot a_{ij}\mathbf{e}^i\mathbf{e}^j \cdot \mathbf{e}_m = \mathbf{e}_k \cdot a^{ij}\mathbf{e}_i\mathbf{e}_j \cdot \mathbf{e}_m.$$

This gives

$$a_{ij}(\mathbf{e}_k \cdot \mathbf{e}^i)(\mathbf{e}^j \cdot \mathbf{e}_m) = a^{ij}(\mathbf{e}_k \cdot \mathbf{e}_i)(\mathbf{e}_j \cdot \mathbf{e}_m),$$

hence

$$a_{km} = a^{ij}g_{ki}g_{jm}.$$

We see that the components of the metric tensor are encountered in this transformation.

Now we can construct the formulas for transforming the tensor components when the change of basis takes the general form

$$\mathbf{e}_i = A_i^j\tilde{\mathbf{e}}_j.$$

From

$$\tilde{a}^{ij}\tilde{\mathbf{e}}_i\tilde{\mathbf{e}}_j = a^{km}\mathbf{e}_k\mathbf{e}_m = a^{km}A_k^p\tilde{\mathbf{e}}_p A_m^q\tilde{\mathbf{e}}_q$$

we obtain

$$\tilde{a}^{ij} = a^{km}A_k^i A_m^j. \tag{3.8}$$

Similarly, the inverse transformation

$$\tilde{\mathbf{e}}_i = \tilde{A}_i^j\mathbf{e}_j$$

leads to

$$a^{ij} = \tilde{a}^{km}\tilde{A}_k^i \tilde{A}_m^j. \tag{3.9}$$

Equations (3.8) and (3.9) together imply that

$$A_j^k\tilde{A}_k^i = \delta_j^i.$$

Various expressions for **A**,

$$\mathbf{A} = a^{ij}\mathbf{e}_i\mathbf{e}_j = a_{kl}\mathbf{e}^k\mathbf{e}^l = a_i{}^j\mathbf{e}^i\mathbf{e}_j = a^k{}_{.l}\mathbf{e}_k\mathbf{e}^l$$
$$= \tilde{a}^{ij}\tilde{\mathbf{e}}_i\tilde{\mathbf{e}}_j = \tilde{a}_{kl}\tilde{\mathbf{e}}^k\tilde{\mathbf{e}}^l = \tilde{a}_i{}^{.j}\tilde{\mathbf{e}}^i\tilde{\mathbf{e}}_j = \tilde{a}^k{}_{.l}\tilde{\mathbf{e}}_k\tilde{\mathbf{e}}^l,$$

lead to other transformation formulas such as

$$\tilde{a}_{ij} = \tilde{A}_i^k\tilde{A}_j^l a_{kl}, \qquad a_{ij} = A_i^k A_j^l\tilde{a}_{kl},$$

and

$$\tilde{a}_i{}^{.j} = \tilde{A}_i^k A_l^j a_k{}^{.l}, \qquad \tilde{a}^i{}_{.j} = A_k^i\tilde{A}_j^l a^k{}_{.l}.$$

Remembering the terminology of § 2.5, we see why the a_{ij} are called the covariant components of **A** while the a^{ij} are called the contravariant components. The components $a^i{}_{.j}$ and $a_i{}^{.j}$ are called the mixed components.

Quick summary

We have

$$\mathbf{A} = \tilde{a}^{ij}\tilde{\mathbf{e}}_i\tilde{\mathbf{e}}_j = \tilde{a}_{kl}\tilde{\mathbf{e}}^k\tilde{\mathbf{e}}^l = \tilde{a}_i^{\cdot j}\tilde{\mathbf{e}}^i\tilde{\mathbf{e}}_j = \tilde{a}^k_{\cdot l}\tilde{\mathbf{e}}_k\tilde{\mathbf{e}}^l$$
$$= a^{ij}\mathbf{e}_i\mathbf{e}_j = a_{kl}\mathbf{e}^k\mathbf{e}^l = a_i^{\cdot j}\mathbf{e}^i\mathbf{e}_j = a^k_{\cdot l}\mathbf{e}_k\mathbf{e}^l$$

where

$$\tilde{a}^{ij} = A_k^i A_l^j a^{kl}, \qquad\qquad a^{ij} = \tilde{A}_k^i \tilde{A}_l^j \tilde{a}^{kl},$$
$$\tilde{a}_{ij} = \tilde{A}_i^k \tilde{A}_j^l a_{kl}, \qquad\qquad a_{ij} = A_i^k A_j^l \tilde{a}_{kl},$$
$$\tilde{a}_{\cdot j}^i = A_k^i \tilde{A}_j^l a_{\cdot l}^k, \qquad\qquad a_{\cdot j}^i = \tilde{A}_k^i A_j^l \tilde{a}_{\cdot l}^k,$$
$$\tilde{a}_i^{\cdot j} = \tilde{A}_i^k A_l^j a_k^{\cdot l}, \qquad\qquad a_i^{\cdot j} = A_i^k \tilde{A}_l^j \tilde{a}_k^{\cdot l}.$$

Exercise 3.4. (a) Express the transformation law

$$\tilde{b}^{ij} = A_k^i A_m^j b^{km}$$

in matrix notation. (b) Repeat for a transformation law of the form

$$\tilde{b}_{ij} = \tilde{A}_i^k \tilde{A}_j^m b_{km}.$$

Exercise 3.5. Our A_i^j values give the transformation from one basis to another; they define a transformation of the space that is an operator, and hence a second-order tensor. Write out the tensor for which the A_i^j are components. Repeat for the inverse transformation.

3.4 More Dyadic Operations

The dot product of two dyads **ab** and **cd** is defined by

$$\mathbf{ab} \cdot \mathbf{cd} = (\mathbf{b} \cdot \mathbf{c})\mathbf{ad}.$$

The result is again a dyad, with a coefficient $\mathbf{b} \cdot \mathbf{c}$. Extensions of this and the formulas (3.2) and (3.3) to operations with sums of dyads and vectors (using (3.1) and the vectorial rules) gives us a number of rules which the dot product obeys. Let **A** and **B** be dyads, **a** and **b** be vectors, and λ and μ be any real numbers. Then

$$\mathbf{A} \cdot (\lambda\mathbf{a} + \mu\mathbf{b}) = \lambda\mathbf{A} \cdot \mathbf{a} + \mu\mathbf{A} \cdot \mathbf{b},$$
$$(\lambda\mathbf{A} + \mu\mathbf{B}) \cdot \mathbf{a} = \lambda\mathbf{A} \cdot \mathbf{a} + \mu\mathbf{B} \cdot \mathbf{a}.$$

Similar identities hold for dot products taken in the opposite orders. These results show that linearity may be assumed in working with these operations.

Now let us pursue the close analogy between the dot product and matrix multiplication. We begin with the simple case of a Cartesian frame. We take a dyad $\mathbf{A}$ and a vector $\mathbf{b}$ and express these relative to a basis $\mathbf{i}_k$:

$$\mathbf{A} = a^{km}\mathbf{i}_k\mathbf{i}_m, \qquad \mathbf{b} = b^j\mathbf{i}_j.$$

Denoting

$$\mathbf{c} = \mathbf{A} \cdot \mathbf{b} \tag{3.10}$$

we have

$$c^k\mathbf{i}_k = a^{km}\mathbf{i}_k\mathbf{i}_m \cdot b^j\mathbf{i}_j$$

so that

$$c^k = \sum_{j=1}^{3} a^{kj}b^j.$$

(We have inserted the summation symbol because j stands in the upper position twice and the summation convention would not apply.) Written out, this is the system of three equations

$$\begin{aligned}
c^1 &= a^{11}b^1 + a^{12}b^2 + a^{13}b^3, \\
c^2 &= a^{21}b^1 + a^{22}b^2 + a^{23}b^3, \\
c^3 &= a^{31}b^1 + a^{32}b^2 + a^{33}b^3,
\end{aligned}$$

or

$$\begin{pmatrix} c^1 \\ c^2 \\ c^3 \end{pmatrix} = \begin{pmatrix} a^{11} & a^{12} & a^{13} \\ a^{21} & a^{22} & a^{23} \\ a^{31} & a^{32} & a^{33} \end{pmatrix} \begin{pmatrix} b^1 \\ b^2 \\ b^3 \end{pmatrix}.$$

Here we have a matrix equation of the form

$$c = Ab \tag{3.11}$$

where c and b are column vectors and A is a 3×3 matrix. The analogy between the dot product and matrix multiplication is evident from (3.10) and (3.11). This analogy extends beyond the confines of Cartesian frames. Let us write, for example,

$$\mathbf{A} = a^{km}\mathbf{e}_k\mathbf{e}_m, \qquad \mathbf{b} = b_j\mathbf{e}^j.$$

This time $\mathbf{c} = \mathbf{A} \cdot \mathbf{b}$ gives

$$c^k \mathbf{e}_k = a^{km} \mathbf{e}_k \mathbf{e}_m \cdot b_j \mathbf{e}^j = a^{km} \mathbf{e}_k \delta_m^j b_j,$$

hence

$$c^k = a^{kj} b_j.$$

The corresponding matrix equation is, of course,

$$\begin{pmatrix} c^1 \\ c^2 \\ c^3 \end{pmatrix} = \begin{pmatrix} a^{11} & a^{12} & a^{13} \\ a^{21} & a^{22} & a^{23} \\ a^{31} & a^{32} & a^{33} \end{pmatrix} \begin{pmatrix} b_1 \\ b_2 \\ b_3 \end{pmatrix}.$$

With suitable understanding we could still write this as (3.11). Note what happens when we express both the dyad and the vector in terms of covariant components:

$$\mathbf{A} = a_{km} \mathbf{e}^k \mathbf{e}^m, \qquad \mathbf{b} = b_j \mathbf{e}^j.$$

We obtain

$$c_k = a_{km} g^{mj} b_j$$

and the metric tensor appears. The corresponding matrix form is

$$\begin{pmatrix} c_1 \\ c_2 \\ c_3 \end{pmatrix} = \begin{pmatrix} a_{11} & a_{12} & a_{13} \\ a_{21} & a_{22} & a_{23} \\ a_{31} & a_{32} & a_{33} \end{pmatrix} \begin{pmatrix} g^{11} & g^{12} & g^{13} \\ g^{21} & g^{22} & g^{23} \\ g^{31} & g^{32} & g^{33} \end{pmatrix} \begin{pmatrix} b_1 \\ b_2 \\ b_3 \end{pmatrix}.$$

Because the metric tensor can raise an index on a vector component, we may also write these equations in the forms

$$c_k = a_{km} b^m$$

and

$$\begin{pmatrix} c_1 \\ c_2 \\ c_3 \end{pmatrix} = \begin{pmatrix} a_{11} & a_{12} & a_{13} \\ a_{21} & a_{22} & a_{23} \\ a_{31} & a_{32} & a_{33} \end{pmatrix} \begin{pmatrix} b^1 \\ b^2 \\ b^3 \end{pmatrix}.$$

Let us examine the dot product between two dyads. There are various possibilities for the components of the dyad

$$\mathbf{C} = \mathbf{A} \cdot \mathbf{B},$$

depending on how we choose to express $\mathbf{A}$ and $\mathbf{B}$. If we use all contravariant components and write

$$\mathbf{A} = a^{km} \mathbf{e}_k \mathbf{e}_m, \qquad \mathbf{B} = b^{km} \mathbf{e}_k \mathbf{e}_m,$$

then

$$C = c^{kn}e_k e_n$$

where

$$c^{kn} = a^{km}g_{mj}b^{jn}.$$

Similarly, the use of all covariant components as in

$$A = a_{km}e^k e^m, \qquad B = b_{km}e^k e^m,$$

leads to

$$C = c_{kn}e^k e^n$$

where

$$c_{kn} = a_{km}g^{mj}b_{jn}.$$

Mixed components appear when we express

$$A = a^{km}e_k e_m, \qquad B = b_{km}e^k e^m.$$

Then

$$C = A \cdot B = a^{km}e_k e_m \cdot b_{jn}e^j e^n = a^{km}\delta^j_m b_{jn}e_k e^n = a^{kj}b_{jn}e_k e^n.$$

Defining

$$c^k_{\cdot n} = a^{kj}b_{jn}$$

we have

$$C = c^k_{\cdot n}e_k e^n.$$

We leave other possibilities to the reader as

Exercise 3.6. (a) Discuss how the formulation $c_k^{\cdot n} = a_{kj}b^{jn}$ arises. (b) Show how all the forms above correspond to matrix multiplication. (c) What happens if mixed components are used on the right-hand sides to express A and B?

Another useful operation that can be performed between tensors is *double dot multiplication*. If ab and cd are dyads, we define

$$ab \cdot\cdot cd = (b \cdot c)(a \cdot d).$$

That is, we first dot multiply the near standing vectors, then the remaining vectors, and thereby obtain a scalar as the result.

Exercise 3.7. (a) Calculate $\mathbf{A} \cdot\!\cdot\, \mathbf{E}$ if $\mathbf{A}$ is a tensor of order two. How does this relate to the trace of the matrix that represents $\mathbf{A}$ in mixed components? (b) Let $\mathbf{A}$ and $\mathbf{B}$ be tensors of order two. Write down several different component forms for the quantity $\mathbf{A} \cdot\!\cdot\, \mathbf{B}$.

Yet another operation is the *scalar product* of two second-order tensors $\mathbf{A}$ and $\mathbf{B}$, denoted by $\mathbf{A} \bullet \mathbf{B}$. This represents a natural extension of the operation

$$\mathbf{ab} \bullet \mathbf{cd} = (\mathbf{a} \cdot \mathbf{c})(\mathbf{b} \cdot \mathbf{d})$$

between two dyads $\mathbf{ab}$ and $\mathbf{cd}$.

3.5 Properties of Second-Order Tensors

Now we would like to consider in more detail those tensors that occur most frequently in applications: tensors of order two. First we recall that such a tensor is represented in dyadic form as

$$\mathbf{A} = a^{ij}\mathbf{e}_i\mathbf{e}_j.$$

Those who work in the applied sciences are probably more accustomed to the matrix representation

$$\begin{pmatrix} a^{11} & a^{12} & a^{13} \\ a^{21} & a^{22} & a^{23} \\ a^{31} & a^{32} & a^{33} \end{pmatrix}. \tag{3.12}$$

When we use the matrix form (3.12) the dyadic basis of the tensor remains implicit. Of course when we use a unique, say Cartesian, frame for the space of vectors, then it does not matter whether we show the dyads. The correspondence between the dyadic and matrix representations suggests that we can introduce many familiar ideas from the theory of matrices.

The tensor transpose

Let us begin with the notion of *transposition*. For a matrix $A = (a^{ij})$ the transposed matrix A^T is

$$A^T = (a^{ji}).$$

Similarly we introduce the transpose operation for the tensor $\mathbf{A}$:

$$\mathbf{A}^T = a^{ji}\mathbf{e}_i\mathbf{e}_j. \tag{3.13}$$

This operation yields a new tensor, in each representation of which the corresponding indices appear in reverse order:

$$\mathbf{A}^T = a^{ji}\mathbf{e}_i\mathbf{e}_j = a_{ji}\mathbf{e}^i\mathbf{e}^j = a^j_{\cdot i}\mathbf{e}^i\mathbf{e}_j = a^{\cdot i}_j\mathbf{e}_i\mathbf{e}^j.$$

A useful relation for any second-order tensor $\mathbf{A}$ and any vector $\mathbf{x}$ is

$$\mathbf{A}\cdot\mathbf{x} = \mathbf{x}\cdot\mathbf{A}^T. \tag{3.14}$$

This follows when we write $\mathbf{x} = x^k\mathbf{e}_k$ and use (3.13) to see that

$$\mathbf{A}^T = a^{ij}\mathbf{e}_j\mathbf{e}_i.$$

Equation (3.14) can be used to define the transpose. Also note that

$$(\mathbf{A}^T)^T = \mathbf{A}$$

for any second-order tensor $\mathbf{A}$.

Exercise 3.8. Let $\mathbf{A}$ and $\mathbf{B}$ be tensors of order two. Demonstrate that

$$\mathbf{A}\bullet\mathbf{B} = \mathbf{A}\cdot\cdot\mathbf{B}^T = \mathbf{A}^T\cdot\cdot\mathbf{B}.$$

Exercise 3.9. Let $\mathbf{A}$ be a second-order tensor. Find $\mathbf{A}\cdot\cdot\mathbf{A}^T$. Demonstrate that $\mathbf{A}\cdot\cdot\mathbf{A}^T = 0$ if and only if $\mathbf{A} = \mathbf{0}$.

Exercise 3.10. (a) Show that if $\mathbf{A}$ and $\mathbf{B}$ are tensors of order two, then

$$(\mathbf{A}\cdot\mathbf{B})^T = \mathbf{B}^T\cdot\mathbf{A}^T.$$

(b) Let $\mathbf{a}$ and $\mathbf{b}$ be vectors and $\mathbf{C}$ be a tensor of order two. Show that

$$\mathbf{a}\cdot\mathbf{C}^T\cdot\mathbf{b} = \mathbf{b}\cdot\mathbf{C}\cdot\mathbf{a}.$$

(c) Show that if $\mathbf{A}$ is a nonsingular tensor of order two, then the components of the tensor $\mathbf{B} = \mathbf{A}^{-1}$ are given by the formulas

$$b^{\cdot j}_i = \frac{1}{2\det\mathbf{A}}\epsilon_{ikl}\epsilon^{jmn}a^{\cdot k}_m a^{\cdot l}_n.$$

(d) Verify the following relations:

$$\det\mathbf{A}^{-1} = (\det\mathbf{A})^{-1}, \qquad (\mathbf{A}\cdot\mathbf{B})^{-1} = \mathbf{B}^{-1}\cdot\mathbf{A}^{-1},$$
$$(\mathbf{A}^T)^{-1} = (\mathbf{A}^{-1})^T, \qquad (\mathbf{A}^{-1})^{-1} = \mathbf{A}.$$

Tensors raised to powers

By analogy with matrix algebra we may raise a tensor to a positive integer power:

$$\mathbf{A}^2 = \mathbf{A} \cdot \mathbf{A}, \qquad \mathbf{A}^3 = \mathbf{A} \cdot \mathbf{A}^2, \qquad \mathbf{A}^4 = \mathbf{A} \cdot \mathbf{A}^3,$$

and so on. Note that $\mathbf{A}^k$ still represents a linear operator. Negative integer powers are defined by raising $\mathbf{A}^{-1}$ to positive integer powers:

$$\mathbf{A}^{-2} = \mathbf{A}^{-1} \cdot \mathbf{A}^{-1}, \qquad \mathbf{A}^{-3} = \mathbf{A}^{-2} \cdot \mathbf{A}^{-1}, \qquad \mathbf{A}^{-4} = \mathbf{A}^{-3} \cdot \mathbf{A}^{-1},$$

and so on. These operations can be used to construct functions of tensors using Taylor expansions of elementary functions. For example,

$$e^x = 1 + \frac{x}{1!} + \frac{x^2}{2!} + \frac{x^3}{3!} + \cdots .$$

By this we can introduce the exponential of the tensor $\mathbf{A}$:

$$e^{\mathbf{A}} = \mathbf{E} + \frac{\mathbf{A}}{1!} + \frac{\mathbf{A}^2}{2!} + \frac{\mathbf{A}^3}{3!} + \cdots .$$

The issue of convergence of such series is approached in a manner similar to the absolute convergence of usual series, but with use of a norm of the tensor $\mathbf{A}$ (see § 3.12). Note that $e^{\mathbf{A}}$ represents a linear operator. We can introduce other functions similarly. This technique is used in the study of nonlinear elasticity, for example.

Symmetric and antisymmetric tensors

Among the class of all second-order tensors, an important role is played by the *symmetric* tensors. These include the strain and stress tensors of the theory of elasticity. The tensor of inertia is symmetric, as is the metric tensor. All these satisfy the relation

$$\mathbf{A} = \mathbf{A}^T.$$

It follows from (3.14) that

$$\mathbf{A} \cdot \mathbf{x} = \mathbf{x} \cdot \mathbf{A} \tag{3.15}$$

and

$$(\mathbf{A} \cdot \mathbf{x}) \cdot \mathbf{y} = \mathbf{x} \cdot (\mathbf{A} \cdot \mathbf{y}) \tag{3.16}$$

if $\mathbf{A}$ is symmetric. The reader will recall that the unit tensor $\mathbf{E}$ satisfies a relation of the form (3.15). A tensor $\mathbf{A}$ is said to be *antisymmetric* if

$$\mathbf{A} = -\mathbf{A}^T.$$

Exercise 3.11. Give the matrix forms corresponding to the cases of symmetric and antisymmetric tensors. How many components can be independently specified for a symmetric tensor? For an antisymmetric tensor?

Both symmetric and antisymmetric tensors arise naturally in the physical sciences. Their significance is also shown by the following

Theorem 3.1. *Any second-order tensor can be decomposed as a sum of symmetric and antisymmetric tensors:*

$$\mathbf{A} = \mathbf{B} + \mathbf{C}$$

where $\mathbf{B} = \mathbf{B}^T$ *and* $\mathbf{C} = -\mathbf{C}^T$.

Proof. Take

$$\mathbf{B} = \frac{1}{2}\left(\mathbf{A} + \mathbf{A}^T\right), \qquad \mathbf{C} = \frac{1}{2}\left(\mathbf{A} - \mathbf{A}^T\right),$$

and check all the statements. □

The dyad $\mathbf{ab}$ can be decomposed into symmetric and antisymmetric parts as

$$\mathbf{ab} = \frac{1}{2}(\mathbf{ab} + \mathbf{ba}) + \frac{1}{2}(\mathbf{ab} - \mathbf{ba})$$

for example.

Exercise 3.12. Show that if $\mathbf{A}$ is symmetric and $\mathbf{B}$ is antisymmetric then $\mathbf{A} \cdot\cdot \mathbf{B} = 0$.

Exercise 3.13. Demonstrate that the quadratic form $\mathbf{x} \cdot \mathbf{A} \cdot \mathbf{x}$ does not change if the second-order tensor $\mathbf{A}$ is replaced by its symmetric part.

Given an antisymmetric tensor $\mathbf{C} = c^{ij}\mathbf{i}_i\mathbf{i}_j$ in a Cartesian frame, we can construct a vector

$$\boldsymbol{\omega} = \omega^k\mathbf{i}_k$$

according to the formulas

$$\omega^1 = c^{32}, \qquad \omega^2 = c^{13}, \qquad \omega^3 = c^{21}.$$

It is easy to verify directly that

$$\mathbf{C} \cdot \mathbf{x} = \boldsymbol{\omega} \times \mathbf{x}, \qquad \mathbf{x} \cdot \mathbf{C} = \mathbf{x} \times \boldsymbol{\omega},$$

where $\mathbf{x}$ is an arbitrary vector. These formulas are written in non-coordinate form so they hold in any frame. The reader can derive the formulas for $\boldsymbol{\omega}$, which is called the *conjugate vector*, for an arbitrary frame.

The cross-products of a tensor $\mathbf{A} = a^{ij}\mathbf{e}_i\mathbf{e}_j$ and a vector $\mathbf{x}$ are defined by the formulas

$$\mathbf{A} \times \mathbf{x} = a^{ij}\mathbf{e}_i(\mathbf{e}_j \times \mathbf{x}), \qquad \mathbf{x} \times \mathbf{A} = a^{ij}(\mathbf{x} \times \mathbf{e}_i)\mathbf{e}_j.$$

Exercise 3.14. Show that $\mathbf{C} = \mathbf{E} \times \boldsymbol{\omega} = \boldsymbol{\omega} \times \mathbf{E}$.

3.6 Eigenvalues and Eigenvectors of a Second-Order Symmetric Tensor

We now consider the question of which basis yields a tensor of simplest form. As the analogous question in matrix theory relates to eigenvalues and eigenvectors, we extend these notions to tensors. The pair

$$(\lambda, \mathbf{x}) \qquad (\mathbf{x} \neq \mathbf{0})$$

is called an *eigenpair* if the equality

$$\mathbf{A} \cdot \mathbf{x} = \lambda \mathbf{x} \tag{3.17}$$

holds. Hence $\mathbf{x}$ is an eigenvector of $\mathbf{A}$ if $\mathbf{A}$ operates on $\mathbf{x}$ to give a vector proportional to $\mathbf{x}$. Equation (3.17) may also be written in the form

$$(\mathbf{A} - \lambda\mathbf{E}) \cdot \mathbf{x} = \mathbf{0}.$$

Exercise 3.15. Find the eigenpairs of the dyad $\mathbf{ab}$. Now try to position an eigenvector on the left: $\mathbf{x} \cdot \mathbf{ab} = \lambda\mathbf{x}$. You should find that this differs from the previous eigenvector, so it makes sense to introduce left and right eigenvectors. In the case of a symmetric tensor they coincide.

The eigenvalues of a second-order tensor $\mathbf{A}$ are found as solutions of the characteristic equation for $\mathbf{A}$, which is derived as follows. In components (3.17) becomes

$$a^{ij}\mathbf{e}_i\mathbf{e}_j \cdot x^k\mathbf{e}_k = \lambda x^i\mathbf{e}_i$$

or

$$a^{ij}g_{jk}x^k\mathbf{e}_i = \lambda x^k\delta_k^i\mathbf{e}_i.$$

Writing this as

$$(a^i_{.k} - \lambda\delta_k^i)x^k = 0,$$

we have a system of three simultaneous equations in the three variables x^k. A nontrivial solution exists if and only if the determinant of the coefficient matrix vanishes:

$$\begin{vmatrix} a^1_{.1} - \lambda & a^1_{.2} & a^1_{.3} \\ a^2_{.1} & a^2_{.2} - \lambda & a^2_{.3} \\ a^3_{.1} & a^3_{.2} & a^3_{.3} - \lambda \end{vmatrix} = 0.$$

This is the characteristic equation[1] for $\mathbf{A}$. Writing it in the form

$$-\lambda^3 + I_1(\mathbf{A})\lambda^2 - I_2(\mathbf{A})\lambda + I_3(\mathbf{A}) = 0 \qquad (3.18)$$

we note that it is cubic in λ, hence there are at most three distinct eigenvalues λ_1, λ_2, λ_3. The coefficients $I_1(\mathbf{A})$, $I_2(\mathbf{A})$, and $I_3(\mathbf{A})$ are called the *first, second, and third invariants* of $\mathbf{A}$, and are expressed in terms of the eigenvalues by the *Viète formulas*

$$I_1(\mathbf{A}) = \lambda_1 + \lambda_2 + \lambda_3,$$
$$I_2(\mathbf{A}) = \lambda_1\lambda_2 + \lambda_1\lambda_3 + \lambda_2\lambda_3,$$
$$I_3(\mathbf{A}) = \lambda_1\lambda_2\lambda_3.$$

After representing the tensor in diagonal form it will be easy to see that $I_1(\mathbf{A})$ and $I_3(\mathbf{A})$ are, respectively, the trace and determinant of the tensor $\mathbf{A}$. In fact,

$$I_1(\mathbf{A}) = \text{tr}\,\mathbf{A}, \quad I_2(\mathbf{A}) = \frac{1}{2}[\text{tr}^2\,\mathbf{A} - \text{tr}\,\mathbf{A}^2], \quad I_3(\mathbf{A}) = \det \mathbf{A}.$$

In nonlinear elasticity, the invariants and eigenvalues of several tensors play important roles in the formulation of various constitutive laws. See, for example, [Lurie (1990); Lurie (2005); Ogden (1997)].

Exercise 3.16. A tensor $\mathbf{A}$, when referred to a certain Cartesian basis, has matrix

$$\begin{pmatrix} 1 & 0 & 1 \\ 2 & -1 & 0 \\ 0 & 1 & 2 \end{pmatrix}.$$

Find the first, second, and third principal invariants of $\mathbf{A}$.

---

[1] Note that it is expressed in terms of mixed components of the tensor. However, it characterizes the properties of the tensor and has invariant properties since the eigenvalues of a tensor do not depend on the coordinate frame in which they are obtained.

In applications, the most important second-order tensors are the real-valued symmetric tensors. These have special properties. For a real-valued tensor that is considered as an operator in the complex space $\mathbb{C}^3$, we have a formula analogous to (3.16):

$$(\mathbf{A} \cdot \mathbf{x}) \cdot \overline{\mathbf{y}} = \mathbf{x} \cdot \overline{(\mathbf{A} \cdot \mathbf{y})}. \tag{3.19}$$

This will be used below. We recall that a bar over an expression denotes complex conjugation.

Theorem 3.2. *The eigenvalues of a real symmetric tensor are real. Moreover, eigenvectors corresponding to distinct eigenvalues are orthogonal.*

Proof. Let $\mathbf{A}$ be a real symmetric tensor and λ an eigenvalue of $\mathbf{A}$ that corresponds to the eigenvector $\mathbf{x} \neq \mathbf{0}$, so

$$\mathbf{A} \cdot \mathbf{x} = \lambda \mathbf{x}.$$

Dot-multiply both sides of this equality by $\overline{\mathbf{x}}$. It follows that

$$\lambda = \frac{(\mathbf{A} \cdot \mathbf{x}) \cdot \overline{\mathbf{x}}}{\mathbf{x} \cdot \overline{\mathbf{x}}}.$$

Now we prove that λ is real. Indeed, $\mathbf{x} \cdot \overline{\mathbf{x}} = |\mathbf{x}|^2$ is positive. To see that $(\mathbf{A} \cdot \mathbf{x}) \cdot \overline{\mathbf{x}}$ takes a real value, we write out

$$(\mathbf{A} \cdot \mathbf{x}) \cdot \overline{\mathbf{x}} = \overline{\mathbf{x} \cdot \overline{(\mathbf{A} \cdot \mathbf{x})}}$$
$$= \overline{(\mathbf{A} \cdot \mathbf{x}) \cdot \overline{\mathbf{x}}}$$

(we have used a property of the inner product in a complex linear space, and then (3.19)). Because the eigenvalues λ are real, the components of the eigenvectors satisfy a linear system of simultaneous equations having real coefficients; hence they are real as well.

To prove the second part of the theorem, suppose that

$$\mathbf{A} \cdot \mathbf{x}_1 = \lambda_1 \mathbf{x}_1, \qquad \mathbf{A} \cdot \mathbf{x}_2 = \lambda_2 \mathbf{x}_2,$$

where $\lambda_2 \neq \lambda_1$. From these we obtain

$$\lambda_1 \mathbf{x}_1 \cdot \mathbf{x}_2 = (\mathbf{A} \cdot \mathbf{x}_1) \cdot \mathbf{x}_2, \qquad \lambda_2 \mathbf{x}_2 \cdot \mathbf{x}_1 = (\mathbf{A} \cdot \mathbf{x}_2) \cdot \mathbf{x}_1,$$

and subtraction gives

$$(\lambda_1 - \lambda_2) \mathbf{x}_1 \cdot \mathbf{x}_2 = \mathbf{x}_2 \cdot \mathbf{A} \cdot \mathbf{x}_1 - \mathbf{x}_1 \cdot \mathbf{A} \cdot \mathbf{x}_2 = 0.$$

It follows that

$$\mathbf{x}_1 \cdot \mathbf{x}_2 = 0. \tag{3.20}$$

From (3.20) we may obtain another property of the eigenvectors $\mathbf{x}_1, \mathbf{x}_2$, known as *generalized orthogonality*:

$$\mathbf{x}_1 \cdot \mathbf{A} \cdot \mathbf{x}_2 = 0.$$

This holds for the eigenvectors corresponding to different eigenvalues of a real symmetric tensor $\mathbf{A}$, and is useful in applications. $\qquad\square$

Exercise 3.17. Show that for any second-order tensor $\mathbf{A}$ (not necessarily symmetric), eigenvectors corresponding to distinct eigenvalues are linearly independent.

In solving the characteristic equation for λ, we may find that there are three distinct solutions or fewer than three. Note that if $\mathbf{x}$ is an eigenvector of $\mathbf{A}$ then so is $\alpha\mathbf{x}$ for any $\alpha \neq 0$. In other words, an eigenvector is determined up to a constant multiple. So when the eigenvalues are distinct we can compose a Cartesian frame from the orthonormal eigenvectors $\mathbf{x}_k$ and then express the tensor $\mathbf{A}$ in terms of its components a_{ij} as

$$\mathbf{A} = \sum a_{ij}\mathbf{x}_i\mathbf{x}_j.$$

Since the frame $\mathbf{x}_k$ is Cartesian the reciprocal basis is the same, and we may calculate the components of $\mathbf{A}$ from

$$a_{ij} = \mathbf{x}_i \cdot \mathbf{A} \cdot \mathbf{x}_j.$$

Since the $\mathbf{x}_i$ are also eigenvectors we can write

$$\mathbf{A} \cdot \mathbf{x}_j = \lambda_j\mathbf{x}_j,$$

and dot multiplication by $\mathbf{x}_i$ from the left gives

$$\mathbf{x}_i \cdot \mathbf{A} \cdot \mathbf{x}_j = \mathbf{x}_i \cdot \lambda_j\mathbf{x}_j = \lambda_j\delta_i^j.$$

Hence the coefficients of the dyads $\mathbf{x}_i\mathbf{x}_j$ in $\mathbf{A}$ are nonzero only for those coefficients that lie on the main diagonal of the matrix representation of $\mathbf{A}$; moreover, these diagonal entries are the eigenvalues of $\mathbf{A}$. We can therefore write

$$\mathbf{A} = \sum_{i=1}^{3} \lambda_i\mathbf{x}_i\mathbf{x}_i. \tag{3.21}$$

This is called the *orthogonal representation* of $\mathbf{A}$. The eigenvectors composing the coordinate frame give us the *principal axes* of $\mathbf{A}$, and the process of referring the tensor to its principal axes is known as *diagonalization*.

When the characteristic equation of a tensor has fewer than three distinct solutions for λ, then the repeated eigenvalue is said to be *degenerate*. If $\mathbf{A}$ is symmetric, we may represent $\mathbf{A}$ in a Cartesian basis and apply facts from the theory of symmetric matrices. Corresponding to a multiple root of the characteristic equation we have a subspace of eigenvectors. If $\lambda_1 = \lambda_2$ is a double root, then the subspace is two-dimensional and we can select an orthonormal pair that is orthogonal to the third eigenvector (since λ_1 and λ_3 are distinct). Such an eigenvalue corresponding to two linearly independent eigenvectors is regarded as a *multiple eigenvalue* (multiplicity two). If $\lambda_1 = \lambda_2 = \lambda_3$ then any vector is an eigenvector, and choosing a Cartesian frame we would have an eigenvalue of multiplicity three. However, this case arises only when the tensor under consideration is proportional to the unit tensor $\mathbf{E}$. Such a tensor is called a *ball tensor*.

Exercise 3.18. Show directly that a second-order tensor (not necessarily symmetric) having three distinct eigenvalues cannot have more than one linearly independent eigenvector corresponding to each eigenvalue.

3.7 The Cayley–Hamilton Theorem

The *Cayley–Hamilton theorem* states that every square matrix satisfies its own characteristic equation. For example the 2×2 matrix

$$A = \begin{pmatrix} a & b \\ c & d \end{pmatrix}$$

has characteristic equation

$$\begin{vmatrix} a - \lambda & b \\ c & d - \lambda \end{vmatrix} = \lambda^2 - (a+d)\lambda + (ad - bc) = 0,$$

and the Cayley–Hamilton theorem tells us that A itself satisfies

$$A^2 - (a+d)A + (ad - bc)I = 0,$$

where I is the 2×2 identity matrix and the zero on the right side denotes the 2×2 zero matrix. Similarly, the Cayley–Hamilton theorem for a second-order tensor $\mathbf{A}$ whose characteristic equation is given by (3.18) states that $\mathbf{A}$ satisfies the equation

$$-\mathbf{A}^3 + I_1(\mathbf{A})\mathbf{A}^2 - I_2(\mathbf{A})\mathbf{A} + I_3(\mathbf{A})\mathbf{E} = \mathbf{0}. \tag{3.22}$$

This permits us to represent $\mathbf{A}^3$ in terms of lower powers of $\mathbf{A}$. Furthermore, we may dot multiply (3.22) by $\mathbf{A}$ and thereby represent $\mathbf{A}^4$ in terms of lower powers of **A**. It is clear that we could continue in this fashion and eventually express any desired power of **A** in terms of **E**, **A**, and $\mathbf{A}^2$. This is useful in certain applications (e.g., nonlinear elasticity) where functions of tensors are represented approximately by truncated Taylor series.

It is easy to establish the Cayley–Hamilton theorem for the case of a symmetric tensor. Such a tensor **A** has the representation (3.21), where we redenote $\mathbf{i}_i = \mathbf{x}_i$ because the eigenvectors $\mathbf{x}_i$ constitute an orthonormal basis:

$$\mathbf{A} = \sum_{i=1}^{3} \lambda_i \mathbf{i}_i \mathbf{i}_i.$$

We get

$$\mathbf{A}^2 = \mathbf{A} \cdot \mathbf{A} = \sum_{i=1}^{3} \lambda_i \mathbf{i}_i \mathbf{i}_i \cdot \sum_{j=1}^{3} \lambda_j \mathbf{i}_j \mathbf{i}_j = \sum_{i,j=1}^{3} \lambda_i \mathbf{i}_i \lambda_j \mathbf{i}_j \delta_j^i = \sum_{i=1}^{3} \lambda_i^2 \mathbf{i}_i \mathbf{i}_i.$$

Similarly

$$\mathbf{A}^3 = \mathbf{A} \cdot \mathbf{A} \cdot \mathbf{A} = \sum_{i=1}^{3} \lambda_i^3 \mathbf{i}_i \mathbf{i}_i. \tag{3.23}$$

Let us return to equation (3.18) written for the ith eigenvalue and put it into the expression (3.23):

$$\mathbf{A}^3 = \sum_{i=1}^{3} [I_1(\mathbf{A})\lambda_i^2 - I_2(\mathbf{A})\lambda_i + I_3(\mathbf{A})] \mathbf{i}_i \mathbf{i}_i$$

$$= I_1(\mathbf{A}) \sum_{i=1}^{3} \lambda_i^2 \mathbf{i}_i \mathbf{i}_i - I_2(\mathbf{A}) \sum_{i=1}^{3} \lambda_i \mathbf{i}_i \mathbf{i}_i + I_3(\mathbf{A}) \sum_{i=1}^{3} \mathbf{i}_i \mathbf{i}_i$$

$$= I_1(\mathbf{A})\mathbf{A}^2 - I_2(\mathbf{A})\mathbf{A} + I_3(\mathbf{A})\mathbf{E},$$

as desired.

Exercise 3.19. Use the Cayley–Hamilton theorem to express $\mathbf{A}^3$ in terms of $\mathbf{A}^2$, **A**, and **E** if $\mathbf{A} = \mathbf{i}_1 \mathbf{i}_1 + \mathbf{i}_2 \mathbf{i}_1 + \mathbf{i}_2 \mathbf{i}_2 + \mathbf{i}_3 \mathbf{i}_2$.

3.8 Other Properties of Second-Order Tensors

Tensors of rotation

A tensor **Q** of order two is said to be *orthogonal* if it satisfies the equality

$$\mathbf{Q} \cdot \mathbf{Q}^T = \mathbf{Q}^T \cdot \mathbf{Q} = \mathbf{E}.$$

We see that

$$\mathbf{Q}^T = \mathbf{Q}^{-1}$$

for an orthogonal tensor. Furthermore,

$$\det \mathbf{Q} = \pm 1.$$

Indeed, $\det \mathbf{Q}$ is determined by the determinant of the matrix of mixed components of $\mathbf{Q}$. Because of the properties of the determinant of a matrix and the correspondence between tensors and matrices, we have for two tensors $\mathbf{A}$ and $\mathbf{B}$ of order two

$$\det(\mathbf{A} \cdot \mathbf{B}) = \det \mathbf{A} \det \mathbf{B}$$

and

$$\det \mathbf{A} = \det \mathbf{A}^T.$$

Thus

$$1 = \det \mathbf{E} = \det \mathbf{Q} \det \mathbf{Q}^T = (\det \mathbf{Q})^2$$

as desired. We call $\mathbf{Q}$ a *proper* orthogonal tensor if $\det \mathbf{Q} = +1$; we call $\mathbf{Q}$ an *improper* orthogonal tensor if $\det \mathbf{Q} = -1$.

Exercise 3.20. (a) Show that the tensor $\mathbf{Q} = -\mathbf{i}_1\mathbf{i}_1 + \mathbf{i}_2\mathbf{i}_2 + \mathbf{i}_3\mathbf{i}_3$ is orthogonal. Is it proper or improper? (b) Show that

$$q_{ij}q^{kj} = \delta_i^k$$

if $\mathbf{Q}$ is orthogonal. (c) Show that if $\mathbf{Q}$ is orthogonal then so is $\mathbf{Q}^n$ for every integer n.

We now consider the orthogonal tensor $\mathbf{Q}$ as an operator in the space of all vectors.

Theorem 3.3. *The operator defined by the orthogonal tensor $\mathbf{Q}$ preserves the magnitudes of vectors and the angles between them.*

Proof. Consider the result of application of $\mathbf{Q}$ to both the multipliers of the inner product $\mathbf{x} \cdot \mathbf{y}$:

$$(\mathbf{Q} \cdot \mathbf{x}) \cdot (\mathbf{Q} \cdot \mathbf{y}) = (\mathbf{x} \cdot \mathbf{Q}^T) \cdot (\mathbf{Q} \cdot \mathbf{y}) = \mathbf{x} \cdot \mathbf{Q}^T \cdot \mathbf{Q} \cdot \mathbf{y} = \mathbf{x} \cdot \mathbf{E} \cdot \mathbf{y} = \mathbf{x} \cdot \mathbf{y}.$$

First we put $\mathbf{y} = \mathbf{x}$ to see that $\mathbf{Q}$ preserves vector magnitudes: $|\mathbf{Q} \cdot \mathbf{x}|^2 = |\mathbf{x}|^2$. Thus, by definition of the dot product in terms of the cosine, angles are also preserved. $\square$

The action of $\mathbf{Q}$ amounts to a rotation of all vectors of the space. (More precisely, this is the case for a proper orthogonal tensor; an improper tensor also causes an axis reflection that changes the "handedness" of the frame.) The situation is analogous to the case of a solid body where the position of a point is defined by a vector beginning at a fixed origin of some frame and ending at the point. Any motion of the solid with a fixed point is a rotation with respect to some axis by some angle. The equations of physics should often be introduced in such a way that they are invariant under rotation of the coordinate frame whose position is not determined in space. Such invariance under rotation should be verified, and in large part this can be done by showing that the application of $\mathbf{Q}$ to all the vectors of the relation does not change the form of the relation.

Let us note that some quantities are always invariant under rotation. One of them is the first invariant $I_1(\mathbf{A})$ of a tensor. This quantity, also called the *trace* of the tensor and denoted $\operatorname{tr}(\mathbf{A})$, is the sum of the diagonal mixed components of $\mathbf{A}$:

$$\operatorname{tr}(\mathbf{A}) = a_i^{\cdot i}.$$

The trace can be equivalently determined in the non-coordinate form

$$\operatorname{tr}(\mathbf{A}) = \mathbf{E} \cdots \mathbf{A} = \mathbf{E} \bullet \mathbf{A}$$

(the reader should check this). To show invariance we consider the tensor

$$\begin{aligned}
\mathbf{Q} \cdot \mathbf{A} \cdot \mathbf{Q}^T &= \mathbf{Q} \cdot (a^{ij} \mathbf{e}_i \mathbf{e}_j) \cdot \mathbf{Q}^T \\
&= a^{ij} (\mathbf{Q} \cdot \mathbf{e}_i)(\mathbf{e}_j \cdot \mathbf{Q}^T) \\
&= a^{ij} (\mathbf{Q} \cdot \mathbf{e}_i)(\mathbf{Q} \cdot \mathbf{e}_j).
\end{aligned}$$

We see that this is a representation of the "rotated" tensor, derived as the result of applying $\mathbf{Q}$ to each vector of the dyadic components of $\mathbf{A}$. The trace of the rotated tensor is given by

$$\begin{aligned}
\mathbf{E} \cdots (\mathbf{Q} \cdot \mathbf{A} \cdot \mathbf{Q}^T) &= \mathbf{E} \cdots (\mathbf{Q} \cdot a_i^{\cdot j} \mathbf{e}^i \mathbf{e}_j \cdot \mathbf{Q}^T) \\
&= \mathbf{E} \cdots a_i^{\cdot j} (\mathbf{Q} \cdot \mathbf{e}^i)(\mathbf{Q} \cdot \mathbf{e}_j).
\end{aligned}$$

Under the action of $\mathbf{Q}$ the frame $\mathbf{e}_i$ transforms to the frame $\mathbf{Q} \cdot \mathbf{e}_i$, to which the reciprocal basis is $\mathbf{Q} \cdot \mathbf{e}^i$. This means that the unit tensor (the metric tensor!) can be represented, in particular, as

$$\mathbf{E} = (\mathbf{Q} \cdot \mathbf{e}_i)(\mathbf{Q} \cdot \mathbf{e}^i) = (\mathbf{Q} \cdot \mathbf{e}^j)(\mathbf{Q} \cdot \mathbf{e}_j).$$

It follows that

$$
\begin{aligned}
\mathbf{E} \cdot\cdot (\mathbf{Q} \cdot \mathbf{A} \cdot \mathbf{Q}^T) &= \left[(\mathbf{Q} \cdot \mathbf{e}^k)(\mathbf{Q} \cdot \mathbf{e}_k) \right] \cdot\cdot\, a_i^{\cdot j} (\mathbf{Q} \cdot \mathbf{e}^i)(\mathbf{Q} \cdot \mathbf{e}_j) \\
&= a_i^{\cdot j} \left[(\mathbf{Q} \cdot \mathbf{e}_k) \cdot (\mathbf{Q} \cdot \mathbf{e}^i) \right] \left[(\mathbf{Q} \cdot \mathbf{e}^k) \cdot (\mathbf{Q} \cdot \mathbf{e}_j) \right] \\
&= a_i^{\cdot j} \left[(\mathbf{e}_k \cdot \mathbf{Q}^T) \cdot (\mathbf{Q} \cdot \mathbf{e}^i) \right] \left[(\mathbf{e}^k \cdot \mathbf{Q}^T) \cdot (\mathbf{Q} \cdot \mathbf{e}_j) \right] \\
&= a_i^{\cdot j} \left[\mathbf{e}_k \cdot \mathbf{Q}^T \cdot \mathbf{Q} \cdot \mathbf{e}^i \right] \left[\mathbf{e}^k \cdot \mathbf{Q}^T \cdot \mathbf{Q} \cdot \mathbf{e}_j \right] \\
&= a_i^{\cdot j} \left[\mathbf{e}_k \cdot \mathbf{E} \cdot \mathbf{e}^i \right] \left[\mathbf{e}^k \cdot \mathbf{E} \cdot \mathbf{e}_j \right] \\
&= a_i^{\cdot j} \left[\mathbf{e}_k \cdot \mathbf{e}^i \right] \left[\mathbf{e}^k \cdot \mathbf{e}_j \right] \\
&= a_i^{\cdot j} \delta_k^i \delta_j^k \\
&= a_k^{\cdot k} \\
&= \operatorname{tr}(\mathbf{A}).
\end{aligned}
$$

Another example demonstrates that under the transformation $\mathbf{Q}$ the eigenvalues of a tensor $\mathbf{A}$ remain the same but the eigenvectors $\mathbf{x}_i$ rotate. Indeed, let

$$
\mathbf{A} \cdot \mathbf{x}_i = \lambda_i \mathbf{x}_i.
$$

We show that $\mathbf{Q} \cdot \mathbf{x}_i$ is an eigenvector of $\mathbf{Q} \cdot \mathbf{A} \cdot \mathbf{Q}^T$:

$$
\begin{aligned}
(\mathbf{Q} \cdot \mathbf{A} \cdot \mathbf{Q}^T) \cdot (\mathbf{Q} \cdot \mathbf{x}_i) &= \mathbf{Q} \cdot \mathbf{A} \cdot \mathbf{Q}^T \cdot \mathbf{Q} \cdot \mathbf{x}_i \\
&= \mathbf{Q} \cdot \mathbf{A} \cdot \mathbf{E} \cdot \mathbf{x}_i \\
&= \mathbf{Q} \cdot (\mathbf{A} \cdot \mathbf{x}_i) \\
&= \mathbf{Q} \cdot (\lambda_i \mathbf{x}_i) \\
&= \lambda_i (\mathbf{Q} \cdot \mathbf{x}_i).
\end{aligned}
$$

Let $\mathbf{e}$ be an axis of rotation defined by an orthogonal tensor $\mathbf{Q}$, and let ω be the angle of rotation about $\mathbf{e}$. It can be shown that the proper orthogonal tensor has the representation

$$
\mathbf{Q} = \mathbf{E} \cos \omega + (1 - \cos \omega) \mathbf{e}\mathbf{e} - \mathbf{e} \times \mathbf{E} \sin \omega.
$$

Polar decomposition

A second-order tensor $\mathbf{A}$ is nonsingular if

$$
\det A \neq 0
$$

where A is the matrix of mixed components of $\mathbf{A}$. It is possible to express such a tensor as a product of a symmetric tensor and another tensor. A statement of this result, known as the *polar decomposition theorem*, requires that we introduce some additional terminology.

A symmetric tensor is said to be *positive definite* if its eigenvalues are all positive. By orthogonal decomposition such a tensor $\mathbf{S}$ may be written in the form

$$\mathbf{S} = \lambda_1 \mathbf{i}_1 \mathbf{i}_1 + \lambda_2 \mathbf{i}_2 \mathbf{i}_2 + \lambda_3 \mathbf{i}_3 \mathbf{i}_3,$$

where all the $\lambda_i > 0$ and the eigenvectors $\mathbf{i}_k$ constitute an orthonormal basis.

If $\mathbf{A}$ is nonsingular then the tensor $\mathbf{A} \cdot \mathbf{A}^T$ is symmetric and positive definite. Symmetry follows from the equation

$$(\mathbf{A} \cdot \mathbf{A}^T)^T = (\mathbf{A}^T)^T \cdot \mathbf{A}^T = \mathbf{A} \cdot \mathbf{A}^T.$$

To see positive definiteness, we begin with the definition

$$(\mathbf{A} \cdot \mathbf{A}^T) \cdot \mathbf{x} = \lambda \mathbf{x}$$

of an eigenvalue λ and dot with $\mathbf{x}$ from the left to get

$$\lambda = \frac{\mathbf{x} \cdot (\mathbf{A} \cdot \mathbf{A}^T) \cdot \mathbf{x}}{|\mathbf{x}|^2}.$$

The numerator is positive because

$$\mathbf{x} \cdot (\mathbf{A} \cdot \mathbf{A}^T) \cdot \mathbf{x} = (\mathbf{x} \cdot \mathbf{A}) \cdot (\mathbf{A}^T \cdot \mathbf{x}) = (\mathbf{x} \cdot \mathbf{A}) \cdot (\mathbf{x} \cdot \mathbf{A}) = |\mathbf{x} \cdot \mathbf{A}|^2.$$

Hence all the eigenvalues λ of $\mathbf{A} \cdot \mathbf{A}^T$ are positive. The diagonalization of $\mathbf{A} \cdot \mathbf{A}^T$ now shows that it is positive definite.

With these facts in hand we may turn to our main result.

Theorem 3.4. *Any nonsingular tensor* $\mathbf{A}$ *of order two may be written as a product of an orthogonal tensor and a positive definite symmetric tensor. The decomposition may be done in two ways: as a* **left polar decomposition**

$$\mathbf{A} = \mathbf{S} \cdot \mathbf{Q} \tag{3.24}$$

or as a **right polar decomposition**

$$\mathbf{A} = \mathbf{Q} \cdot \mathbf{S}'. \tag{3.25}$$

Here $\mathbf{Q}$ *is an orthogonal tensor of order two, and* $\mathbf{S}$ *and* $\mathbf{S}'$ *are positive definite and symmetric.*

Proof. Because $\mathbf{A} \cdot \mathbf{A}^T$ is positive definite and symmetric, we have

$$\mathbf{A} \cdot \mathbf{A}^T = \lambda_1 \mathbf{i}_1 \mathbf{i}_1 + \lambda_2 \mathbf{i}_2 \mathbf{i}_2 + \lambda_3 \mathbf{i}_3 \mathbf{i}_3,$$

where the $\mathbf{i}_k$ are orthonormal. We define

$$\mathbf{S} \equiv \left(\mathbf{A} \cdot \mathbf{A}^T\right)^{1/2} = \sqrt{\lambda_1} \mathbf{i}_1 \mathbf{i}_1 + \sqrt{\lambda_2} \mathbf{i}_2 \mathbf{i}_2 + \sqrt{\lambda_3} \mathbf{i}_3 \mathbf{i}_3$$

since the λ_i are positive. We see that $\mathbf{S}^{-1}$ exists and is equal to

$$\mathbf{S}^{-1} = \frac{1}{\sqrt{\lambda_1}} \mathbf{i}_1 \mathbf{i}_1 + \frac{1}{\sqrt{\lambda_2}} \mathbf{i}_2 \mathbf{i}_2 + \frac{1}{\sqrt{\lambda_3}} \mathbf{i}_3 \mathbf{i}_3.$$

Now we set

$$\mathbf{Q} \equiv \mathbf{S}^{-1} \cdot \mathbf{A}.$$

To see that $\mathbf{Q}$ is orthogonal, we write

$$
\begin{aligned}
\mathbf{Q} \cdot \mathbf{Q}^T &= (\mathbf{S}^{-1} \cdot \mathbf{A}) \cdot (\mathbf{S}^{-1} \cdot \mathbf{A})^T \\
&= (\mathbf{S}^{-1} \cdot \mathbf{A}) \cdot (\mathbf{A}^T \cdot (\mathbf{S}^{-1})^T) \\
&= \mathbf{S}^{-1} \cdot (\mathbf{A} \cdot \mathbf{A}^T) \cdot (\mathbf{S}^T)^{-1} \\
&= \mathbf{S}^{-1} \cdot \mathbf{S}^2 \cdot \mathbf{S}^{-1} \\
&= \mathbf{E}.
\end{aligned}
$$

Thus we have expressed $\mathbf{A} = \mathbf{S} \cdot \mathbf{Q}$ as in (3.24). The validity of (3.25) follows from defining $\mathbf{S}' \equiv \mathbf{Q}^T \cdot \mathbf{S} \cdot \mathbf{Q}$. $\qquad\square$

Exercise 3.21. We have called a tensor positive definite if its eigenvalues are all positive. An alternative definition is that $\mathbf{A}$ is positive definite if $\mathbf{x} \cdot (\mathbf{A} \cdot \mathbf{x}) > 0$ for all $\mathbf{x} \neq 0$. Explain.

As an application of polar decomposition, let us show that a nonsingular tensor $\mathbf{A}$ operates on the position vectors of the points on the unit sphere to produce position vectors defining the points of an ellipsoid. If $\mathbf{r}$ locates any point on the unit sphere then

$$\mathbf{r} \cdot \mathbf{r} = 1. \tag{3.26}$$

Let $\mathbf{x}$ be the image of $\mathbf{r}$ under $\mathbf{A}$:

$$\mathbf{x} = \mathbf{A} \cdot \mathbf{r}.$$

Then $\mathbf{r} = \mathbf{A}^{-1} \cdot \mathbf{x}$, and substitution into (3.26) along with (3.24) gives

$$
\begin{aligned}
1 &= [(\mathbf{S} \cdot \mathbf{Q})^{-1} \cdot \mathbf{x}] \cdot [(\mathbf{S} \cdot \mathbf{Q})^{-1} \cdot \mathbf{x}] \\
&= \mathbf{x} \cdot \{[(\mathbf{S} \cdot \mathbf{Q})^{-1}]^T \cdot (\mathbf{S} \cdot \mathbf{Q})^{-1}\} \cdot \mathbf{x} \\
&= \mathbf{x} \cdot \{\mathbf{S}^{-1} \cdot \mathbf{S}^{-1}\} \cdot \mathbf{x}
\end{aligned}
$$

(the reader can supply the missing details). Because

$$
\mathbf{S} = \sqrt{\lambda_1}\mathbf{i}_1\mathbf{i}_1 + \sqrt{\lambda_2}\mathbf{i}_2\mathbf{i}_2 + \sqrt{\lambda_3}\mathbf{i}_3\mathbf{i}_3,
$$

expansion in the Cartesian frame $\mathbf{i}_k$ with use of

$$
\mathbf{x} = \sum_{i=1}^{3} x_i \mathbf{i}_i, \qquad \mathbf{S}^{-1} = \sum_{j=1}^{3} \frac{1}{\sqrt{\lambda_j}} \mathbf{i}_j \mathbf{i}_j
$$

reduces the above equation $\mathbf{x} \cdot \{\mathbf{S}^{-1} \cdot \mathbf{S}^{-1}\} \cdot \mathbf{x} = 1$ to the form

$$
\sum_{i=1}^{3} \frac{1}{\lambda_i} x_i^2 = 1.
$$

This is the equation of an ellipsoid.

Polar decomposition provides the background for introducing measures of deformation in nonlinear continuum mechanics.

Deviator and ball tensor representation

For $\mathbf{A}$ we can introduce the representation

$$
\mathbf{A} = \frac{1}{3} I_1(\mathbf{A}) \mathbf{E} + \operatorname{dev} \mathbf{A}.
$$

Such a representation is found useful in the theory of elasticity. Moreover, it is used to formulate constitutive equations in the theories of plasticity, creep, and viscoelasticity.

The tensor $\operatorname{dev} \mathbf{A}$ is defined by the above equality. It has the same eigenvectors as $\mathbf{A}$, but eigenvalues that differ from the eigenvalues of $\mathbf{A}$ by $(1/3)I_1(\mathbf{A})$:

$$
\tilde{\lambda}_i = \lambda_i - \frac{1}{3} \operatorname{tr} \mathbf{A}.
$$

3.9 Extending the Dyad Idea

Third-order tensors can be introduced in a way that parallels the introduction of dyads in §3.1. Using the tensor product as before, we introduce *triad* quantities of the type

$$\mathbf{R} = \mathbf{abc}$$

where $\mathbf{a}$, $\mathbf{b}$, and $\mathbf{c}$ are vectors. Expanding these vectors in terms of a basis $\mathbf{e}_i$ we obtain

$$\mathbf{R} = a^i b^j c^k \mathbf{e}_i \mathbf{e}_j \mathbf{e}_k.$$

We then consider a linear space whose basis is the set of 27 quantities $\mathbf{e}_i \mathbf{e}_j \mathbf{e}_k$ and call it the space of third-order tensors. We continue to refer to the numerical values $a^i b^j c^k$ as the tensor's components. A general element of this space, a tensor $\mathbf{R}$ of order three, has the representation

$$\mathbf{R} = r^{ijk} \mathbf{e}_i \mathbf{e}_j \mathbf{e}_k. \tag{3.27}$$

The property of objectivity of $\mathbf{R}$ remains paramount, leading to the requirement that the components transform appropriately when we change the frame. The now familiar procedure of setting

$$\tilde{r}^{ijk} \tilde{\mathbf{e}}_i \tilde{\mathbf{e}}_j \tilde{\mathbf{e}}_k = r^{mnp} \mathbf{e}_m \mathbf{e}_n \mathbf{e}_p$$

under the change of frame

$$\mathbf{e}_i = A_i^j \tilde{\mathbf{e}}_j$$

gives

$$\tilde{r}^{ijk} = r^{mnp} A_m^i A_n^j A_p^k$$

— a direct extension of (3.8). As an alternative to the representation (3.27) in contravariant components, we could use the covariant-type representation

$$\mathbf{R} = r_{ijk} \mathbf{e}^i \mathbf{e}^j \mathbf{e}^k$$

or either of the mixed representations

$$\mathbf{R} = r^{ij}_{\ \cdot\cdot k} \mathbf{e}_i \mathbf{e}_j \mathbf{e}^k, \qquad \mathbf{R} = r^{i}_{\ \cdot jk} \mathbf{e}_i \mathbf{e}^j \mathbf{e}^k.$$

These necessitate the respective transformation laws

$$\tilde{r}_{ijk} = r_{mnp} \tilde{A}_i^m \tilde{A}_j^n \tilde{A}_k^p$$

and

$$\tilde{r}^{ij}_{\cdot\cdot k} = r^{mn}{}_{\cdot\cdot p} A^i_m A^j_n \tilde{A}^p_k, \qquad \tilde{r}^i{}_{\cdot jk} = r^m{}_{np} A^i_m \tilde{A}^n_j \tilde{A}^p_k,$$

as is easily verified.

Dot products involving triads follow familiar rules. The dot product of a triad with a vector is given by the formula

$$\mathbf{abc} \cdot \mathbf{x} = \mathbf{ab}(\mathbf{c} \cdot \mathbf{x}),$$

while the double dot product of a triad with a dyad is given by

$$\mathbf{abc} \cdot\cdot \mathbf{xy} = \mathbf{a}(\mathbf{c} \cdot \mathbf{x})(\mathbf{b} \cdot \mathbf{y}).$$

One may also define a triple dot product of a triad with another triad:

$$\mathbf{abc} \cdots \mathbf{xyz} = (\mathbf{c} \cdot \mathbf{x})(\mathbf{b} \cdot \mathbf{y})(\mathbf{a} \cdot \mathbf{z}).$$

The scalar product of triads is defined by the rule

$$\mathbf{abc} \bullet \mathbf{xyz} = (\mathbf{a} \cdot \mathbf{x})(\mathbf{b} \cdot \mathbf{y})(\mathbf{c} \cdot \mathbf{z}).$$

The reader has surmised by now that the order of a tensor is always equal to the number of free indices needed to specify its components. A vector, for instance, is a tensor of order one. In Chapter 2 we also met a quantity whose components are specified by three indices: ϵ_{ijk}. This quantity, which arose naturally in our discussion of the vector cross product, is known as the *Levi–Civita tensor* and is given by

$$\mathcal{E} = \epsilon_{ijk}\mathbf{e}^i\mathbf{e}^j\mathbf{e}^k.$$

Exercise 3.22. Verify that $\mathcal{E} = -\mathbf{E} \times \mathbf{E}$.

Exercise 3.23. Verify the following formulas for operations involving the Levi–Civita tensor:

(a) $\mathcal{E} \cdots \mathbf{zyx} = \mathbf{x} \cdot (\mathbf{y} \times \mathbf{z})$,

(b) $\mathcal{E} \cdot\cdot \mathbf{xy} = \mathbf{y} \times \mathbf{x}$,

(c) $\mathcal{E} \cdot \mathbf{x} = -\mathbf{x}\times$.

Note: The notation of (c) may require some explanation. The result of applying the third-order tensor $\mathcal{E}$ to a vector $\mathbf{x}$ is a second-order tensor $\mathcal{E} \cdot \mathbf{x}$. When $\mathcal{E} \cdot \mathbf{x}$ is applied to another vector $\mathbf{y}$, it becomes equivalent to the cross product $\mathbf{y} \times \mathbf{x}$. This is the meaning of the right side of (c). Although the notation is awkward, it is rare that the action of some tensor can be described as the action of two vectors, and the development of a special notation is unwarranted.

3.10　Tensors of the Fourth and Higher Orders

We can obviously extend the present treatment to tensors of any desired order. Let us illustrate the essential points using tensors of order four.

A fourth-order tensor $\mathbf{C}$ can be represented by several types of components:

$$c^{ijkl}, \qquad c_i^{\;jkl}, \qquad c_{ij}^{\;\;kl}, \qquad c_{ijk}^{\;\;\;l}, \qquad c_{ijkl}.$$

The first and last are purely contravariant and purely covariant, respectively, while the other three are mixed with indices in various positions. As before these components represent $\mathbf{C}$ with respect to various bases; for instance,

$$\mathbf{C} = c^{ijkl}\mathbf{e}_i\mathbf{e}_j\mathbf{e}_k\mathbf{e}_l.$$

Dot products with vectors can be taken as before: the rule is that we simply dot multiply the basis vectors positioned nearest to the dot. Carrying out such operations we may obtain results of various kinds. A dot product of a fourth-order tensor with a vector gives, for example,

$$\begin{aligned}
\mathbf{R} &= \mathbf{C} \cdot \mathbf{x} \\
&= c^{ijkl}\mathbf{e}_i\mathbf{e}_j\mathbf{e}_k\mathbf{e}_l \cdot x_m\mathbf{e}^m \\
&= c^{ijkl}\mathbf{e}_i\mathbf{e}_j\mathbf{e}_k\delta_l^m x_m \\
&= c^{ijkl}x_l\mathbf{e}_i\mathbf{e}_j\mathbf{e}_k
\end{aligned}$$

— a tensor of order three. Similarly, a dot product between a third-order tensor and a vector gives a second-order tensor. Wherever the dot product is utilized, it continues to enjoy the linearity properties stated earlier.

Double dot products also appear in applications. For example, in elasticity one encounters double dot products between the tensor of elastic constants and the strain tensor. In generalized form Hooke's law becomes

$$\boldsymbol{\sigma} = \mathbf{C} \cdot\cdot \,\boldsymbol{\varepsilon}$$

where $\boldsymbol{\sigma}$ is the stress tensor, $\mathbf{C}$ is the tensor of elastic constants, and $\boldsymbol{\varepsilon}$ is the strain tensor. The density of the function of internal (elastic) energy in linear elasticity is

$$\frac{1}{2}\boldsymbol{\sigma} \cdot\cdot\, \boldsymbol{\varepsilon} = \frac{1}{2}(\mathbf{C} \cdot\cdot\, \boldsymbol{\varepsilon}) \cdot\cdot\, \boldsymbol{\varepsilon}.$$

Because $\boldsymbol{\sigma}$ and $\boldsymbol{\varepsilon}$ are symmetric, the last expression can be put in a more symmetrical form

$$\frac{1}{2}\boldsymbol{\varepsilon}\cdot\cdot\,\mathbf{C}\cdot\cdot\,\boldsymbol{\varepsilon}$$

in which the result does not depend on the order of operations.

Isotropic tensors

In engineering, isotropic materials play an important role. These are the materials whose properties are the same in all directions. Air, for example, is isotropic: it is equally transparent in all directions. It is impossible to tell whether a ball made of isotropic material has been rotated through some angle. In mechanics, material properties are expressed via constitutive relations. From a mathematical point of view, a material is isotropic when its constitutive equations are invariant with respect to certain transformations: the rotations and mirror reflections.

First we consider the question of when various tensorial quantities can be isotropic. We say that a tensor is *isotropic* if its individual components are invariant under all possible rotations and mirror reflections in $\mathbb{R}^3$.

Any scalar quantity is isotropic. Clearly, the only isotropic vector is $\mathbf{0}$. Let $\mathbf{A}$ be a second-order tensor so that in a basis $\mathbf{e}_1, \mathbf{e}_2, \mathbf{e}_3$ we have

$$\mathbf{A} = a^{ij}\mathbf{e}_i\mathbf{e}_j.$$

Recall that any rotation or mirror reflection of $\mathbb{R}^3$ is uniquely defined by an orthogonal tensor $\mathbf{Q}$. Let us apply $\mathbf{Q}$ to each vector of the basis:

$$\tilde{\mathbf{e}}_k = \mathbf{Q}\cdot\mathbf{e}_k.$$

In the new basis we have

$$\mathbf{A} = \tilde{a}^{ij}\tilde{\mathbf{e}}_i\tilde{\mathbf{e}}_j.$$

Let $\mathbf{A}$ be isotropic. By the above definition we must have $a^{ij} = \tilde{a}^{ij}$. So

$$\mathbf{A} = \tilde{a}^{ij}\tilde{\mathbf{e}}_i\tilde{\mathbf{e}}_j = a^{ij}\mathbf{Q}\cdot\mathbf{e}_i\mathbf{Q}\cdot\mathbf{e}_j = \mathbf{Q}\cdot(a^{ij}\mathbf{e}_i\mathbf{e}_j)\cdot\mathbf{Q}^T.$$

This means that $\mathbf{A}$ is isotropic if and only if the equation

$$\mathbf{A} = \mathbf{Q}\cdot\mathbf{A}\cdot\mathbf{Q}^T$$

holds for any orthogonal tensor $\mathbf{Q}$.

Common sense tells us that the metric tensor $\mathbf{E}$, which is the unit tensor as well, should be isotropic. Let us demonstrate this. For any orthogonal $\mathbf{Q}$ we have

$$\mathbf{Q} \cdot \mathbf{Q}^T = \mathbf{E}.$$

This can be rewritten as

$$\mathbf{E} = \mathbf{Q} \cdot \mathbf{E} \cdot \mathbf{Q}^T.$$

Hence $\mathbf{E}$ is isotropic. If λ is a scalar, then clearly the ball tensor $\lambda\mathbf{E}$ is isotropic as well.

The following is, unfortunately, *not* a trivial exercise. It asserts that any isotropic second-order tensor takes the form $\lambda\mathbf{E}$.

Exercise 3.24. Show that $\mathbf{A}$ is an isotropic tensor of order two if and only if it is a ball tensor: that is, $\mathbf{A} = \lambda\mathbf{E}$ for some scalar λ.

Under an orthogonal transformation $\mathbf{Q}$ of $\mathbb{R}^3$, a fourth-order tensor

$$\mathbf{C} = c^{ijmn}\mathbf{e}_i\mathbf{e}_j\mathbf{e}_m\mathbf{e}_n$$

takes the form

$$\mathbf{C} = \tilde{c}^{ijmn}(\mathbf{Q} \cdot \mathbf{e}_i)(\mathbf{Q} \cdot \mathbf{e}_j)(\mathbf{Q} \cdot \mathbf{e}_m)(\mathbf{Q} \cdot \mathbf{e}_n).$$

By the general definition, it is isotropic if $\tilde{c}^{ijmn} = c^{ijmn}$ for all $\mathbf{Q}$.

In a Cartesian frame, the general form of the fourth-order isotropic tensor is

$$\alpha\mathbf{E}\mathbf{E} + \beta\mathbf{e}_k\mathbf{E}\mathbf{e}^k + \gamma\mathbf{I}, \tag{3.28}$$

where α, β, γ are arbitrary scalars [Jeffreys (1931)]. The proof is cumbersome and we omit it. The properties of the tensors in the representation are exhibited in Exercises 3.27, 3.28, and 3.29. This fact can be applied to the tensor of elastic constants for an isotropic material. The quantities $\mathbf{C}$, $\boldsymbol{\sigma}$, and $\boldsymbol{\varepsilon}$ will be considered further in Chapter 6.

3.11   Functions of Tensorial Arguments

The reader is familiar with the notion of a function $f(x_1, \ldots, x_n)$ in n variables. If we regard x_k as a Cartesian component of a vector

$$\mathbf{x} = (x_1, \ldots, x_n),$$

then we can regard f as a function of the vectorial argument $\mathbf{x}$:

$$f = f(\mathbf{x}).$$

But a logically good definition of such a function dictated by physics should require that f be independent of the representation of $\mathbf{x}$ in a basis.

Similarly, we can consider a function of one or more tensorial arguments. In a fixed basis, such a function reduces to a function in many variables, the components of the tensorial arguments. Again, however, a true function of a tensorial argument cannot depend on the basis representations of its arguments.

To extend this notion further, we can consider functions that take values in the set of vectors — or even tensors. Such functions arise in applications. For example, a force vector $\mathbf{f} = \mathbf{f}(t)$ can be given as a function of time t. Later, we will encounter other functions that take values in the set of tensors of some order. Depending on this latter set, the function may be termed scalar-valued, vector-valued, or tensor-valued.

As for any function in many variables, we can apply the tools of calculus to tensor-valued functions. These include the notion of continuity, the first differential, and derivatives. We will consider these topics later.

Linear functions

In linear elasticity and linear shell theory, linear relations and quadratic functions (such as occur in strain energy expressions) play central roles. We define a linear function of a tensorial variable as a function f which, for any tensors $\mathbf{A}, \mathbf{B}$ and scalars λ, μ, satisfies the relation

$$f(\lambda \mathbf{A} + \mu \mathbf{B}) = \lambda f(\mathbf{A}) + \mu f(\mathbf{B}).$$

This mimics the definition of a linear matrix operator. From this point of view, the equation $y = kx + b$ represents a linear function only if $b = 0$.

Theorem 3.5. *Let f be a scalar-valued function of a vectorial argument $\mathbf{x}$. There is a unique $\mathbf{c}$ such that for all $\mathbf{x}$,*

$$f(\mathbf{x}) = \mathbf{c} \cdot \mathbf{x}. \tag{3.29}$$

Proof. We expand $\mathbf{x} = x^k \mathbf{e}_k$ with respect to the basis $\mathbf{e}_k$. By linearity, $f(\mathbf{x}) = x^k f(\mathbf{e}_k)$. Equation (3.29) holds with $\mathbf{c} = f(\mathbf{e}_k) \mathbf{e}^k$. Supposing the existence of two vectors $\mathbf{c}_1$ and $\mathbf{c}_2$ such that $\mathbf{c}_1 \cdot \mathbf{x} = \mathbf{c}_2 \cdot \mathbf{x}$, and putting $\mathbf{x} = \mathbf{c}_2 - \mathbf{c}_1$, we get $\mathbf{c}_1 = \mathbf{c}_2$. $\qquad\square$

Now let us consider a scalar-valued function whose argument is a second-order tensor $\mathbf{A}$. Clearly, for a second-order tensor $\mathbf{B}$, the function

$$f(\mathbf{A}) = \mathrm{tr}(\mathbf{B} \cdot \mathbf{A}) = \mathbf{B} \cdot\cdot \mathbf{A}^T$$

is linear.

Exercise 3.25. Show that $\mathbf{B} \cdot\cdot \mathbf{A}^T = \mathbf{A} \cdot\cdot \mathbf{B}^T$.

Theorem 3.6. *Let f be a scalar-valued function of a second-order tensor $\mathbf{A}$. There is a unique second-order tensor $\mathbf{B}$ such that for all $\mathbf{A}$,*

$$f(\mathbf{A}) = \mathrm{tr}(\mathbf{B} \cdot \mathbf{A}^T). \tag{3.30}$$

The proof is left to the reader.

The representation of any tensor-valued function of a tensorial argument is similar. We use it to introduce the tensor of elastic constants in linear elasticity.

Theorem 3.7. *Let $\mathbf{F} = \mathbf{F}(\mathbf{A})$ be a linear function from the set of second-order tensors $\mathbf{A}$ to the same set of second-order tensors. There is a unique fourth-order tensor $\mathbf{C}$ such that for all $\mathbf{A}$,*

$$\mathbf{F}(\mathbf{A}) = \mathbf{C} \cdot\cdot \mathbf{A}^T. \tag{3.31}$$

Proof.　Write $\mathbf{A} = a^{mn}\mathbf{e}_m\mathbf{e}_n$ and introduce $\mathbf{C}$ by the formula

$$\mathbf{C} = \mathbf{F}(\mathbf{e}_m\mathbf{e}_n)\mathbf{e}^m\mathbf{e}^n.$$

Then

$$\mathbf{C} \cdot\cdot \mathbf{A}^T = \mathbf{F}(\mathbf{e}_m\mathbf{e}_n)\mathbf{e}^m\mathbf{e}^n \cdot\cdot a^{ij}\mathbf{e}_j\mathbf{e}_i = \mathbf{F}(\mathbf{e}_m\mathbf{e}_n)a^{mn} = \mathbf{F}(\mathbf{A}).$$

So the representation is valid. Proof of uniqueness is left to the reader. $\square$

Exercise 3.26. An operator on the set of vectors $\mathbf{x}$ is given by the formula $\mathbf{y} = \mathbf{B} \cdot \mathbf{x}$ where $\mathbf{B}$ is a second-order tensor. This can be extended to the set of second-order tensors by the equation $\mathbf{Y} = \mathbf{B} \cdot \mathbf{X}^T$. Show that by using the fourth-order tensor $\mathbf{C} = \mathbf{B} \cdot \mathbf{e}^n\mathbf{E}\mathbf{e}_n$, we get $\mathbf{C} \cdot\cdot \mathbf{X}^T = \mathbf{B} \cdot \mathbf{X}^T$ for all $\mathbf{X}$. Note that we cannot represent Hooke's law using only this operation $\mathbf{B} \cdot \mathbf{X}$.

Looking back, we note that the general form of a fourth-order isotropic tensor contains three independent isotropic tensors. Their properties are exhibited in the following exercises.

Exercise 3.27. Show that the identity operator from the representation (3.31) is $\mathbf{I} = \mathbf{e}_k\mathbf{e}_m\mathbf{e}^k\mathbf{e}^m$; that is, for all $\mathbf{A}$ we have $\mathbf{I} \cdot\cdot \mathbf{A}^T = \mathbf{A}$.

Exercise 3.28. A linear function is defined by the equality $\mathbf{F}(\mathbf{A}) = \mathbf{A}^T$. Show that the corresponding fourth-order tensor is $\mathbf{e}_k \mathbf{E} \mathbf{e}^k$, i.e., show that

$$\mathbf{e}_k \mathbf{E} \mathbf{e}^k \cdot\cdot \mathbf{A}^T = \mathbf{A}^T.$$

Exercise 3.29. A linear function is defined by $\mathbf{F}(\mathbf{A}) = (\operatorname{tr}\mathbf{A})\mathbf{E}$. Show that the corresponding fourth-order tensor is $\mathbf{E}\mathbf{E}$, i.e., show that

$$\mathbf{E}\mathbf{E} \cdot\cdot \mathbf{A}^T = (\operatorname{tr}\mathbf{A})\mathbf{E}.$$

Isotropic scalar-valued functions

A scalar function of a tensor is said to be *isotropic* if it retains its form under any orthogonal transformation of the space (or equivalently, of its basis).

A scalar-valued function $f(\mathbf{A})$ of a second-order argument $\mathbf{A}$ is isotropic if and only if for any orthogonal tensor $\mathbf{Q}$ we have

$$f(\mathbf{A}) = f\left(\mathbf{Q}\cdot\mathbf{A}\cdot\mathbf{Q}^T\right).$$

Let us demonstrate that any eigenvalue λ of a second-order tensor $\mathbf{A}$, when considered as a scalar-valued function of $\mathbf{A}$ (i.e., $\lambda = f(\mathbf{A})$), is isotropic. Indeed, an eigenvalue satisfies the characteristic equation

$$\det(\mathbf{A} - \lambda\mathbf{E}) = 0.$$

The eigenvalues of $\mathbf{Q}\cdot\mathbf{A}\cdot\mathbf{Q}^T$ satisfy

$$\det(\mathbf{Q}\cdot\mathbf{A}\cdot\mathbf{Q}^T - \lambda\mathbf{E}) = 0.$$

The equality

$$
\begin{aligned}
\det(\mathbf{Q}\cdot\mathbf{A}\cdot\mathbf{Q}^T - \lambda\mathbf{E}) &= \det(\mathbf{Q}\cdot\mathbf{A}\cdot\mathbf{Q}^T - \lambda\mathbf{Q}\cdot\mathbf{E}\cdot\mathbf{Q}^T)\\
&= \det\left[\mathbf{Q}\cdot(\mathbf{A} - \lambda\mathbf{E})\cdot\mathbf{Q}^T\right]\\
&= (\det\mathbf{Q})^2 \det(\mathbf{A} - \lambda\mathbf{E})\\
&= \det(\mathbf{A} - \lambda\mathbf{E})
\end{aligned}
$$

shows that the eigenvalues of $\mathbf{A}$ and $\mathbf{Q}\cdot\mathbf{A}\cdot\mathbf{Q}^T$ satisfy the same equation. Hence they coincide.

Exercise 3.30. Show that the invariants $I_1(\mathbf{A})$, $I_2(\mathbf{A})$, $I_3(\mathbf{A})$ are isotropic functions of $\mathbf{A}$.

It can be shown that any scalar-valued isotropic function of a second-order tensor is a function of its invariants [Lurie (1990); Ogden (1997); Truesdell and Noll (2004)]. This is used in nonlinear elasticity to introduce the constitutive equations for isotropic bodies.

By Theorem 3.6, a scalar-valued linear function has the representation

$$f(\mathbf{A}) = \mathbf{B} \cdot\cdot \mathbf{A}^T$$

for some second-order tensor $\mathbf{B}$.

Theorem 3.8. *The function $f(\mathbf{A}) = \mathbf{B} \cdot\cdot \mathbf{A}^T$ is isotropic if and only if $\mathbf{B}$ is an isotropic tensor of order two, and hence a ball tensor: $\mathbf{B} = \lambda\mathbf{E}$.*

Proof.   Because f is isotropic, we have

$$\mathbf{B} \cdot\cdot \mathbf{A}^T = \mathbf{B} \cdot\cdot (\mathbf{Q} \cdot \mathbf{A} \cdot \mathbf{Q}^T)^T$$

for any orthogonal $\mathbf{Q}$. Using

$$\mathbf{B} \cdot\cdot \mathbf{A}^T = \operatorname{tr}(\mathbf{B} \cdot \mathbf{A})$$

we get

$$\begin{aligned}
\mathbf{B} \cdot\cdot (\mathbf{Q} \cdot \mathbf{A} \cdot \mathbf{Q}^T)^T &= \operatorname{tr}(\mathbf{B} \cdot \mathbf{Q} \cdot \mathbf{A} \cdot \mathbf{Q}^T) \\
&= \operatorname{tr}(\mathbf{Q}^T \cdot \mathbf{B} \cdot \mathbf{Q} \cdot \mathbf{A}) \\
&= (\mathbf{Q}^T \cdot \mathbf{B} \cdot \mathbf{Q}) \cdot\cdot \mathbf{A}^T.
\end{aligned}$$

Hence for any $\mathbf{A}$ we have

$$\mathbf{B} \cdot\cdot \mathbf{A}^T = (\mathbf{Q}^T \cdot \mathbf{B} \cdot \mathbf{Q}) \cdot\cdot \mathbf{A}^T.$$

But this occurs if and only if the relation

$$\mathbf{B} = \mathbf{Q}^T \cdot \mathbf{B} \cdot \mathbf{Q}$$

holds for any $\mathbf{Q}$. Therefore $\mathbf{B}$ is isotropic. $\square$

Exercise 3.31. Show that a scalar-valued, linear, isotropic function of a second-order tensor $\mathbf{A}$ is a linear function of $\operatorname{tr} \mathbf{A}$: that is, $f(\mathbf{X}) = \lambda \operatorname{tr} \mathbf{X}$.

If a scalar function maintains its form under some subgroup of orthogonal transformations, then one can find functions that are invariant under these. The reader will find applications of this idea in books on crystallography and elasticity.

Isotropic tensor-valued functions

Now we consider a function $\mathbf{F}$ whose domain and range are the set of second-order tensors. We say that $\mathbf{F}(\mathbf{A})$ is *isotropic* if the components of its image value do not change from one Cartesian basis to another. So $\mathbf{F}$ is isotropic if

$$\mathbf{F}\left(\mathbf{Q} \cdot \mathbf{A} \cdot \mathbf{Q}^T\right) = \mathbf{Q} \cdot \mathbf{F}(\mathbf{A}) \cdot \mathbf{Q}^T$$

holds for any orthogonal tensor $\mathbf{Q}$. An example of an isotropic tensor-valued function is $\mathbf{F} = \lambda \mathbf{A}$.

As with a scalar-valued function, it can be shown that a tensor-valued linear function $\mathbf{F}$ represented in terms of a fourth-order tensor $\mathbf{C}$,

$$\mathbf{F} = \mathbf{C} \cdot\cdot \, \mathbf{A}^T,$$

is isotropic if and only if $\mathbf{C}$ is isotropic and therefore is given by (3.28). So $\mathbf{F}$ takes the form

$$\mathbf{F}(\mathbf{A}) = \alpha \mathbf{E} \operatorname{tr} \mathbf{A} + \beta \mathbf{A}^T + \gamma \mathbf{A}.$$

Mechanicists employ functions whose domains and ranges can be sets of symmetric tensors. This imposes certain additional restrictions on the form of $\mathbf{C}$. Indeed, let $\mathbf{A} = \mathbf{A}^T$ and $\mathbf{F}(\mathbf{A}) = \mathbf{F}(\mathbf{A})^T$. Using the representation

$$\mathbf{F}(\mathbf{A}) = c^{ijmn} a_{mn} \mathbf{i}_i \mathbf{i}_j$$

with a Cartesian basis $\mathbf{i}_i$, we see that

$$c^{ijmn} = c^{jimn} = c^{ijnm}.$$

In the general case, $\mathbf{C}$ has 81 independent components. But, in view of symmetry, $\mathbf{C}$ has only 36 independent components. An isotropic fourth-order tensor $\mathbf{C}$ satisfying the symmetry conditions takes the form

$$\alpha \mathbf{E} \mathbf{E} + \beta(\mathbf{e}_k \mathbf{E} \mathbf{e}^k + \mathbf{I}).$$

So the general form of an isotropic linear function satisfying the symmetry condition is

$$\mathbf{F}(\mathbf{A}) = \alpha \mathbf{E} \operatorname{tr} \mathbf{A} + 2\beta \mathbf{A}.$$

We will introduce the elements of calculus for functions of tensorial arguments. First, we require a way to gauge the magnitude of a vector or tensor. Suitable norms exist for this purpose.

3.12 Norms for Tensors, and Some Spaces

In § 2.7 we introduced a norm in the space $\mathbb{R}^3$. It can be immediately extended to $\mathbb{R}^k$ for any k as

$$\|\mathbf{x}\| = (\mathbf{x} \cdot \mathbf{x})^{1/2},$$

where the inner product of $\mathbf{x}$ and $\mathbf{y}$ in the space is given by $\mathbf{x} \cdot \mathbf{y}$.

Similarly, we can introduce a norm and inner product in the set of second-order tensors. We denote the inner product by $(\mathbf{A}, \mathbf{B})$ and define it using the dot product as

$$\begin{aligned}
(\mathbf{A}, \mathbf{B}) &= \mathbf{A} \cdot\cdot \, \mathbf{B}^T \\
&= a^{ij} \mathbf{e}_i \mathbf{e}_j \cdot\cdot \, b^{ts} \mathbf{e}_s \mathbf{e}_t \\
&= a^{ij} b^{ts} g_{js} g_{it} \\
&= a_{ij} b_{ts} g^{js} g^{it} \\
&= a^{ij} \mathbf{e}_i \mathbf{e}_j \cdot\cdot \, b_{ts} \mathbf{e}^s \mathbf{e}^t \\
&= a^{ij} b_{ts} \delta_j^s \delta_i^t \\
&= a^{ij} b_{ij}.
\end{aligned}$$

It is clear that in a Cartesian frame $(\mathbf{A}, \mathbf{A})$ is the sum of all the squared components of $\mathbf{A}$, so this is quite similar to the scalar product of vectors. Using the same reasoning as above, we can show that the axioms of the scalar product hold here as well (note that using only a Cartesian frame, we could regard the components of a tensor as those of a nine-dimensional vector, so this is another reason why the axioms hold). In this case the inner product axioms are written for arbitrary second-order tensors $\mathbf{A}, \mathbf{B}, \mathbf{C}$ as

(i) $(\mathbf{A}, \mathbf{A}) \geq 0$, and $(\mathbf{A}, \mathbf{A}) = 0$ if and only if $\mathbf{A} = \mathbf{0}$;

(ii) $(\mathbf{A}, \mathbf{B}) = (\mathbf{B}, \mathbf{A})$;

(iii) $(\alpha \mathbf{A} + \beta \mathbf{B}, \mathbf{C}) = \alpha(\mathbf{A}, \mathbf{C}) + \beta(\mathbf{B}, \mathbf{C})$ for any real α, β.

By linear algebra, the expression

$$\|\mathbf{A}\| = (\mathbf{A}, \mathbf{A})^{1/2}$$

is a norm in the set of second-order tensors. It satisfies the following axioms.

(i) $\|\mathbf{A}\| \geq 0$, with $\|\mathbf{A}\| = 0$ if and only if $\mathbf{A} = \mathbf{0}$;

(ii) $\|\alpha \mathbf{A}\| = |\alpha| \, \|\mathbf{A}\|$ for any real α;

(iii) $\|\mathbf{A} + \mathbf{B}\| \leq \|\mathbf{A}\| + \|\mathbf{B}\|$.

We still have the Schwarz inequality

$$\|(\mathbf{A}, \mathbf{B})\| \le \|\mathbf{A}\|\,\|\mathbf{B}\|\,.$$

In linear algebra, many particular implementations of the vector and tensor norms can be introduced. We can do the same in any fixed frame, but if we wish to change the frames under consideration we must remember that these norms should change in accordance with the tensor transformation rules.

Note that the representation of a linear function in the previous section used $\mathbf{A}^T$ as an argument. This reflects the form of the inner product on the set of second-order tensors.

It is worth noting two important properties of the norms we have introduced, as these are used in analysis. Let $\mathbf{A}$ and $\mathbf{B}$ be second-order tensors and $\mathbf{x}$ a vector. Then the relations

$$\|\mathbf{A} \cdot \mathbf{x}\| \le \|\mathbf{A}\|\,\|\mathbf{x}\|$$

and

$$\|\mathbf{A} \cdot \mathbf{B}\| \le \|\mathbf{A}\|\,\|\mathbf{B}\|$$

hold. The latter implies that

$$\left\|\mathbf{A}^k\right\| \le \|\mathbf{A}\|^k\,. \tag{3.32}$$

Using this property, we can justify the introduction of tensor-valued functions like $e^{\mathbf{A}}$.

Exercise 3.32. Using (3.32), prove convergence of the series

$$e^{\mathbf{A}} = \mathbf{E} + \frac{1}{1!}\mathbf{A} + \cdots + \frac{1}{k!}\mathbf{A}^k + \cdots\,.$$

It is easy to extend the notion of inner product to tensors of any order; we introduce the inner product

$$(\mathbf{A}, \mathbf{B}) = a^{i_1 i_2 \cdots i_n} b_{i_1 i_2 \cdots i_n}\,.$$

The reader can represent this in all its particular forms and verify the inner product axioms.

Some elements of calculus

As in ordinary calculus, we can introduce the notion of a function in one or many variables that takes values in a set of vectors or tensors. Such a function will be given on a certain set. Consider, for example, a function on the segment $[a, b]$ with values in $\mathbb{R}^3$. This mapping pairs each point of $[a, b]$ with at most one vector from $\mathbb{R}^3$. We may similarly construct a function from $[a, b]$ to the set of tensors of some order. Furthermore, we can introduce the notions of limit and continuity at a point t_0 in the same way as in calculus:

> *The function $\mathbf{f}\colon [a, b] \to \mathbb{R}^3$ has **limit** $\mathbf{a}$ at $t = t_0 \in [a, b]$ if for any $\varepsilon > 0$ there is a $\delta > 0$, dependent on ε, such that for any $t \neq t_0$ and $|t - t_0| \leq \delta$ we have $\|\mathbf{f}(t) - \mathbf{a}\| < \varepsilon$. When $\mathbf{a} = \mathbf{f}(t_0)$, we say the function is **continuous** at t_0.*

Various norms can be used on $\mathbb{R}^3$. However, in linear algebra it is shown that on a finite-dimensional space all norms are equivalent. Equivalence of two norms $\|\cdot\|_1$ and $\|\cdot\|_2$ means that there exist positive constants c_1 and c_2 such that for any element $\mathbf{x}$ of the space we have

$$0 < c_1 \leq \frac{\|\mathbf{x}\|_1}{\|\mathbf{x}\|_2} \leq c_2 < \infty,$$

where c_1 and c_2 do not depend on $\mathbf{x}$. Consequently, either norm can be used in the definition of limit: the limit will exist or not, independently of the form taken by the norm.

As in calculus, we say that a function $\mathbf{f}(t)$ is continuous on $[a, b]$ if it is continuous at each point of $[a, b]$. When a function takes values in $\mathbb{R}^3$, the ordinary definitions of limit, derivative, integral, etc., can be modified by replacing the absolute value with a suitable norm. The derivative of $\mathbf{f}(t)$ at t_0 is given by

$$\mathbf{f}'(t_0) = \frac{d\mathbf{f}}{dt}\bigg|_{t=t_0} = \lim_{t \to t_0} \frac{\mathbf{f}(t) - \mathbf{f}(t_0)}{t - t_0}.$$

The derivative of a vector-valued function has many properties familiar from ordinary calculus. For example, if $\mathbf{f}(t)$ and $\mathbf{g}(t)$ are both differentiable at t, then the product rule holds in the form

$$(\mathbf{f}(t) \cdot \mathbf{g}(t))' = \mathbf{f}'(t) \cdot \mathbf{g}(t) + \mathbf{f}(t) \cdot \mathbf{g}'(t).$$

If we expand $\mathbf{f}$ in a basis $\mathbf{e}_1, \mathbf{e}_2, \mathbf{e}_3$ so that $\mathbf{f}(t) = f^k(t)\mathbf{e}_k$, then

$$\frac{d\mathbf{f}(t)}{dt} = \frac{df^k(t)}{dt}\mathbf{e}_k.$$

Here we have assumed that the $\mathbf{e}_k$ do not depend on t; otherwise an application of the product rule would have been required (we will encounter this situation later). Definite integration can also be carried out in component-wise fashion:

$$\int_a^b \mathbf{f}(t)\,dt = \left(\int_a^b f^k(t)\,dt\right)\mathbf{e}_k.$$

The integral has all the properties familiar from calculus.

Clearly, all this can be extended to the case of a function in one scalar variable taking values in the set of tensors of some order. Moreover, the theory of functions in many variables is generalized in a similar way to the tensorial functions. The reader should remember that formally, in all the definitions of ordinary calculus, we should change the absolute value to the norm. We record here only the formula for the derivative of a tensorial function $\mathbf{F}(t) = f^{mn}(t)\mathbf{e}_m\mathbf{e}_n$:

$$\frac{d\mathbf{F}(t)}{dt} = \frac{df^{mn}(t)}{dt}\mathbf{e}_m\mathbf{e}_n.$$

Some normed spaces

In textbooks on functional analysis, the norm of a vector function is usually introduced in a Cartesian frame. For example, the norm on the space $C(V)$ of continuous vector functions given on a compact region V is

$$\|\mathbf{f}(\mathbf{x})\|_C = \|f_k(\mathbf{x})\mathbf{i}_k\|_C = \max_k\left(\max_V\|f_k(\mathbf{x})\|\right). \qquad (3.33)$$

This formula depends on the Cartesian frame $\mathbf{i}_k$. If we use a similar norm involving vector components in a frame having singular points in V — as may be the case with spherical coordinates — we obtain a norm that is not equivalent to (3.33). This means that (3.33) is an improper way to characterize the intensity of a *vector* field.

However, the proper norm for a function given in curvilinear coordinates is based on the above norm:

$$\|\mathbf{f}(\mathbf{x})\| = \|f^i(\mathbf{x})\mathbf{r}_i\| = \max_V\left[f^i(\mathbf{x})f_i(\mathbf{x})\right]^{1/2}. \qquad (3.34)$$

If we would like to use a norm of the type (3.33), we need to remember that during transformation of the frame we must change the form of the norm accordingly.

On the set of second-order tensor functions continuous on a compact region V, we can introduce a norm similar to (3.34):

$$\|\mathbf{A}(\mathbf{x})\| = \left\|a^{ij}(\mathbf{x})\mathbf{r}_i\mathbf{r}_j\right\| = \max_V \left[a^{ij}(\mathbf{x})a_{ij}(\mathbf{x})\right]^{1/2}.$$

Finally, note that instead of the norms for continuous vector and tensor functions we can introduce the norms and scalar products corresponding to the space of scalar functions $L^2(V)$. The inner product is then

$$(\mathbf{A}, \mathbf{B}) = \int_V a^{ij}b_{ij}\, dV. \tag{3.35}$$

The reader should verify all the axioms of the inner product for this, and introduce all forms of the inner product and norm corresponding to (3.35).

Exercise 3.33. Let $\mathbf{A}$ be a tensor of order two with $\|\mathbf{A}\| = q < 1$. Demonstrate that $\mathbf{E} - \mathbf{A}$ has the inverse $(\mathbf{E} - \mathbf{A})^{-1}$ that is equal to

$$\mathbf{E} + \mathbf{A} + \mathbf{A}^2 + \mathbf{A}^3 + \cdots + \mathbf{A}^n + \cdots.$$

Exercise 3.34. Let $\mathbf{A}$ be a tensor of order two with $\|\mathbf{A}\| = q < 1$. What is the inverse to $\mathbf{E} + \mathbf{A}$?

3.13 Differentiation of Tensorial Functions

In the linear theory of elasticity, there are functions that relate two tensors. The stress and strain tensors, for example, are related though the generalized form of Hooke's law. More complex relations occur in nonlinear elasticity or plasticity. Here one must differentiate tensorial functions. In elasticity, for example, the stress tensor can be found as the derivative of the strain energy with respect to the strain tensor.

We recall that for an ordinary function $f(x)$, the derivative is

$$f'(x) = \lim_{\Delta x \to 0} \frac{f(x + \Delta x) - f(x)}{\Delta x}.$$

The first differential is given by the formula

$$df = f'(x)\, dx. \tag{3.36}$$

For a function in n variables we have

$$df(x_1, \ldots, x_n) = \sum_{k=1}^{n} \frac{\partial f}{\partial x_k}\, dx_k.$$

Using a Cartesian basis $\mathbf{i}_1, \ldots, \mathbf{i}_n$, we can formally represent this as

$$df(x_1, \ldots, x_n) = \left(\sum_{k=1}^{n} \frac{\partial f}{\partial x_k} \mathbf{i}_k \right) \cdot \sum_{m=1}^{n} dx_m \mathbf{i}_m.$$

Let us regard

$$\mathbf{x} = \sum_{m=1}^{n} x_m \mathbf{i}_m$$

as a vector, write

$$f(x_1, \ldots, x_m) = f(\mathbf{x}),$$

and consider this as a function in a vectorial variable. In the same way we introduce

$$d\mathbf{x} = \sum_{m=1}^{n} dx_m \mathbf{i}_m$$

where the dx_m are some quantities that are not necessarily infinitesimal. Let ε be a real variable. For fixed $\mathbf{x}$ and $d\mathbf{x}$, the function $f(\mathbf{x} + \varepsilon\, d\mathbf{x})$ is a function in one variable ε. The chain rule formally applied to this function gives us

$$\left. \frac{df(\mathbf{x} + \varepsilon\, d\mathbf{x})}{d\varepsilon} \right|_{\varepsilon=0} = \left(\sum_{k=1}^{n} \frac{\partial f}{\partial x_k} \mathbf{i}_k \right) \cdot \sum_{m=1}^{n} dx_m \mathbf{i}_m, \tag{3.37}$$

hence

$$df(x_1, \ldots, x_n) = \left. \frac{df(\mathbf{x} + \varepsilon\, d\mathbf{x})}{d\varepsilon} \right|_{\varepsilon=0}. \tag{3.38}$$

The right-hand side of this equality is termed the *Gâteaux derivative* of f at the point $\mathbf{x}$ in the direction $d\mathbf{x}$. For the intermediate expression, we introduce the notation

$$f_{,\mathbf{x}} = \sum_{k=1}^{n} \frac{\partial f}{\partial x_k} \mathbf{i}_k$$

and call it the *derivative* of f with respect to $\mathbf{x}$. Later we will refer to this vector quantity as the *gradient* of f.

We began with a function in n variables, but could in fact consider a function of a vectorial argument $f(\mathbf{x})$ and present the same operations in non-component form. From (3.38) and (3.37) it follows that

$$df(x_1, \ldots, x_n) = \left. \frac{df(\mathbf{x} + \varepsilon\, d\mathbf{x})}{d\varepsilon} \right|_{\varepsilon=0} = f_{,\mathbf{x}} \cdot d\mathbf{x}.$$

When we derive a relation in Cartesian coordinates but present it in non-component form, it becomes valid for any basis. The reader may wish to verify this by direct calculation. Note that in any basis, the components of $f_{,\mathbf{x}}$ are partial derivatives of f with respect to the components of the expansion of $\mathbf{x}$ in the basis. When we wish to have this in a form that does not include the basis vectors, we must use the expansion of $d\mathbf{x}$ in the dual basis.

These ideas can be extended to tensorial functions of tensorial arguments in a straightforward manner.

First we consider a scalar-valued function $f(\mathbf{X})$ whose argument $\mathbf{X}$ belongs to the second-order tensors. In a fixed basis, it can be considered as a function in $3 \times 3 = 9$ variables, the components of $\mathbf{X}$. As above, in a Cartesian basis in $\mathbb{R}^3$ we can introduce the first differential df. Then we introduce the Gâteaux derivative:

$$\frac{\partial}{\partial \varepsilon} f(\mathbf{X} + \varepsilon \, d\mathbf{X}) \bigg|_{\varepsilon=0} \equiv \lim_{\varepsilon \to 0} \frac{f(\mathbf{X} + \varepsilon \, d\mathbf{X}) - f(\mathbf{X})}{\varepsilon} = f_{,\mathbf{x}} \cdot\cdot \, d\mathbf{X}^T \qquad (3.39)$$

for any tensor $d\mathbf{X}$, not necessarily infinitesimal. Here, in the Cartesian basis we have

$$f_{,\mathbf{x}} = \frac{\partial f}{\partial x_{mn}} \mathbf{i}_m \mathbf{i}_n,$$

where the x_{mn} are the Cartesian components of $\mathbf{X}$. Then

$$df = f_{,\mathbf{x}} \cdot\cdot \, d\mathbf{X}^T.$$

The expression $f_{,\mathbf{x}}$ is called the derivative of f with respect to the tensor argument $\mathbf{X}$. Although the last formula was derived in Cartesian coordinates, it holds in any basis.

For a function that maps values $\mathbf{X}$ in the set of second-order tensors into the same set, the derivative $\mathbf{F}_{,\mathbf{x}}(\mathbf{X})$ is defined as

$$\mathbf{F}_{,\mathbf{x}} \cdot\cdot \, d\mathbf{X}^T = \frac{\partial}{\partial \varepsilon} \mathbf{F}(\mathbf{X} + \varepsilon d\mathbf{X}) \bigg|_{\varepsilon=0} \equiv \lim_{\varepsilon \to 0} \frac{\mathbf{F}(\mathbf{X} + \varepsilon d\mathbf{X}) - \mathbf{F}(\mathbf{X})}{\varepsilon}. \qquad (3.40)$$

This is a particular case of the Gâteaux derivative. Again, we can repeat the method used above for the first differential written in Cartesian components. The order of $\mathbf{F}_{,\mathbf{x}}(\mathbf{X})$ is four. In component form it is

$$\mathbf{F}_{,\mathbf{x}} = \frac{\partial f_{ij}}{\partial x_{mn}} \mathbf{i}_i \mathbf{i}_j \mathbf{i}_m \mathbf{i}_n.$$

(We recall that in Cartesian bases $\mathbf{i}_k = \mathbf{i}^k$, and hence the summation convention applies in situations such as this.) For

$$f_{,\mathbf{x}}(\mathbf{X}) \qquad \text{and} \qquad \mathbf{F}_{,\mathbf{x}}(\mathbf{X}),$$

the notations

$$\frac{df(\mathbf{X})}{d\mathbf{X}} \quad \text{and} \quad \frac{d\mathbf{F}(\mathbf{X})}{d\mathbf{X}}$$

are also used.

In a similar way, we can define the derivative of a tensor-valued function for tensors of any order.

Clearly, when calculating $d\mathbf{F}$ we get a linear function in $d\mathbf{X}$. This is why we studied linear functions earlier.

Now we introduce a partial derivative for a scalar-valued function $f(\mathbf{X}_1, \ldots, \mathbf{X}_m)$ in several tensorial arguments. Let the $\mathbf{X}_i$ be second-order tensors. The partial derivative of f with respect to $\mathbf{X}_i$, denoted by $\partial f/\partial \mathbf{X}_i$, is defined by the equality

$$\frac{\partial f}{\partial \mathbf{X}_i} \cdot\cdot \, \mathbf{Y}^T = \frac{\partial}{\partial \varepsilon} f(\mathbf{X}_1, \ldots, \mathbf{X}_i + \varepsilon \mathbf{Y}, \ldots, \mathbf{X}_m) \Big|_{\varepsilon = 0}$$

for any second-order tensor $\mathbf{Y}$. When function $\mathbf{F}$ takes values in the set of second-order tensors, the partial derivative $\partial \mathbf{F}/\partial \mathbf{X}_i$ is similarly defined by equality

$$\frac{\partial \mathbf{F}}{\partial \mathbf{X}_i} \cdot\cdot \, \mathbf{Y}^T = \frac{\partial}{\partial \varepsilon} \mathbf{F}(\mathbf{X}_1, \ldots, \mathbf{X}_i + \varepsilon \mathbf{Y}, \ldots, \mathbf{X}_m) \Big|_{\varepsilon = 0}.$$

As in case of a tensorial function of one variable, the components of the above partial derivatives can be expressed as ordinary partial derivatives of the components. We describe this representation in Cartesian coordinates. Let

$$f(\mathbf{X}_1, \ldots, \mathbf{X}_m) = f(x_{ij}^{(1)}, \ldots, x_{ij}^{(m)})$$

be a function in $9m$ variables, the components $x_{ij}^{(k)}$ of $\mathbf{X}_k$. Then

$$\frac{\partial f}{\partial \mathbf{X}_i} = \frac{\partial f}{\partial x_{jk}^{(i)}} \mathbf{i}_j \mathbf{i}_k.$$

Exercise 3.35. Find the derivative of $f(\mathbf{X}) = I_1(\mathbf{X}) \equiv \operatorname{tr} \mathbf{X}$ with respect to $\mathbf{X}$.

Exercise 3.36. Find the derivative of $f(\mathbf{X}) = \operatorname{tr} \mathbf{X}^2$.

Exercise 3.37. Using the method applied to the previous exercise, show that the derivative of $f(\mathbf{X}) = \operatorname{tr} \mathbf{X}^3$ with respect to $\mathbf{X}$ is $3(\mathbf{X}^T)^2$.

Exercise 3.38. Show that the derivative of

$$f(\mathbf{X}) = I_2(\mathbf{X}) \equiv \frac{1}{2}[\text{tr}^2\,\mathbf{X} - \text{tr}\,\mathbf{X}^2]$$

with respect to $\mathbf{X}$ is

$$I_2(\mathbf{X}),_{\mathbf{X}} = I_1(\mathbf{X})\mathbf{E} - \mathbf{X}^T.$$

Exercise 3.39. Show that the derivative of

$$f(\mathbf{X}) = I_3(\mathbf{X}) \equiv \det \mathbf{X}$$

with respect to $\mathbf{X}$ is

$$[\mathbf{X}^2 - I_1(\mathbf{X})\mathbf{X} + I_2(\mathbf{X})\mathbf{E}]^T.$$

The formulas for differentiating the invariants of the strain tensor are used to write down the constitutive equation for a nonlinear elastic isotropic material under finite deformation.

Exercise 3.40. The strain energy of an isotropic elastic medium is a function of the invariants I_k of the strain tensor, $f = f(I_1, I_2, I_3)$. Using the results of Exercises 3.35, 3.36, and 3.39, demonstrate that its derivative is

$$f,_{\mathbf{X}} = \left[\frac{\partial f}{\partial I_1} + I_1 \frac{\partial f}{\partial I_2} + I_2 \frac{\partial f}{\partial I_3}\right]\mathbf{E} - \left(\frac{\partial f}{\partial I_2} + I_1 \frac{\partial f}{\partial I_3}\right)\mathbf{X}^T + \frac{\partial f}{\partial I_3}\mathbf{X}^{T2}.$$

We have presented examples of the derivatives of scalar-valued functions. The derivatives of tensor-valued functions are more complicated. Note that the derivative of a linear function

$$\mathbf{F}(\mathbf{X}) = \mathbf{C} \cdot\cdot\, \mathbf{X}^T$$

is $\mathbf{C}$:

$$\mathbf{F},_{\mathbf{X}} = \mathbf{C}.$$

Exercise 3.41. Using Exercises 3.27 and 3.28, verify that if $\mathbf{F}(\mathbf{X}) = \mathbf{X}$ then $\mathbf{F},_{\mathbf{X}} = \mathbf{I}$. If $\mathbf{F}(\mathbf{X}) = \mathbf{X}^T$, then $\mathbf{F},_{\mathbf{X}} = \mathbf{e}_k \mathbf{E} \mathbf{e}^k$.

By the exercise above,

$$\mathbf{X},_{\mathbf{X}} = \mathbf{I}.$$

That is, the derivative of a second-order tensor $\mathbf{X}$ with respect to $\mathbf{X}$ is $\mathbf{I}$, a tensor of order four.

On symmetric tensor functions

The principal tensors of linear elasticity, the stress tensor $\boldsymbol{\sigma}$ and strain tensor $\boldsymbol{\varepsilon}$, are symmetric (cf., Chapter 6). So we consider the problem of differentiating a tensor-valued function of a symmetric tensorial argument, as this specific case has some peculiarities.

Let us consider the derivative of a scalar-valued function $f(\mathbf{X})$ of a symmetric second-order tensor $\mathbf{X}$. We modify the definition (3.39) as follows. The derivative $f_{,\mathbf{X}}$ is a second-order symmetric tensor that satisfies the condition

$$\frac{\partial}{\partial \varepsilon} f(\mathbf{X} + \varepsilon \, d\mathbf{X}) \bigg|_{\varepsilon=0} = f_{,\mathbf{X}} \cdot\cdot \, d\mathbf{X} \tag{3.41}$$

for any symmetric tensor $d\mathbf{X}$. The component representation of $f_{,\mathbf{X}}$ in a Cartesian basis is

$$f_{,\mathbf{X}} = \frac{\partial f}{\partial x_{mn}} \, \mathbf{i}_m \mathbf{i}_n.$$

So we have brought the symmetry of $f_{,\mathbf{X}}$ into the definition:

$$(f_{,\mathbf{X}})^T = f_{,\mathbf{X}}.$$

Why do we require $f_{,\mathbf{X}}$ to be symmetric? The sets of symmetric and antisymmetric second-order tensors are subspaces of the space of all second-order tensors. Let $\mathbf{A}$ be a symmetric tensor and $\mathbf{B}$ an antisymmetric second-order tensor. It is easy to see that $\mathbf{A} \cdot\cdot \mathbf{B} = 0$. So the subspaces of symmetric and antisymmetric tensors are mutually orthogonal. In (3.41), $d\mathbf{X}$ is arbitrary but symmetric. If in this definition we do not require $f_{,\mathbf{X}}$ to be symmetric, then (3.41), holding only for all the symmetric tensors $d\mathbf{X}$, defines $f_{,\mathbf{X}}$ non-uniquely up to an additive term $\mathbf{B}$ that can be any antisymmetric second-order tensor.

Note that the definition of derivative for a function with respect to a symmetric tensor argument is closely related to the problem of representing a linear function acting on the subspace of symmetric tensors, which differs slightly from the general case (see Problems 3.52, 3.53, and 3.54).

As an example, we find the derivative of the function

$$f(\mathbf{X}) = \mathbf{C} \cdot\cdot \mathbf{X},$$

where $\mathbf{C}$ is a second-order tensor and $\mathbf{X}^T = \mathbf{X}$. By definition,

$$f_{,\mathbf{X}} \cdot\cdot \, d\mathbf{X} = \mathbf{C} \cdot\cdot \, d\mathbf{X}$$

for all $d\mathbf{X}$ such that $d\mathbf{X} = d\mathbf{X}^T$. At first glance it seems that $f_{,\mathbf{X}} = \mathbf{C}$, and this does hold for an arbitrary argument $\mathbf{X}$. But for a symmetric argument $\mathbf{X}$ the answer changes. Indeed the derivative must be a symmetric tensor, so $f_{,\mathbf{X}} = \mathbf{C}$ if and only if $\mathbf{C} = \mathbf{C}^T$. Now we consider the case when $\mathbf{C}$ is not symmetric. We said that $f_{,\mathbf{X}}$ must be symmetric. A consequence is that $f_{,\mathbf{X}}$ becomes

$$f_{,\mathbf{X}} = \frac{1}{2}(\mathbf{C} + \mathbf{C}^T).$$

This expression is uniquely defined. Indeed, we can represent $\mathbf{C}$ as a sum of symmetric and antisymmetric tensors:

$$\mathbf{C} = \frac{1}{2}(\mathbf{C} + \mathbf{C}^T) + \frac{1}{2}(\mathbf{C} - \mathbf{C}^T)$$

But

$$\frac{1}{2}(\mathbf{C} - \mathbf{C}^T) \cdot\cdot\, d\mathbf{X} = \frac{1}{2}(\mathbf{C} - \mathbf{C}) \cdot\cdot\, d\mathbf{X} = 0$$

and thus

$$\mathbf{C} \cdot\cdot\, d\mathbf{X} = \frac{1}{2}(\mathbf{C} + \mathbf{C}^T) \cdot\cdot\, d\mathbf{X}$$

as claimed.

In a similar fashion, we should modify the definition of the derivative of a tensor-valued function of a second-order symmetric argument. If $\mathbf{F}$ takes values in the set of tensors of order n, the derivative with respect to $\mathbf{X}$ is denoted by $\mathbf{F}_{,\mathbf{X}}$; it takes values in the subset of tensors of order $n + 2$ which are symmetric in the two last indices and satisfy the equality

$$\mathbf{F}_{,\mathbf{X}} \cdot\cdot\, d\mathbf{X} = \left.\frac{\partial}{\partial \varepsilon}\mathbf{F}(\mathbf{X} + \varepsilon\, d\mathbf{X})\right|_{\varepsilon=0} \tag{3.42}$$

for all symmetric second-order tensors $d\mathbf{X}$. The symmetry of $\mathbf{F}_{,\mathbf{X}}$ in the last two indices means that for the components $F_{ij...pt}$ of $\mathbf{F}_{,\mathbf{X}}$ we have

$$F_{ij...pt} = F_{ij...tp}.$$

Similar changes permit the definition of partial derivative for tensorial functions in many symmetric tensorial arguments.

Exercise 3.42. Let

$$\mathbf{F}(\mathbf{X}) = \mathbf{C} \cdot\cdot\, \mathbf{X},$$

where $\mathbf{C} = c_{mnpt}\mathbf{i}_m\mathbf{i}_n\mathbf{i}_p\mathbf{i}_t$ is a fourth-order tensor and $\mathbf{X}$ is a second-order symmetric tensor. Demonstrate that

$$\mathbf{F}_{,\mathbf{X}} = \frac{1}{2}(\mathbf{C} + \mathbf{C}'),$$

where $\mathbf{C}'$ is derived from $\mathbf{C}$ by transposing the last two indices: $\mathbf{C}' = c_{mnpt}\mathbf{i}_m\mathbf{i}_n\mathbf{i}_t\mathbf{i}_p$.

Exercise 3.43. Let

$$f(\mathbf{X}) = \frac{1}{2}\mathbf{X} \cdot\cdot\, \mathbf{C} \cdot\cdot\, \mathbf{X},$$

where $\mathbf{C} = c_{mnpt}\mathbf{i}_m\mathbf{i}_n\mathbf{i}_p\mathbf{i}_t$ and $\mathbf{X}$ is a second-order symmetric tensor. Demonstrate that

$$f_{,\mathbf{X}} = \frac{1}{4}(\mathbf{C} \cdot\cdot\, \mathbf{X} + \mathbf{C}'' \cdot\cdot\, \mathbf{X} + \mathbf{X} \cdot\cdot\, \mathbf{C} + \mathbf{X} \cdot\cdot\, \mathbf{C}'),$$

where $\mathbf{C}'$ is derived from $\mathbf{C}$ by transposing the last two indices of the components of $\mathbf{C}$, i.e., $\mathbf{C}' = c_{mnpt}\mathbf{i}_m\mathbf{i}_n\mathbf{i}_t\mathbf{i}_p$, and $\mathbf{C}''$ by transposing the first two indices of the components, i.e., $\mathbf{C}'' = c_{mnpt}\mathbf{i}_n\mathbf{i}_m\mathbf{i}_p\mathbf{i}_t$.

Exercise 3.44. Let

$$f(\mathbf{X}) = \frac{1}{2}\mathbf{X} \cdot\cdot\, \mathbf{C} \cdot\cdot\, \mathbf{X},$$

where $\mathbf{C}$ is a fourth-order tensor and $\mathbf{X}$ is a second-order symmetric tensor. Suppose $\mathbf{C} = \mathbf{C}' = \mathbf{C}''$ and $\mathbf{C} \cdot\cdot\, \mathbf{X} = \mathbf{X} \cdot\cdot\, \mathbf{C}$ for any symmetric tensor $\mathbf{X}$; in terms of components, this means the equalities $c_{mnpt} = c_{nmpt}$ and $c_{mnpt} = c_{ptmn}$ hold for any sets of indices. Using the solution of Exercise 3.43, demonstrate that $f_{,\mathbf{X}} = \mathbf{C} \cdot\cdot\, \mathbf{X}$.

3.14 Problems

In this section, unless otherwise stated, we use $\mathbf{A}, \mathbf{B}, \mathbf{C}, \mathbf{X}$ to denote second-order tensors, λ a scalar, $\boldsymbol{\Omega}$ an antisymmetric second-order tensor, and $\mathbf{Q}$ an orthogonal second-order tensor.

3.1 Write out the components of the dyad $\mathbf{i}_1\mathbf{i}_2$ in a Cartesian basis $(\mathbf{i}_1, \mathbf{i}_2, \mathbf{i}_3)$.

3.2 Find the components of the tensor $\mathbf{i}_1\mathbf{i}_2 - \mathbf{i}_2\mathbf{i}_1 + 2\mathbf{i}_3\mathbf{i}_3$ in the Cartesian basis $(\mathbf{i}_1, \mathbf{i}_2, \mathbf{i}_3)$.

3.3 Write out the components of the tensor that is the dyad $\mathbf{a}_1 \mathbf{a}_2$ in the Cartesian basis $(\mathbf{i}_1, \mathbf{i}_2, \mathbf{i}_3)$, where $\mathbf{a}_1 = (-1, 2, -2)$ and $\mathbf{a}_2 = (1, -1, 2)$.

3.4 Show that $\mathbf{a}0 = 0\mathbf{b} = 0$.

3.5 Determine the symmetric and antisymmetric parts of the following tensors:

(a) $\mathbf{i}_1 \mathbf{i}_2$;
(b) $\mathbf{i}_1 \mathbf{i}_2 - \mathbf{i}_2 \mathbf{i}_1 + 2\mathbf{i}_3 \mathbf{i}_3$;
(c) $\mathbf{i}_1 \mathbf{i}_2 - 2\mathbf{i}_2 \mathbf{i}_1 + \mathbf{i}_1 \mathbf{i}_3$;
(d) $\mathbf{i}_1 \mathbf{i}_2 + \mathbf{i}_2 \mathbf{i}_3 + \mathbf{i}_1 \mathbf{i}_3$;
(e) $\mathbf{i}_1 \mathbf{i}_1 + 2\mathbf{i}_1 \mathbf{i}_2 + 2\mathbf{i}_2 \mathbf{i}_1 + \mathbf{i}_3 \mathbf{i}_1 + \mathbf{i}_1 \mathbf{i}_3$.

3.6 Determine the ball and deviator parts of the following tensors:

(a) $\mathbf{i}_1 \mathbf{i}_2$;
(b) $\mathbf{i}_1 \mathbf{i}_2 + \mathbf{i}_2 \mathbf{i}_1$;
(c) $\mathbf{i}_1 \mathbf{i}_1$;
(d) $\mathbf{a}\mathbf{a}$;
(e) $\mathbf{i}_1 \mathbf{i}_1 + 2\mathbf{i}_1 \mathbf{i}_2 + 2\mathbf{i}_2 \mathbf{i}_1 + \mathbf{i}_3 \mathbf{i}_1 + \mathbf{i}_1 \mathbf{i}_3$.

3.7 Show that

$$I_1(\mathbf{X}) = \operatorname{tr} \mathbf{X}, \quad I_2(\mathbf{X}) = \frac{1}{2}[\operatorname{tr}^2 \mathbf{X} - \operatorname{tr} \mathbf{X}^2], \quad I_3(\mathbf{X}) = \det \mathbf{X}.$$

3.8 Find the invariants of the following tensors:

(a) $\mathbf{a}\mathbf{a}$;
(b) $\mathbf{i}_1 \mathbf{i}_2$;
(c) $\mathbf{i}_1 \mathbf{i}_1 + \mathbf{i}_2 \mathbf{i}_2$;
(d) $\lambda \mathbf{E}$;
(e) $2\mathbf{i}_1 \mathbf{i}_1 + 3\mathbf{i}_2 \mathbf{i}_2 + 4\mathbf{i}_3 \mathbf{i}_3$.

3.9 Let $\mathbf{A}$ be a symmetric, invertible tensor. Show that

$$I_1(\mathbf{A}^{-1}) = \frac{I_2(\mathbf{A})}{I_3(\mathbf{A})}, \quad I_2(\mathbf{A}^{-1}) = \frac{I_1(\mathbf{A})}{I_3(\mathbf{A})}, \quad I_3(\mathbf{A}^{-1}) = I_3^{-1}(\mathbf{A}).$$

3.10 Find the left and right polar decompositions of the following tensors:

(a) $\lambda \mathbf{E}$;
(b) $\mathbf{a}\mathbf{a} + \mathbf{b}\mathbf{b} + \mathbf{c}\mathbf{c}$ if $\mathbf{a}, \mathbf{b}, \mathbf{c}$ are mutually orthogonal;

(c) $\lambda \mathbf{E} + a\mathbf{i}_1\mathbf{i}_1$;

(d) $\lambda \mathbf{E} + a\mathbf{i}_1\mathbf{i}_1 + b\mathbf{i}_2\mathbf{i}_2$;

(e) $a\mathbf{i}_1\mathbf{i}_1 + b\mathbf{i}_2\mathbf{i}_2 + c\mathbf{i}_3\mathbf{i}_3$.

3.11 Suppose $\operatorname{tr}\mathbf{A}^2 \neq 0$, a is an arbitrary scalar, and $\mathbf{Y}$ is an arbitrary second-order tensor. Demonstrate that

$$\mathbf{X} = a\mathbf{Y} - a\frac{\operatorname{tr}(\mathbf{A}\cdot\mathbf{Y})}{\operatorname{tr}\mathbf{A}^2}\mathbf{A}$$

satisfies the equation $\operatorname{tr}(\mathbf{A}\cdot\mathbf{X}) = 0$.

3.12 Let a nonzero scalar a and second-order tensors $\mathbf{A}, \mathbf{B}$ be given. Find a solution $\mathbf{X}$ of the equation

$$a\mathbf{X} + \operatorname{tr}(\mathbf{A}\cdot\mathbf{X})\mathbf{E} = \mathbf{B}.$$

3.13 For the equations

(a) $\mathbf{X} + \operatorname{tr}(\mathbf{A}\cdot\mathbf{X})\mathbf{B} = \mathbf{C}$,

(b) $\mathbf{X}^T + \operatorname{tr}(\mathbf{A}\cdot\mathbf{X})\mathbf{B} = \mathbf{C}$,

(c) $\mathbf{X} + a(\operatorname{tr}\mathbf{X})\mathbf{A} = \mathbf{B}$,

find a solution $\mathbf{X}$ and establish conditions for uniqueness.

3.14 Under what conditions will the equation

$$a\mathbf{X} + \mathbf{E}\operatorname{tr}\mathbf{X} = \mathbf{0}$$

have a nonzero solution $\mathbf{X}$?

3.15 Under what conditions will the equation

$$a\mathbf{X} + \mathbf{A}\operatorname{tr}\mathbf{X} = \mathbf{0}$$

have a nontrivial solution?

3.16 Demonstrate that in a Cartesian basis

$$\mathcal{E} = \mathbf{i}_1\mathbf{i}_2\mathbf{i}_3 + \mathbf{i}_2\mathbf{i}_3\mathbf{i}_1 + \mathbf{i}_3\mathbf{i}_1\mathbf{i}_2 - \mathbf{i}_1\mathbf{i}_3\mathbf{i}_2 - \mathbf{i}_3\mathbf{i}_2\mathbf{i}_1 - \mathbf{i}_2\mathbf{i}_1\mathbf{i}_3.$$

3.17 Let $\mathbf{A}$ be a symmetric tensor. Show that $I_2(\operatorname{dev}\mathbf{A}) \leq 0$.

3.18 Let $\mathbf{A}$ be a nonsingular tensor and let $\mathbf{a}$ and $\mathbf{b}$ be arbitrary vectors. Prove that $\mathbf{A} + \mathbf{ab}$ is a nonsingular tensor if and only if $1 + \mathbf{b}\cdot\mathbf{A}^{-1}\cdot\mathbf{a} \neq 0$, and

$$(\mathbf{A} + \mathbf{ab})^{-1} = \mathbf{A}^{-1} - \frac{1}{1 + \mathbf{b}\cdot\mathbf{A}^{-1}\cdot\mathbf{a}}(\mathbf{A}^{-1}\cdot\mathbf{a})(\mathbf{b}\cdot\mathbf{A}^{-1}).$$

3.19 Show that $(\mathbf{E} \times \boldsymbol{\omega})^2 = \boldsymbol{\omega}\boldsymbol{\omega} - \mathbf{E}\boldsymbol{\omega} \cdot \boldsymbol{\omega}$.

3.20 Show that

(a) $(\mathbf{E} \times \boldsymbol{\omega})^{2n} = (-1)^{n-1}(\boldsymbol{\omega} \cdot \boldsymbol{\omega})^{n-1}(\boldsymbol{\omega}\boldsymbol{\omega} - \mathbf{E}\boldsymbol{\omega} \cdot \boldsymbol{\omega})$,

(b) $(\mathbf{E} \times \boldsymbol{\omega})^{2n+1} = (-1)^n(\boldsymbol{\omega} \cdot \boldsymbol{\omega})^n \mathbf{E} \times \boldsymbol{\omega}$,

(c) $\mathbf{a} \times \mathbf{E} \times \mathbf{b} = \mathbf{b}\mathbf{a} - \mathbf{a} \cdot \mathbf{b}\mathbf{E}$.

3.21 For a second-order tensor $\mathbf{A} = a_{mn}\mathbf{e}^m\mathbf{e}^n$, the *vectorial invariant* was introduced by J.W. Gibbs (1839–1903) as

$$\mathbf{A}_\times = a_{mn}\mathbf{e}^m \times \mathbf{e}^n.$$

For example, $(\mathbf{ab})_\times = \mathbf{a} \times \mathbf{b}$. Find the vectorial invariants of the following tensors:

(a) $\mathbf{aa}$;

(b) $\mathbf{E} \times \boldsymbol{\omega}$;

(c) $\mathbf{ab} - \mathbf{ba}$;

(d) $\mathbf{E}$;

(e) $\mathbf{A}$ if $\mathbf{A} = \mathbf{A}^T$.

3.22 Show that $\mathbf{A}_\times = \frac{1}{2}(\mathbf{A} - \mathbf{A}^T)_\times$.

3.23 Show that $(\mathbf{a} \times \mathbf{A})_\times = \mathbf{A} \cdot \mathbf{a} - \mathbf{a}\,\text{tr}\,\mathbf{A}$.

3.24 Show that $(\mathbf{A} \times \mathbf{B}) \cdot\cdot\, \mathbf{B}^T = 0$ holds for arbitrary second-order tensors $\mathbf{A}$ and $\mathbf{B}$.

3.25 Show that $\text{tr}\,[\mathbf{b} \times (\mathbf{a} \times \mathbf{A})] = \mathbf{b} \cdot \mathbf{A} \cdot \mathbf{a} - \mathbf{a} \cdot \mathbf{b}\,\text{tr}\,\mathbf{A}$.

3.26 Show that

$$\mathbf{A} \times \boldsymbol{\omega} = -(\boldsymbol{\omega} \times \mathbf{A}^T)^T, \qquad \boldsymbol{\omega} \times \mathbf{A} = -(\mathbf{A}^T \times \boldsymbol{\omega})^T.$$

3.27 Two second-order tensors $\mathbf{A}$ and $\mathbf{B}$ are called *commutative* if $\mathbf{A} \cdot \mathbf{B} = \mathbf{B} \cdot \mathbf{A}$. Show that symmetrical tensors $\mathbf{A}$ and $\mathbf{B}$ are commutative if their sets of eigenvectors coincide.

3.28 Let $\mathbf{a} \cdot \mathbf{b} = 0$. Demonstrate that the dyads $\mathbf{aa}$ and $\mathbf{bb}$ are commutative.

3.29 Let the symmetric tensors $\mathbf{A}$ and $\mathbf{B}$ be commutative. Demonstrate that $(\mathbf{A} \cdot \mathbf{B})_\times = \mathbf{0}$.

3.30 Show that the tensor $\mathbf{Q} = \mathbf{i}_1 \mathbf{i}_2 - \mathbf{i}_2 \mathbf{i}_1 + \mathbf{i}_3 \mathbf{i}_3$ is orthogonal.

3.31 Show that the principal invariants of an orthogonal tensor satisfy the relations $I_1 I_3 = I_2$ and $I_3^2 = 1$.

3.32 Let $\mathbf{e} \cdot \mathbf{e} = 1$. Show that the tensor $\mathbf{Q} = \mathbf{E} - 2\mathbf{e}\mathbf{e}$ is orthogonal.

3.33 Show that if $\boldsymbol{\Omega}$ is an antisymmetric tensor, then the tensor

$$\mathbf{Q} = (\mathbf{E} + \boldsymbol{\Omega}) \cdot (\mathbf{E} - \boldsymbol{\Omega})^{-1}$$

is orthogonal.

3.34 Show that the tensor $\mathbf{Q} = e^{\boldsymbol{\Omega}}$ is orthogonal if $\boldsymbol{\Omega}$ is antisymmetric.

3.35 Demonstrate that the tensor

$$\mathbf{Q} = \frac{1}{4 + \theta^2} \left[(4 - \theta^2)\mathbf{E} + 2\boldsymbol{\theta}\boldsymbol{\theta} - 4\mathbf{E} \times \boldsymbol{\theta} \right],$$

where $\theta^2 = \boldsymbol{\theta} \cdot \boldsymbol{\theta}$, is orthogonal. We call $\boldsymbol{\theta}$ a *finite rotation vector*.

3.36 Let $\mathbf{Q}$ be an orthogonal tensor. Establish the identity

$$(\mathbf{Q} - \mathbf{E}) \cdots (\mathbf{Q}^T - \mathbf{E}) = 6 - 2\operatorname{tr}\mathbf{Q}.$$

3.37 Let $\mathbf{e}_k$ $(k = 1, 2, 3)$ and $\mathbf{d}_m$ $(m = 1, 2, 3)$ be two orthonormal bases. Verify that $\mathbf{Q} = \mathbf{e}_k \mathbf{d}_k$ is an orthogonal tensor.

3.38 Let $\mathbf{Q}$ be a proper orthogonal tensor. Establish the identity

$$\mathbf{Q}_\times = 2\sin\omega\,\mathbf{e},$$

where ω and $\mathbf{e}$ are the rotation angle and axis, respectively.

3.39 Find the principal invariants of the tensor $\alpha\mathbf{E} + \beta\mathbf{e}\mathbf{e}$.

3.40 Find the principal invariants of the tensor $\mathbf{E} \times \boldsymbol{\omega}$.

3.41 Find the principal invariants of the tensor $(\mathbf{E} \times \boldsymbol{\omega})^2$.

3.42 Show that

$$I_3(\mathbf{A}) = \frac{1}{6} \left(\operatorname{tr}^3 \mathbf{A} - 3\operatorname{tr}\mathbf{A}\operatorname{tr}\mathbf{A}^2 + 2\operatorname{tr}\mathbf{A}^3 \right).$$

3.43 From the Cayley–Hamilton theorem for a non-degenerate tensor, it follows that

$$\mathbf{A}^{-1} = \frac{1}{\det \mathbf{A}} \left(\mathbf{A}^2 - I_1(\mathbf{A})\mathbf{A} + I_2(\mathbf{A})\mathbf{E} \right).$$

Use this to find the inverses of the following tensors:

(a) $i_1 i_1 + 2 i_2 i_2 + i_3 i_3 + i_1 i_2 + i_2 i_1$;

(b) $a i_1 i_3 + b i_2 i_2 + c i_3 i_1$, $a, b, c \neq 0$;

(c) $a i_1 i_1 + i_2 i_2 + i_3 i_3 + b i_1 i_2$;

(d) $a \mathbf{E} + b i_1 i_2$.

3.44 Find the derivative with respect to the tensor $\mathbf{X}$ of the following scalar-valued functions:

(a) $\operatorname{tr} \mathbf{X}^4$;

(b) $\mathbf{a} \cdot \mathbf{X} \cdot \mathbf{a}$;

(c) $\mathbf{a} \cdot \mathbf{X} \cdot \mathbf{b}$;

(d) $\operatorname{tr}(\mathbf{B} \cdot \mathbf{X})$.

3.45 Let $\mathbf{A}$ be a second-order tensor. A tensor $\mathbf{X}$ is called the *cofactor* of $\mathbf{A}$ and denoted by $\mathbf{X} = \operatorname{cof} \mathbf{A}$ if $\mathbf{X}$ satisfies the equation

$$\mathbf{A} \cdot \mathbf{X}^T = \mathbf{X}^T \cdot \mathbf{A} = (\det \mathbf{A}) \mathbf{E}.$$

If $\mathbf{A}$ is nonsingular, then $\operatorname{cof} \mathbf{A} = (\det \mathbf{A}) \mathbf{A}^{-T}$. Check the following properties of cofactor:

(a) $\operatorname{cof}(\mathbf{A}^T) = (\operatorname{cof} \mathbf{A})^T$;

(b) $\operatorname{cof}(\mathbf{A} \cdot \mathbf{B}) = \operatorname{cof} \mathbf{A} \cdot \operatorname{cof} \mathbf{B}$;

(c) $\operatorname{cof} \lambda \mathbf{A} = \lambda^2 \operatorname{cof} \mathbf{A}$;

(d) $I_2(\mathbf{A}) = \operatorname{tr} \operatorname{cof} \mathbf{A}$.

3.46 Let $\mathbf{A} = \lambda_1 i_1 i_1 + \lambda_2 i_2 i_2 + \lambda_3 i_3 i_3$. Demonstrate that

$$\operatorname{cof} \mathbf{A} = \lambda_2 \lambda_3 i_1 i_1 + \lambda_1 \lambda_3 i_2 i_2 + \lambda_1 \lambda_2 i_3 i_3.$$

3.47 Let $f(x)$ be an analytic function in an open ball $x < r$ centered at $x = 0$; that is, suppose the representation

$$f(x) = \sum_{k=0}^{\infty} \frac{f^{(k)}(0)}{k!} x^k$$

holds and the series converges for $x < r$. In the ball, the relation

$$f'(x) = \sum_{k=1}^{\infty} \frac{f^{(k)}(0)}{(k-1)!} x^{k-1}$$

holds. Let

$$\mathbf{F}(\mathbf{X}) = \sum_{k=0}^{\infty} \frac{f^{(k)}(0)}{k!} \mathbf{X}^k.$$

Demonstrate that

$$[\mathrm{tr}(\mathbf{F}(\mathbf{X}))]_{,\mathbf{X}} = \sum_{k=1}^{\infty} \frac{f^{(k)}(0)}{(k-1)!} \mathbf{X}^{T^{k-1}} = f'(\mathbf{X}^T)$$

for $\|\mathbf{X}\| < r$.

3.48 Let $f(\mathbf{X})$ be a scalar-valued function. Show that the derivative of $f(\mathbf{X})$ with respect to $\mathbf{X}^T$ is equal to the transpose of its derivative with respect to $\mathbf{X}$, i.e.,

$$f(\mathbf{X})_{,\mathbf{X}^T} = (f(\mathbf{X})_{,\mathbf{x}})^T.$$

3.49 Find the derivative with respect to $\mathbf{X}$ of the following tensor-valued functions:

(a) $\mathrm{tr}(\mathbf{A} \cdot \mathbf{X}^T)\mathbf{E}$;
(b) $\mathrm{tr}(\mathbf{A} \cdot \mathbf{X})\mathbf{E}$;
(c) $\mathbf{a} \cdot \mathbf{X} \cdot \mathbf{bcd}$.

3.50 For a nonsingular tensor $\mathbf{X}$, demonstrate that

$$(\det \mathbf{X})_{,\mathbf{X}} = (\det \mathbf{X})\mathbf{X}^{-T}.$$

3.51 Let S be the set of all symmetric second-order tensors. Show that S is a linear subspace of the linear space of all the second-order tensors. On S the inner product $\mathbf{A} \cdot\cdot \, \mathbf{B}^T$ of the whole space takes the form

$$\mathbf{A} \cdot\cdot \, \mathbf{B}^T = \mathbf{A} \cdot\cdot \, \mathbf{B}.$$

3.52 Let f be a scalar-valued function of a symmetric second-order tensor $\mathbf{A}$. Show that there is a unique symmetric second-order tensor $\mathbf{B}$ such that

$$f(\mathbf{A}) = \mathrm{tr}(\mathbf{B} \cdot \mathbf{A})$$

for any symmetric $\mathbf{A}$, that is $\mathbf{A} = \mathbf{A}^T$.

3.53 Let $\mathbf{F} = \mathbf{F}(\mathbf{A})$ be a linear function from the set of second-order symmetric tensors $\mathbf{A}$ to the set of second-order tensors. Show that there is a unique fourth-order tensor $\mathbf{C}$ such that

$$\mathbf{F}(\mathbf{A}) = \mathbf{C} \cdot\cdot \, \mathbf{A}, \qquad \mathbf{C} = c_{mnpt}\mathbf{i}_m\mathbf{i}_n\mathbf{i}_p\mathbf{i}_t, \qquad c_{mnpt} = c_{mntp}$$

for all $\mathbf{A} = \mathbf{A}^T$.

3.54 Let $\mathbf{F} = \mathbf{F}(\mathbf{A})$ be a linear function from the set of second-order symmetric tensors $\mathbf{A}$ to the same set of symmetric second-order tensors $(\mathbf{F}(\mathbf{A}) = \mathbf{F}(\mathbf{A})^T)$. Show that there is a unique fourth-order tensor $\mathbf{C}$ such that

$$\mathbf{F}(\mathbf{A}) = \mathbf{C} \cdot\cdot \, \mathbf{A}, \quad \mathbf{C} = c_{mnpt}\mathbf{i}_m\mathbf{i}_n\mathbf{i}_p\mathbf{i}_t, \quad c_{mnpt} = c_{mntp}, \quad c_{mnpt} = c_{nmpt}$$

whenever $\mathbf{A} = \mathbf{A}^T$.

3.55 Let Q be the set of all antisymmetric second-order tensors. Show that Q is a linear subspace of the linear space of all second-order tensors. On Q the inner product $\mathbf{A} \cdot\cdot \, \mathbf{B}^T$ of the whole space takes the form

$$\mathbf{A} \cdot\cdot \, \mathbf{B}^T = -\mathbf{A} \cdot\cdot \, \mathbf{B} = -\mathbf{A}^T \cdot\cdot \, \mathbf{B}^T.$$

3.56 Let $\mathbf{C}$ be a tensor of order n that is antisymmetric in the two last components: $c_{mn\ldots pt} = -c_{mn\ldots tp}$. Let $\mathbf{A}$ be an arbitrary symmetric second-order tensor. Demonstrate that $\mathbf{C} \cdot\cdot \, \mathbf{A} = \mathbf{0}$, where $\mathbf{0}$ is the zero tensor of order $n - 2$.

Chapter 4

Tensor Fields

4.1 Vector Fields

The state of a point (more precisely of an infinitely small volume) within a natural object is frequently characterized by a vector or a tensor. Hence inside a spatial object there arises what we call a *vector* or *tensor field*. As a rule these fields are governed by simultaneous partial differential equations. Such equations are usually derived using a Cartesian space frame. In this frame, the operations of calculus closely parallel those of one dimensional analysis: the differentiation of a vector function proceeds on a component by component basis, for example. However, it is often convenient to introduce curvilinear coordinates in the body, in terms of which the problem formulation is simpler. In this way we get a frame that changes from point to point, and component-wise differentiation is not enough to characterize the change of the vector function. Thus we need to develop the apparatus of calculus for vector and tensor functions when the frame in the object is changeable. Another reason for introducing these tools is the objectivity of the laws of nature: we must be able to formulate frame independent statements of these laws. Finally, there is an aesthetic reason: in non-coordinate form many statements of mathematical physics look much less cumbersome than their counterparts stated in terms of coordinates. It is said that beauty governs the world; although this is not absolutely true, most students would prefer a short and beautiful statement to a nightmarish formula taking half a page.

The position of a point in space is characterized by three numbers called coordinates of the point. Coordinate systems in common use include the Cartesian, cylindrical, and spherical systems. The first of these differs from the latter two in an important respect: the frame vectors of a Cartesian system are unique, while those for the curvilinear systems change from point to point. In a general coordinate system the position of a point in space is determined uniquely by three numbers q^1, q^2, q^3. These are referred to as *curvilinear coordinates* if the frame is not Cartesian. See Fig. 4.1.

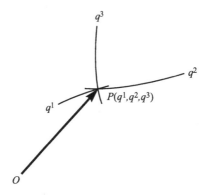

Fig. 4.1   Curvilinear coordinates.

If we fix two of the coordinates and change the third, we get a line in space called a *coordinate line*. The Cartesian coordinates x^1, x^2, x^3 of a point can be determined through the general curvilinear coordinates by relations of the form

$$x^i = x^i(q^1, q^2, q^3) \qquad (i = 1, 2, 3). \tag{4.1}$$

Except for some set of singular points in space, the correspondence (4.1) is one to one. We suppose the functions $x^i = x^i(q^1, q^2, q^3)$ are smooth (continuously differentiable). In this case the local one-to-one correspondence is provided by the requirement that the Jacobian

$$\left| \frac{\partial x^i}{\partial q^j} \right| \neq 0. \tag{4.2}$$

Fixing some origin O, we characterize the position of a point $P(q^1, q^2, q^3)$ by a vector $\mathbf{r}$ connecting the points O and P:

$$\mathbf{r} = \mathbf{r}(q^1, q^2, q^3).$$

When a point moves along a coordinate line, along the q^1 line for instance, the end of its position vector moves along this line and the difference vector

$$\Delta \mathbf{r} = \mathbf{r}(q^1 + \Delta q^1, q^2, q^3) - \mathbf{r}(q^1, q^2, q^3)$$

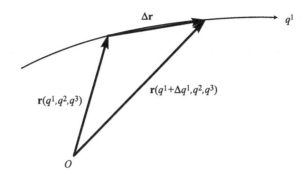

Fig. 4.2   Generation of a local tangent vector to a coordinate line.

is directed along the chord (Fig. 4.2). The smaller the value of Δq^1, the closer $\Delta \mathbf{r}$ is to being tangent to the q^1 line. The limit as $\Delta q^1 \to 0$ of the ratio $\Delta \mathbf{r}/\Delta q^1$ is a vector

$$\frac{\partial \mathbf{r}(q^1, q^2, q^3)}{\partial q^1} = \mathbf{r}_1$$

tangent to the q^1 coordinate line. At the same point (q^1, q^2, q^3) we can introduce two other vectors

$$\mathbf{r}_2 = \frac{\partial \mathbf{r}(q^1, q^2, q^3)}{\partial q^2}, \qquad \mathbf{r}_3 = \frac{\partial \mathbf{r}(q^1, q^2, q^3)}{\partial q^3},$$

tangent to the q^2 and q^3 coordinate lines, respectively. If the coordinate point (q^1, q^2, q^3) is not singular then the vectors $\mathbf{r}_i$ are non-coplanar and therefore constitute a frame triad. The mixed product $\mathbf{r}_1 \cdot (\mathbf{r}_2 \times \mathbf{r}_3)$ is the volume of the frame parallelepiped; renumbering the coordinates if necessary, we obtain from this the same expression for the Jacobian (4.2). Let us denote it

$$\sqrt{g} = \mathbf{r}_1 \cdot (\mathbf{r}_2 \times \mathbf{r}_3) = \left| \frac{\partial x^i}{\partial q^j} \right|.$$

Unlike the previous chapters in which all frame vectors were constant, we now deal with frame vectors that change from point to point. However, at each point we can repeat our prior reasoning. In particular, let us introduce the reciprocal basis $\mathbf{r}^i$ by the relation

$$\mathbf{r}^i \cdot \mathbf{r}_j = \delta^i_j.$$

By the previous chapter we define the metric coefficients

$$g_{ij} = \mathbf{r}_i \cdot \mathbf{r}_j, \qquad g^{ij} = \mathbf{r}^i \cdot \mathbf{r}^j, \qquad g^j_i = \mathbf{r}_i \cdot \mathbf{r}^j = \delta^j_i,$$

which are the components of the metric tensor $\mathbf{E}$ at each point in space.

Above we differentiated a vector function $\mathbf{r}(q^1, q^2, q^3)$. Note that the rules for differentiating vector functions are quite similar to those for differentiating ordinary functions. For brevity we consider the case of fields depending on one variable t. Let $\mathbf{e}_1(t)$ and $\mathbf{e}_2(t)$ be continuously differentiable at some finite t, which means that for $i = 1, 2$ there exist

$$\frac{d\mathbf{e}_i(t)}{dt} = \lim_{\Delta t \to 0} \frac{\mathbf{e}_i(t + \Delta t) - \mathbf{e}_i(t)}{\Delta t}.$$

It is easily seen that

$$\frac{d(\mathbf{e}_1(t) + \mathbf{e}_2(t))}{dt} = \frac{d\mathbf{e}_1(t)}{dt} + \frac{d\mathbf{e}_2(t)}{dt}.$$

Indeed,

$$\frac{d(\mathbf{e}_1(t) + \mathbf{e}_2(t))}{dt} = \lim_{\Delta t \to 0} \frac{\mathbf{e}_1(t + \Delta t) - \mathbf{e}_1(t) + \mathbf{e}_2(t + \Delta t) - \mathbf{e}_2(t)}{\Delta t}$$

$$= \lim_{\Delta t \to 0} \frac{\mathbf{e}_1(t + \Delta t) - \mathbf{e}_1(t)}{\Delta t} + \lim_{\Delta t \to 0} \frac{\mathbf{e}_2(t + \Delta t) - \mathbf{e}_2(t)}{\Delta t}$$

$$= \frac{d\mathbf{e}_1(t)}{dt} + \frac{d\mathbf{e}_2(t)}{dt}.$$

Similarly, for a constant c we have

$$\frac{d(c\mathbf{e}_1(t))}{dt} = c\frac{d\mathbf{e}_1(t)}{dt}.$$

Finally, the product of a scalar function $f(t)$ and a vector function $\mathbf{e}(t)$ is differentiated by a rule similar to the formula for differentiating a product of scalar functions:

$$\frac{d(f(t)\mathbf{e}(t))}{dt} = \frac{df(t)}{dt}\mathbf{e}(t) + f(t)\frac{d\mathbf{e}(t)}{dt}.$$

The reader can adapt the proof for the scalar case almost word for word. The rules for partial differentiation of vector functions in several variables look quite similar and we leave their formulation to the reader. Representing vectors in Cartesian coordinates it is easy to see the validity of the following formulas:

$$\frac{d}{dt}(\mathbf{e}_1(t) \cdot \mathbf{e}_2(t)) = \mathbf{e}_1'(t) \cdot \mathbf{e}_2(t) + \mathbf{e}_1(t) \cdot \mathbf{e}_2'(t),$$

$$\frac{d}{dt}(\mathbf{e}_1(t) \times \mathbf{e}_2(t)) = \mathbf{e}_1'(t) \times \mathbf{e}_2(t) + \mathbf{e}_1(t) \times \mathbf{e}_2'(t).$$

Exercise 4.1. (a) Differentiate $\mathbf{e}_1(t) \cdot \mathbf{e}_2(t)$ and $\mathbf{e}_1(t) \times \mathbf{e}_2(t)$ with respect to t if

$$\mathbf{e}_1(t) = \mathbf{i}_1 e^{-t} + \mathbf{i}_2, \qquad \mathbf{e}_2(t) = -\mathbf{i}_1 \sin^2 t + \mathbf{i}_2 e^{-t}.$$

(b) Show that $[\mathbf{e}(t) \times \mathbf{e}'(t)]' = \mathbf{e}(t) \times \mathbf{e}''(t)$ for any differentiable vector function $\mathbf{e}(t)$.

Exercise 4.2. For the mixed product

$$[\mathbf{e}_1(t), \mathbf{e}_2(t), \mathbf{e}_3(t)] = (\mathbf{e}_1(t) \times \mathbf{e}_2(t)) \cdot \mathbf{e}_3(t)$$

show that

$$\frac{d}{dt}[\mathbf{e}_1(t), \mathbf{e}_2(t), \mathbf{e}_3(t)]$$
$$= [\mathbf{e}_1'(t), \mathbf{e}_2(t), \mathbf{e}_3(t)] + [\mathbf{e}_1(t), \mathbf{e}_2'(t), \mathbf{e}_3(t)] + [\mathbf{e}_1(t), \mathbf{e}_2(t), \mathbf{e}_3'(t)].$$

Cylindrical coordinates

In the cylindrical coordinate system we have

$$(q^1, q^2, q^3) = (\rho, \phi, z),$$

where ρ is the radial distance from the z-axis and ϕ is the azimuthal angle (Fig. 4.3).

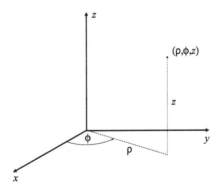

Fig. 4.3   Cylindrical coordinate system.

Using the expression

$$\mathbf{r} = \hat{\mathbf{x}}x + \hat{\mathbf{y}}y + \hat{\mathbf{z}}z$$

for the position vector in rectangular coordinates and the coordinate transformation formulas

$$x = \rho \cos \phi, \qquad y = \rho \sin \phi, \qquad z = z,$$

we have

$$\mathbf{r}(\rho, \phi, z) = \hat{\mathbf{x}} \rho \cos \phi + \hat{\mathbf{y}} \rho \sin \phi + \hat{\mathbf{z}} z.$$

Then

$$\mathbf{r}_1 = \frac{\partial \mathbf{r}(\rho, \phi, z)}{\partial \rho} = \hat{\mathbf{x}} \cos \phi + \hat{\mathbf{y}} \sin \phi.$$

Similarly we compute

$$\mathbf{r}_2 = \frac{\partial \mathbf{r}(\rho, \phi, z)}{\partial \phi} = -\hat{\mathbf{x}} \rho \sin \phi + \hat{\mathbf{y}} \rho \cos \phi, \qquad \mathbf{r}_3 = \frac{\partial \mathbf{r}(\rho, \phi, z)}{\partial z} = \hat{\mathbf{z}}.$$

Then

$$\begin{aligned}
\sqrt{g} &= \mathbf{r}_1 \cdot (\mathbf{r}_2 \times \mathbf{r}_3) \\
&= (\hat{\mathbf{x}} \cos \phi + \hat{\mathbf{y}} \sin \phi) \cdot [(-\hat{\mathbf{x}} \rho \sin \phi + \hat{\mathbf{y}} \rho \cos \phi) \times \hat{\mathbf{z}}] \\
&= (\hat{\mathbf{x}} \cos \phi + \hat{\mathbf{y}} \sin \phi) \cdot (\hat{\mathbf{y}} \rho \sin \phi + \hat{\mathbf{x}} \rho \cos \phi) \\
&= \rho(\cos^2 \phi + \sin^2 \phi) \\
&= \rho.
\end{aligned}$$

It is easy to verify that the frame vectors $\mathbf{r}_1, \mathbf{r}_2, \mathbf{r}_3$ are mutually perpendicular. For instance,

$$\begin{aligned}
\mathbf{r}_1 \cdot \mathbf{r}_2 &= (\hat{\mathbf{x}} \cos \phi + \hat{\mathbf{y}} \sin \phi) \cdot (-\hat{\mathbf{x}} \rho \sin \phi + \hat{\mathbf{y}} \rho \cos \phi) \\
&= -\rho \cos \phi \sin \phi + \rho \cos \phi \sin \phi = 0.
\end{aligned}$$

By direct computation we also find that

$$|\mathbf{r}_1| = 1, \qquad |\mathbf{r}_2| = \rho, \qquad |\mathbf{r}_3| = 1.$$

These facts may be used to construct the reciprocal basis as

$$\mathbf{r}^1 = \mathbf{r}_1, \qquad \mathbf{r}^2 = \mathbf{r}_2/\rho^2, \qquad \mathbf{r}^3 = \mathbf{r}_3.$$

The various metric coefficients for the cylindrical frame are easily computed, and are

$$(g_{ij}) = \begin{pmatrix} 1 & 0 & 0 \\ 0 & \rho^2 & 0 \\ 0 & 0 & 1 \end{pmatrix}, \qquad (g^{ij}) = \begin{pmatrix} 1 & 0 & 0 \\ 0 & 1/\rho^2 & 0 \\ 0 & 0 & 1 \end{pmatrix}.$$

Commonly the *unit basis* vectors $\hat{\rho}$, $\hat{\phi}$, and $\hat{z}$ are introduced in cylindrical coordinates. These are unit vectors along the directions of $\mathbf{r}_1$, $\mathbf{r}_2$, and $\mathbf{r}_3$, respectively, at the point of interest. Given any vector $\mathbf{v}$ at a point (ρ, ϕ, z) in space, it is conventional in applications to write

$$\mathbf{v} = \hat{\rho}v_\rho + \hat{\phi}v_\phi + \hat{z}v_z.$$

The quantities v_ρ, v_ϕ, and v_z are known as the *physical components* of $\mathbf{v}$. (For a vector, the physical components are the projections of the vector on the directions of the corresponding frame vectors.[1]) But since we can also express $\mathbf{v}$ in the forms

$$\mathbf{v} = v^1\mathbf{r}_1 + v^2\mathbf{r}_2 + v^3\mathbf{r}_3 = v_1\mathbf{r}^1 + v_2\mathbf{r}^2 + v_3\mathbf{r}^3,$$

we can identify the contravariant and covariant components of $\mathbf{v}$ from our previous expressions. They are

$$(v^1, v^2, v^3) = (v_\rho, v_\phi/\rho, v_z), \qquad (v_1, v_2, v_3) = (v_\rho, v_\phi\rho, v_z).$$

We see that expression of $\mathbf{v}$ in terms of the basis $(\hat{\rho}, \hat{\phi}, \hat{z})$ yields components (v_ρ, v_ϕ, v_z) that are neither contravariant nor covariant.

Exercise 4.3. Show that the acceleration of a particle in plane polar coordinates (ρ, ϕ) is given by

$$\frac{d^2\mathbf{r}}{dt^2} = \hat{\rho}\left[\frac{d^2\rho}{dt^2} - \rho\left(\frac{d\phi}{dt}\right)^2\right] + \hat{\phi}\left(\rho\frac{d^2\phi}{dt^2} + 2\frac{d\rho}{dt}\frac{d\phi}{dt}\right).$$

Spherical coordinates

In the spherical coordinate system we have

$$(q^1, q^2, q^3) = (r, \theta, \phi),$$

---

[1]Mathematicians often prefer to deal with dimensionless quantities. But in applications many quantities have physical dimensions (e.g., N/m^2 for a stress). When we introduce coordinates in space some can also have dimensions; in spherical coordinates, for instance, r has the dimension of length whereas θ and ϕ are dimensionless. Since $\mathbf{r}$ has the length dimension, the frame vectors corresponding to θ and ϕ must have the length dimension, whereas the frame vector for r is dimensionless. In this way some components of the metric tensor may have dimensions and, when involved in transformation formulas, may thus introduce not only numerical changes but also dimensional changes in the terms so that it becomes possible to add such quantities. When one uses physical components of vectors and tensors they take the dimension of the corresponding tensor in full, and the frame vectors become dimensionless.

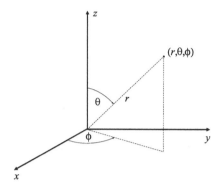

Fig. 4.4   Spherical coordinate system.

where r is the radial distance from the origin, ϕ is the azimuthal angle, and θ is the polar angle (Fig. 4.4). The coordinate transformation formulas

$$x = r \sin \theta \cos \phi, \qquad y = r \sin \theta \sin \phi, \qquad z = r \cos \theta$$

give

$$\mathbf{r}(r, \theta, \phi) = \hat{\mathbf{x}} r \sin \theta \cos \phi + \hat{\mathbf{y}} r \sin \theta \sin \phi + \hat{\mathbf{z}} r \cos \theta.$$

Then

$$\mathbf{r}_1 = \frac{\partial \mathbf{r}(r, \theta, \phi)}{\partial r} = \hat{\mathbf{x}} \sin \theta \cos \phi + \hat{\mathbf{y}} \sin \theta \sin \phi + \hat{\mathbf{z}} \cos \theta,$$

$$\mathbf{r}_2 = \frac{\partial \mathbf{r}(r, \theta, \phi)}{\partial \theta} = \hat{\mathbf{x}} r \cos \theta \cos \phi + \hat{\mathbf{y}} r \cos \theta \sin \phi - \hat{\mathbf{z}} r \sin \theta,$$

$$\mathbf{r}_3 = \frac{\partial \mathbf{r}(r, \theta, \phi)}{\partial \phi} = -\hat{\mathbf{x}} r \sin \theta \sin \phi + \hat{\mathbf{y}} r \sin \theta \cos \phi.$$

In this case

$$\sqrt{g} = \begin{vmatrix} \sin \theta \cos \phi & \sin \theta \sin \phi & \cos \theta \\ r \cos \theta \cos \phi & r \cos \theta \sin \phi & -r \sin \theta \\ -r \sin \theta \sin \phi & r \sin \theta \cos \phi & 0 \end{vmatrix} = r^2 \sin \theta.$$

In this system we find that

$$|\mathbf{r}_1| = 1, \qquad |\mathbf{r}_2| = r, \qquad |\mathbf{r}_3| = r \sin \theta,$$

and the frame vectors are again mutually orthogonal. Hence we have

$$\mathbf{r}^1 = \mathbf{r}_1, \qquad \mathbf{r}^2 = \mathbf{r}_2/r^2, \qquad \mathbf{r}^3 = \mathbf{r}_3/r^2 \sin^2 \theta$$

for the vectors of the reciprocal basis. The metric coefficients are

$$(g_{ij}) = \begin{pmatrix} 1 & 0 & 0 \\ 0 & r^2 & 0 \\ 0 & 0 & r^2\sin^2\theta \end{pmatrix}, \qquad (g^{ij}) = \begin{pmatrix} 1 & 0 & 0 \\ 0 & 1/r^2 & 0 \\ 0 & 0 & 1/r^2\sin^2\theta \end{pmatrix}.$$

Exercise 4.4. The unit basis vectors of spherical coordinates are denoted $\hat{\mathbf{r}}$, $\hat{\boldsymbol{\theta}}$, and $\hat{\boldsymbol{\phi}}$. Give expressions relating the various sets of components (v^1, v^2, v^3), (v_1, v_2, v_3), and (v_r, v_θ, v_ϕ) of a vector $\mathbf{v}$.

We see that both the spherical and cylindrical frames are orthogonal. In this case the reciprocal basis vectors have the same directions as the vectors of the main basis, but have reciprocal lengths so that $|\mathbf{r}_i||\mathbf{r}^i| = 1$ for each i.

In § 2.4 we obtained the transformation laws that apply to the components of a vector under a change of frame. These laws are a consequence of the change of the old curvilinear coordinates q^k to other coordinates $\tilde{q}^j$, so

$$\tilde{q}^j = \tilde{q}^j(q^1, q^2, q^3), \quad q^k = q^k(\tilde{q}^1, \tilde{q}^2, \tilde{q}^3) \qquad (j, k = 1, 2, 3).$$

In the coordinates $\tilde{q}^j$, we denote all the quantities in the same manner as for the coordinates q^k, but with the tilde above: $\tilde{\mathbf{r}}^k$, $\tilde{\mathbf{r}}_k$, and so on. We now extend the transformation laws to the case of a vector field in general coordinates. Because

$$\mathbf{r}_i = \frac{\partial \mathbf{r}}{\partial q^i} = \frac{\partial \mathbf{r}}{\partial \tilde{q}^j}\frac{\partial \tilde{q}^j}{\partial q^i} = \tilde{\mathbf{r}}_j \frac{\partial \tilde{q}^j}{\partial q^i}$$

we can write

$$\mathbf{r}_i = A_i^j \tilde{\mathbf{r}}_j, \qquad A_i^j = \frac{\partial \tilde{q}^j}{\partial q^i}, \tag{4.3}$$

to describe the change of frame. For the inverse transformation

$$\tilde{\mathbf{r}}_i = \tilde{A}_i^j \mathbf{r}_j, \qquad \tilde{A}_i^j = \frac{\partial q^j}{\partial \tilde{q}^i}. \tag{4.4}$$

The results of § 2.4 now apply, and we can write immediately

$$\tilde{f}^i = A_j^i f^j, \qquad \tilde{f}_i = \tilde{A}_i^j f_j, \qquad f^i = \tilde{A}_j^i \tilde{f}^j, \qquad f_i = A_i^j \tilde{f}_j,$$

for the transformation laws pertaining to the components of a vector field $\mathbf{f}$. The f^i are still termed contravariant components of $\mathbf{f}$, while the f_i are still termed covariant components — the main difference is that now A_i^j and $\tilde{A}_i^j$ can change from point to point in space. The transformation laws for the components of higher-order tensor fields are written similarly.

Exercise 4.5. A set of *oblique rectilinear coordinates* (u, v) in the plane is related to a set of Cartesian coordinates (x, y) by the transformation equations

$$x = u \cos \alpha + v \cos \beta, \qquad y = u \sin \alpha + v \sin \beta,$$

where α, β are constants. (a) Sketch a few u and v coordinate curves superimposed on the xy-plane. (b) Find the basis vectors $\mathbf{r}_i$ $(i = 1, 2)$ and the reciprocal basis vectors $\mathbf{r}^i$ in the (u, v) system. Use these to calculate the metric coefficients. (c) Let $\mathbf{z}$ be a given vector. In the oblique system find the covariant components of $\mathbf{z}$ in terms of the contravariant components of $\mathbf{z}$.

4.2   Differentials and the Nabla Operator

Let us consider the infinitesimal vector extending from point (q^1, q^2, q^3) to point $(q^1 + dq^1, q^2 + dq^2, q^3 + dq^3)$, denoted by $d\mathbf{r}$:

$$d\mathbf{r} = \frac{\partial \mathbf{r}}{\partial q^i} \, dq^i = \mathbf{r}_i \, dq^i.$$

Hence we can define dq^i as

$$dq^i = \mathbf{r}^i \cdot d\mathbf{r} = d\mathbf{r} \cdot \mathbf{r}^i. \tag{4.5}$$

(Here there is no summation over i. Note that we only introduce spatial coordinates having superscripts. However, for Cartesian frames it is common to see subscripts used exclusively.) The length of this infinitesimal vector is defined by

$$(ds)^2 = d\mathbf{r} \cdot d\mathbf{r} = \mathbf{r}_i \, dq^i \cdot \mathbf{r}_j \, dq^j = g_{ij} \, dq^i \, dq^j. \tag{4.6}$$

On the right we have a quadratic form with respect to the variables dq^i; the coefficients of this quadratic form are the covariant components of the metric tensor. In a Cartesian frame (4.6) takes the familiar form

$$(ds)^2 = (dx^1)^2 + (dx^2)^2 + (dx^3)^2.$$

In cylindrical and spherical frames we have

$$(ds)^2 = (d\rho)^2 + \rho^2 (d\phi)^2 + (dz)^2,$$
$$(ds)^2 = (dr)^2 + r^2 (d\theta)^2 + r^2 \sin^2 \theta (d\phi)^2,$$

respectively.

Exercise 4.6. Find $(ds)^2$ for the oblique system of Exercise 4.5.

Let $f(q^1, q^2, q^3)$ be a scalar differentiable function of the variables q^i. Its differential is

$$df(q^1, q^2, q^3) = \frac{\partial f(q^1, q^2, q^3)}{\partial q^i} \, dq^i.$$

Here and in similar situations the i in the denominator stands in the lower position and so we must sum over i. Using (4.5) we can write

$$df = \mathbf{r}^i \frac{\partial f}{\partial q^i} \cdot d\mathbf{r}.$$

The first multiplier on the right is a vector $\mathbf{r}^i \partial f / \partial q^i$. Let us introduce a symbolic vector

$$\nabla = \mathbf{r}^i \frac{\partial}{\partial q^i}$$

called the *nabla operator*, whose action on a function f is as given above:

$$\nabla f = \mathbf{r}^i \frac{\partial f}{\partial q^i}$$

(we repeat that there is summation over i here). The nabla operator is often referred to as the gradient operator.

Exercise 4.7. (a) Show that the unit normal to the surface $\varphi(q^1, q^2, q^3) = c = \text{constant}$ is given by

$$\mathbf{n} = \frac{g^{ij} \dfrac{\partial \varphi}{\partial q^i}}{\sqrt{g^{mn} \dfrac{\partial \varphi}{\partial q^m} \dfrac{\partial \varphi}{\partial q^n}}} \mathbf{r}_j.$$

(b) Show that the angle at a point of intersection of the surfaces

$$\varphi(q^1, q^2, q^3) = c_1, \qquad \psi(q^1, q^2, q^3) = c_2,$$

is given by

$$\cos \theta = \frac{g^{ij} \dfrac{\partial \varphi}{\partial q^i} \dfrac{\partial \psi}{\partial q^j}}{\sqrt{g^{mn} \dfrac{\partial \varphi}{\partial q^m} \dfrac{\partial \varphi}{\partial q^n} g^{rt} \dfrac{\partial \psi}{\partial q^r} \dfrac{\partial \psi}{\partial q^t}}}$$

(c) Show that the angle between the coordinate surfaces $q^1 = c_1, q^2 = c_2$ is given by

$$\cos \theta_{12} = \frac{g^{12}}{\sqrt{g^{11} g^{22}}}.$$

(d) Derive the condition for orthogonality of surfaces

$$g^{ij} \frac{\partial \varphi}{\partial q^i} \frac{\partial \psi}{\partial q^j} = 0.$$

Exercise 4.8. Show that the gradient operation in the cylindrical and spherical frames is given by the formulas

$$\nabla f = \hat{\boldsymbol{\rho}} \frac{\partial f}{\partial \rho} + \hat{\boldsymbol{\phi}} \frac{1}{\rho} \frac{\partial f}{\partial \phi} + \hat{\mathbf{z}} \frac{\partial f}{\partial z},$$

$$\nabla f = \hat{\mathbf{r}} \frac{\partial f}{\partial r} + \hat{\boldsymbol{\theta}} \frac{1}{r} \frac{\partial f}{\partial \theta} + \hat{\boldsymbol{\phi}} \frac{1}{r \sin \theta} \frac{\partial f}{\partial \phi},$$

respectively. What are the expressions for these when using the components connected with the triads $\mathbf{r}_i$ and $\mathbf{r}^i$?

Let us find a formula for the differential of a vector function $\mathbf{f}(q^1, q^2, q^3)$:

$$d\mathbf{f}(q^1, q^2, q^3) = \frac{\partial \mathbf{f}(q^1, q^2, q^3)}{\partial q^i} \, dq^i.$$

We use (4.5) again. Then

$$d\mathbf{f}(q^1, q^2, q^3) = \frac{\partial \mathbf{f}(q^1, q^2, q^3)}{\partial q^i} (\mathbf{r}^i \cdot d\mathbf{r})$$

$$= (d\mathbf{r} \cdot \mathbf{r}^i) \frac{\partial \mathbf{f}(q^1, q^2, q^3)}{\partial q^i}$$

$$= d\mathbf{r} \cdot \left(\mathbf{r}^i \frac{\partial \mathbf{f}(q^1, q^2, q^3)}{\partial q^i} \right).$$

Thus we can represent this as

$$d\mathbf{f} = d\mathbf{r} \cdot \nabla \mathbf{f}.$$

The quantity $\nabla \mathbf{f}$, known as the *gradient* of $\mathbf{f}$, is clearly a tensor of order two. With the aid of transposition we can present $d\mathbf{f}$ in the form

$$d\mathbf{f} = \nabla \mathbf{f}^T \cdot d\mathbf{r}.$$

Sometimes $\nabla \mathbf{f}^T$ is called the gradient of $\mathbf{f}$; it is also called the derivative of $\mathbf{f}$ in the direction of $\mathbf{r}$ and denoted as

$$\frac{d\mathbf{f}}{d\mathbf{r}} = \nabla \mathbf{f}^T.$$

An application of the gradient to a tensor brings a new tensor whose order is one higher than that of the original tensor.

For a pair of vectors, we introduced two types of multiplication: the dot and cross product operations. We can apply these to the pair consisting of the nabla operator (which is regarded as a formal vector) and a vector function $\mathbf{f} = \mathbf{f}(q^1, q^2, q^3)$. In this way we get two operations: the *divergence*

$$\operatorname{div} \mathbf{f} = \nabla \cdot \mathbf{f} = \mathbf{r}^i \cdot \frac{\partial \mathbf{f}}{\partial q^i},$$

and the *rotation*

$$\operatorname{rot} \mathbf{f} = \nabla \times \mathbf{f} = \mathbf{r}^i \times \frac{\partial \mathbf{f}}{\partial q^i}. \tag{4.7}$$

The vector

$$\boldsymbol{\omega} = \frac{1}{2} \operatorname{rot} \mathbf{f}$$

is called the *curl* of $\mathbf{f}$. In terms of the curl of $\mathbf{f}$ we can introduce a tensor $\boldsymbol{\Omega}$ as

$$\boldsymbol{\Omega} = \mathbf{E} \times \boldsymbol{\omega}.$$

This tensor is antisymmetric (the reader should verify this), and is called the tensor of spin. It can be shown that

$$\boldsymbol{\Omega} = \frac{1}{2} \left(\nabla \mathbf{f}^T - \nabla \mathbf{f} \right).$$

Similarly we can introduce the divergence and rotation operations for a tensor $\mathbf{A}$ of any order:

$$\nabla \cdot \mathbf{A} = \mathbf{r}^i \cdot \frac{\partial}{\partial q^i} \mathbf{A}, \qquad \nabla \times \mathbf{A} = \mathbf{r}^i \times \frac{\partial}{\partial q^i} \mathbf{A}.$$

Let us see what happens when we apply the nabla operator to the radius vector:

$$\nabla \mathbf{r} = \mathbf{r}^i \frac{\partial}{\partial q^i} \mathbf{r} = \mathbf{r}^i \mathbf{r}_i = \mathbf{E},$$

$$\nabla \cdot \mathbf{r} = \mathbf{r}^i \cdot \frac{\partial}{\partial q^i} \mathbf{r} = \mathbf{r}^i \cdot \mathbf{r}_i = 3,$$

and

$$\nabla \times \mathbf{r} = \mathbf{r}^i \times \frac{\partial}{\partial q^i} \mathbf{r} = \mathbf{r}^i \times \mathbf{r}_i = \mathbf{0}.$$

Exercise 4.9. Calculate $\nabla \mathbf{E}$, $\nabla \cdot \mathbf{E}$, $\nabla \times \mathbf{E}$, $\nabla \boldsymbol{\mathcal{E}}$, $\nabla \cdot \boldsymbol{\mathcal{E}}$, and $\nabla \times \boldsymbol{\mathcal{E}}$.

Exercise 4.10. Let f and g be functions, let $\mathbf{f}$ be a vector, and let $\mathbf{Q}$ be a tensor of order two. Verify the following identities:

$$\nabla(fg) = g\nabla f + f\nabla g,$$
$$\nabla(f\mathbf{f}) = (\nabla f)\mathbf{f} + f\nabla\mathbf{f},$$
$$\nabla(f\mathbf{Q}) = (\nabla f)\mathbf{Q} + f\nabla\mathbf{Q}.$$

Exercise 4.11. Show that the following identities hold ($\mathbf{f}$ and $\mathbf{g}$ are any vectors):

$$\nabla(\mathbf{f} \cdot \mathbf{g}) = (\nabla\mathbf{f}) \cdot \mathbf{g} + \mathbf{f} \cdot \nabla\mathbf{g}^T = (\nabla\mathbf{f}) \cdot \mathbf{g} + (\nabla\mathbf{g}) \cdot \mathbf{f},$$
$$\nabla(\mathbf{f} \times \mathbf{g}) = (\nabla\mathbf{f}) \times \mathbf{g} - (\nabla\mathbf{g}) \times \mathbf{f},$$
$$\nabla \times (\mathbf{f} \times \mathbf{g}) = \mathbf{g} \cdot \nabla\mathbf{f} - \mathbf{g}\nabla \cdot \mathbf{f} - \mathbf{f} \cdot \nabla\mathbf{g} + \mathbf{f}\nabla \cdot \mathbf{g},$$
$$\nabla \cdot (\mathbf{f}\mathbf{g}) = (\nabla \cdot \mathbf{f})\mathbf{g} + \mathbf{f}\nabla \cdot \mathbf{g},$$
$$\nabla \cdot (\mathbf{f} \times \mathbf{g}) = \mathbf{g} \cdot (\nabla \times \mathbf{f}) - \mathbf{f} \cdot (\nabla \times \mathbf{g}).$$

4.3   Differentiation of a Vector Function

We have differentiated vector functions with respect to q^i, with the tacit understanding that the formulas necessary to do this resemble those from ordinary calculus. Moreover, the reader certainly knows that to differentiate a vector function

$$\mathbf{f}(x_1, x_2, x_3) = f^k(x_1, x_2, x_3)\mathbf{i}_k$$

with respect to a Cartesian variable x_i it is enough to differentiate each component of $\mathbf{f}$ with respect to this variable:

$$\frac{\partial}{\partial x_i}\mathbf{f}(x_1, x_2, x_3) = \frac{\partial f^k(x_1, x_2, x_3)}{\partial x_i}\mathbf{i}_k. \tag{4.8}$$

Let us consider how to differentiate a vector function $\mathbf{f}$ that is written out in curvilinear coordinates:

$$\mathbf{f}(q^1, q^2, q^3) = f^i(q^1, q^2, q^3)\mathbf{r}_i.$$

In this case the frame vectors $\mathbf{r}_i$ depend on q^k as well, which means that simple differentiation of the components of $\mathbf{f}$ does not result in the needed formula. To understand this let us consider a constant function $\mathbf{f}(q^1, q^2, q^3) = \mathbf{c}$. By the general definition of the partial derivative we have $\partial\mathbf{f}(q^1, q^2, q^3)/\partial q^i = \mathbf{0}$. But the components of $\mathbf{f}$ are not constant since the

$\mathbf{r}_k$ are variable, hence the derivatives of the components of this function are not zero.

The derivative of the product of a simple function and a vector function is taken by the product rule:

$$\frac{\partial}{\partial q^k}\mathbf{f}(q^1, q^2, q^3) = \frac{\partial}{\partial q^k}\left[f^i(q^1, q^2, q^3)\mathbf{r}_i\right]$$

$$= \frac{\partial f^i(q^1, q^2, q^3)}{\partial q^k}\mathbf{r}_i + f^i(q^1, q^2, q^3)\frac{\partial \mathbf{r}_i}{\partial q^k}.$$

So the derivative of a vector function written in component form consists of two terms. The first is the same as for the derivative in the Cartesian frame (4.8), and the other contains the derivatives of the frame vectors. Thus we need to find the latter.

4.4 Derivatives of the Frame Vectors

We would like to find the value of the derivative

$$\frac{\partial}{\partial q^j}\mathbf{r}_i = \frac{\partial}{\partial q^j}\left(\frac{\partial}{\partial q^i}\mathbf{r}\right) = \frac{\partial^2 \mathbf{r}}{\partial q^j \partial q^i} = \frac{\partial^2 \mathbf{r}}{\partial q^i \partial q^j} = \frac{\partial}{\partial q^i}\mathbf{r}_j. \qquad (4.9)$$

Of course if we have the expression for the Cartesian components of the frame vector we can compute these derivatives (and will do so in the exercises); however in the general case we exploit only the fact that the partial derivative of a frame vector is a vector as well, hence it can be expanded in the same frame $\mathbf{r}_i$ ($i = 1, 2, 3$). Let us denote the coefficients of the expansion by Γ^k_{ij}:

$$\frac{\partial}{\partial q^j}\mathbf{r}_i = \Gamma^k_{ij}\mathbf{r}_k. \qquad (4.10)$$

The quantities Γ^k_{ij} are called *Christoffel coefficients of the second kind* and are often denoted by

$$\Gamma^k_{ij} = \left\{\begin{array}{c} k \\ ij \end{array}\right\}. \qquad (4.11)$$

By (4.9) there is symmetry in the subscripts of the Christoffel symbols:

$$\Gamma^k_{ij} = \Gamma^k_{ji}. \qquad (4.12)$$

Now we can write out the formula for the derivative of a vector function in full:

$$\frac{\partial}{\partial q^k}\mathbf{f}(q^1,q^2,q^3) = \frac{\partial f^i(q^1,q^2,q^3)}{\partial q^k}\mathbf{r}_i + \Gamma^j_{ki}f^i(q^1,q^2,q^3)\mathbf{r}_j$$

$$= \frac{\partial f^i(q^1,q^2,q^3)}{\partial q^k}\mathbf{r}_i + \Gamma^i_{kt}f^t(q^1,q^2,q^3)\mathbf{r}_i$$

$$= \left(\frac{\partial f^i}{\partial q^k} + \Gamma^i_{kt}f^t\right)\mathbf{r}_i. \tag{4.13}$$

This is called *covariant differentiation*. The coefficients of $\mathbf{r}_i$ are called the *covariant derivatives* of the contravariant components of $\mathbf{f}$, and are denoted by

$$\nabla_k f^i = \frac{\partial f^i}{\partial q^k} + \Gamma^i_{kt}f^t.$$

Let us discuss some properties of the Christoffel coefficients.

4.5   Christoffel Coefficients and their Properties

In a Cartesian frame the Christoffel symbols are zero, so it is impossible to obtain them for another frame by the usual transformation rules for tensor components. This is easy to understand: the Christoffel symbols depend not only on the frame vectors themselves but on their rates of change from point to point, and these rates do not appear in the transformation rules for tensors. So the Christoffel symbols are not the components of a tensor, despite their notation. This is why many authors prefer the notation shown on the right side of equation (4.11).

It is important to have formulas for computing the Christoffel symbols. Let us introduce the notation

$$\mathbf{r}_{ij} = \frac{\partial}{\partial q^j}\mathbf{r}_i = \frac{\partial^2 \mathbf{r}}{\partial q^i \partial q^j}.$$

Relation (4.10) can be written as

$$\mathbf{r}_{ij} = \Gamma^k_{ij}\mathbf{r}_k \tag{4.14}$$

from which it follows that

$$\mathbf{r}_{ij} \cdot \mathbf{r}_t = \Gamma^k_{ij}\mathbf{r}_k \cdot \mathbf{r}_t = \Gamma^k_{ij}g_{kt}.$$

The left-hand side of this can be expressed in terms of the components of the metric tensor. Indeed

$$\frac{\partial}{\partial q^j} g_{it} = \frac{\partial}{\partial q^j} \mathbf{r}_i \cdot \mathbf{r}_t = \mathbf{r}_{ij} \cdot \mathbf{r}_t + \mathbf{r}_i \cdot \mathbf{r}_{tj}. \tag{4.15}$$

Similarly

$$\frac{\partial}{\partial q^t} g_{ji} = \frac{\partial}{\partial q^t} \mathbf{r}_j \cdot \mathbf{r}_i = \mathbf{r}_{jt} \cdot \mathbf{r}_i + \mathbf{r}_j \cdot \mathbf{r}_{it}, \tag{4.16}$$

$$\frac{\partial}{\partial q^i} g_{tj} = \frac{\partial}{\partial q^i} \mathbf{r}_t \cdot \mathbf{r}_j = \mathbf{r}_{ti} \cdot \mathbf{r}_j + \mathbf{r}_t \cdot \mathbf{r}_{ji}. \tag{4.17}$$

Now we obtain $\mathbf{r}_{ij} \cdot \mathbf{r}_t$ by subtracting (4.16) from the sum of (4.15) and (4.17) and dividing by 2:

$$\mathbf{r}_{ij} \cdot \mathbf{r}_t = \frac{1}{2} \left(\frac{\partial g_{it}}{\partial q^j} + \frac{\partial g_{tj}}{\partial q^i} - \frac{\partial g_{ji}}{\partial q^t} \right) = \Gamma_{ijt}. \tag{4.18}$$

The quantities Γ_{ijk} are called *Christoffel coefficients of the first kind*. They are denoted frequently as

$$[ij, k] = \Gamma_{ijk}.$$

It is clear that they are symmetric in the first two subscripts: $\Gamma_{ijk} = \Gamma_{jik}$. Thus we have

$$\Gamma_{ij}^k g_{kt} = \Gamma_{ijt} \tag{4.19}$$

and it follows that

$$\Gamma_{ij}^k = g^{kt} \Gamma_{ijt}. \tag{4.20}$$

Using these we can obtain

$$\frac{\partial \mathbf{r}^j}{\partial q^i} = -\Gamma_{it}^j \mathbf{r}^t. \tag{4.21}$$

Indeed

$$0 = \frac{\partial}{\partial q^t} \delta_j^i = \frac{\partial}{\partial q^t} \left(\mathbf{r}^i \cdot \mathbf{r}_j \right) = \frac{\partial \mathbf{r}^i}{\partial q^t} \cdot \mathbf{r}_j + \mathbf{r}^i \cdot \mathbf{r}_{jt} = \frac{\partial \mathbf{r}^i}{\partial q^t} \cdot \mathbf{r}_j + \mathbf{r}^i \cdot \Gamma_{jt}^k \mathbf{r}_k,$$

so

$$\frac{\partial \mathbf{r}^i}{\partial q^t} \cdot \mathbf{r}_j = -\mathbf{r}^i \cdot \Gamma_{jt}^k \mathbf{r}_k = -\Gamma_{jt}^k \delta_k^i = -\Gamma_{jt}^i$$

and we have

$$\frac{\partial \mathbf{r}^i}{\partial q^t} = -\Gamma_{jt}^i \mathbf{r}^j.$$

It is also useful to have formulas for transformation of the Christoffel symbols under change of coordinates. Suppose the new coordinates $\tilde{q}^i$ are determined by the old coordinates q^i through relations of the form $\tilde{q}^i = \tilde{q}^i(q^1, q^2, q^3)$, and refer back to equations (4.3)–(4.4). Now (4.14) applied in the new system gives

$$\tilde{\Gamma}^k_{ij} \tilde{\mathbf{r}}_k = \tilde{\mathbf{r}}_{ij}. \tag{4.22}$$

But

$$\tilde{\mathbf{r}}_{ij} = \frac{\partial}{\partial \tilde{q}^j} \tilde{\mathbf{r}}_i = \frac{\partial q^m}{\partial \tilde{q}^j} \frac{\partial}{\partial q^m} \tilde{\mathbf{r}}_i = \tilde{A}^m_j \frac{\partial}{\partial q^m} (\tilde{A}^n_i \mathbf{r}_n),$$

and expansion by the product rule gives

$$\tilde{\mathbf{r}}_{ij} = \tilde{A}^m_j \left(\frac{\partial \tilde{A}^n_i}{\partial q^m} \mathbf{r}_n + \tilde{A}^n_i \frac{\partial \mathbf{r}_n}{\partial q^m} \right)$$

$$= \tilde{A}^m_j \frac{\partial \tilde{A}^n_i}{\partial q^m} \mathbf{r}_n + \tilde{A}^m_j \tilde{A}^n_i \Gamma^p_{nm} \mathbf{r}_p$$

$$= \tilde{A}^m_j \frac{\partial \tilde{A}^n_i}{\partial q^m} A^k_n \tilde{\mathbf{r}}_k + \tilde{A}^m_j \tilde{A}^n_i \Gamma^p_{nm} A^k_p \tilde{\mathbf{r}}_k.$$

Comparison with (4.22) shows that

$$\tilde{\Gamma}^k_{ij} = \tilde{A}^m_j \frac{\partial \tilde{A}^n_i}{\partial q^m} A^k_n + \tilde{A}^m_j \tilde{A}^n_i A^k_p \Gamma^p_{nm}.$$

The presence of the first term on the right means that the Christoffel coefficient is not a tensor of order three. This confirms our earlier statement of this fact, which was based on different reasoning.

Exercise 4.12. Show that the only nonzero Christoffel coefficients of the first kind for cylindrical coordinates are

$$\Gamma_{221} = -\rho, \qquad \Gamma_{122} = \Gamma_{212} = \rho.$$

Show that the nonzero Christoffel coefficients of the first kind for spherical coordinates are

$$\Gamma_{221} = -r, \qquad\qquad \Gamma_{122} = \Gamma_{212} = r,$$
$$\Gamma_{331} = -r \sin^2 \theta, \qquad\qquad \Gamma_{332} = -r^2 \sin \theta \cos \theta,$$
$$\Gamma_{313} = \Gamma_{133} = r \sin^2 \theta, \qquad\qquad \Gamma_{233} = \Gamma_{323} = r^2 \sin \theta \cos \theta.$$

Exercise 4.13. Show that the only nonzero Christoffel coefficients of the second kind for cylindrical coordinates are

$$\Gamma_{22}^1 = -\rho, \qquad \Gamma_{12}^2 = \Gamma_{21}^2 = 1/\rho.$$

Show that the nonzero Christoffel coefficients of the second kind for spherical coordinates are

$$\Gamma_{22}^1 = -r, \qquad\qquad \Gamma_{12}^2 = \Gamma_{21}^2 = 1/r,$$
$$\Gamma_{33}^1 = -r\sin^2\theta, \qquad\qquad \Gamma_{33}^2 = -\sin\theta\cos\theta,$$
$$\Gamma_{31}^3 = \Gamma_{13}^3 = 1/r, \qquad\qquad \Gamma_{23}^3 = \Gamma_{32}^3 = \cot\theta.$$

Exercise 4.14. A system of plane elliptic coordinates (u, v) is introduced according to the transformation formulas

$$x = c\cosh u \cos v, \qquad y = c\sinh u \sin v,$$

where c is a constant. Find the Christoffel coefficients.

Euclidean vs. non-Euclidean spaces

When we derive the length of the elementary vector $d\mathbf{r}$ we write

$$(ds)^2 = d\mathbf{r} \cdot d\mathbf{r} = \mathbf{r}_i \, dq^i \cdot \mathbf{r}_j \, dq^j = g_{ij} \, dq^i \, dq^j. \tag{4.23}$$

With respect to the variables dq^i and dq^j we have, by construction, a positive definite quadratic form. This is one of the main properties of the metric tensor for a real coordinate space.

If a space is *Euclidean*, that is, if it can be described by a set of Cartesian coordinates x^t ($t = 1, 2, 3$), then the line element $(ds)^2$ can be written as

$$(ds)^2 = \sum_{t=1}^{3} (dx^t)^2. \tag{4.24}$$

Given a set of admissible transformation equations $x^t = x^t(q^n)$, the same line element can be expressed in terms of general coordinates q^n. Since

$$dx^t = \frac{\partial x^t}{\partial q^n} \, dq^n$$

we have

$$(ds)^2 = \sum_{t=1}^{3} \left(\frac{\partial x^t}{\partial q^n} \, dq^n \right)^2,$$

and comparison with (4.23) gives

$$g_{ij} = \sum_{t=1}^{3} \frac{\partial x^t}{\partial q^i} \frac{\partial x^t}{\partial q^j} \qquad (i, j = 1, 2, 3), \qquad (4.25)$$

for the metric coefficients in the q^n system. We now pose the following question. Suppose a space is originally described in terms of a set of general coordinates q^n. Such a description must include a set of metric coefficients g_{ij} having (4.23) positive definite at each point. Will it be possible to introduce a set of Cartesian coordinates x^t that also describe this space? That is, are we guaranteed the existence of functions $x^t(q^n)$ such that in the resulting coordinates $x^t = x^t(q^n)$ the line element takes the form (4.24)? Put still another way, *is the space Euclidean?* Not necessarily. Given the g_{ij} in the q^n system, (4.25) provides us with six equations for the three unknown functions $x^t(q^n)$. This implies that some additional conditions must be fulfilled by the given metric coefficients. It is possible to formulate these restrictions neatly in terms of a certain fourth-order tensor $R^p_{.ijk}$ as the equality

$$R^p_{.ijk} = 0. \qquad (4.26)$$

The tensor $R^p_{.ijk}$ is known as the *Riemann–Christoffel tensor*, and its associated tensor

$$R_{nijk} = g_{np} R^p_{.ijk}$$

is called the *curvature tensor* for the space. It can be shown that

$$R^p_{.ijk} = \frac{\partial \Gamma^p_{ij}}{\partial q^k} - \frac{\partial \Gamma^p_{ik}}{\partial q^j} - \left(\Gamma^p_{mj}\Gamma^m_{ik} - \Gamma^p_{mk}\Gamma^m_{ij} \right).$$

Moreover, $R^p_{.ijk}$ actually has only six independent components, so (4.26) represents six conditions on the g_{ij} that must hold for the space described by the g_{ij} to be Euclidean [Sokolnikoff (1994)].

We have included this information only for the reader's background, as such considerations are important in certain application areas. Further details and more rigorous formulations can be found in some of the more comprehensive references.

4.6 Covariant Differentiation

Let us return to the problem of differentiation of a vector function. We obtained the formula for the derivative (4.13)

$$\frac{\partial}{\partial q^k}\mathbf{f} = \frac{\partial}{\partial q^k}(f^i\mathbf{r}_i) = \left(\frac{\partial f^i}{\partial q^k} + \Gamma^i_{kt}f^t\right)\mathbf{r}_i$$

and observed that the coefficients of this expansion denoted by

$$\nabla_k f^i = \frac{\partial f^i}{\partial q^k} + \Gamma^i_{kt}f^t \tag{4.27}$$

are called the covariant derivatives of contravariant components of vector function $\mathbf{f}$. Let us express the same derivatives in terms of the covariant components:

$$\frac{\partial}{\partial q^k}\mathbf{f} = \frac{\partial}{\partial q^k}(f_i\mathbf{r}^i) = \frac{\partial f_i}{\partial q^k}\mathbf{r}^i + f_i\frac{\partial}{\partial q^k}\mathbf{r}^i.$$

Using (4.21) we get

$$\frac{\partial}{\partial q^k}\mathbf{f} = \frac{\partial f_i}{\partial q^k}\mathbf{r}^i - f_i\Gamma^i_{kt}\mathbf{r}^t = \left(\frac{\partial f_i}{\partial q^k} - \Gamma^j_{ki}f_j\right)\mathbf{r}^i.$$

The coefficients of this expansion are called covariant derivatives of the covariant components and are denoted by

$$\nabla_k f_i = \frac{\partial f_i}{\partial q^k} - \Gamma^j_{ki}f_j. \tag{4.28}$$

In these notations we can write out the formula for differentiation in the form

$$\frac{\partial \mathbf{f}}{\partial q^i} = \mathbf{r}^k\nabla_i f_k = \mathbf{r}_j\nabla_i f^j. \tag{4.29}$$

Because

$$\nabla \mathbf{f} = \mathbf{r}^i\frac{\partial}{\partial q^i}(f_j\mathbf{r}^j) = \mathbf{r}^i(\nabla_i f_j)\mathbf{r}^j = \mathbf{r}^i\mathbf{r}^j\nabla_i f_j,$$

we see that $\nabla_i f_j$ is a covariant component of $\nabla\mathbf{f}$. Similarly,

$$\nabla \mathbf{f} = \mathbf{r}^i\frac{\partial}{\partial q^i}(f^j\mathbf{r}_j) = \mathbf{r}^i\mathbf{r}_j\nabla_i f^j$$

shows that $\nabla_i f^j$ is a mixed component of $\nabla\mathbf{f}$.

Exercise 4.15. Write out the covariant derivatives of the contravariant components of a vector $\mathbf{f}$ in plane polar coordinates.

Quick summary

The formulas for differentiation of a vector are

$$\frac{\partial \mathbf{f}}{\partial q^i} = \mathbf{r}^k \nabla_i f_k = \mathbf{r}_j \nabla_i f^j$$

where

$$\nabla_k f_i = \frac{\partial f_i}{\partial q^k} - \Gamma_{ki}^j f_j, \qquad \nabla_k f^i = \frac{\partial f^i}{\partial q^k} + \Gamma_{kt}^i f^t.$$

The formulas

$$\nabla \mathbf{f} = \mathbf{r}^i \mathbf{r}^j \nabla_i f_j = \mathbf{r}^i \mathbf{r}_j \nabla_i f^j$$

show that $\nabla_i f_j$ is a covariant component of $\nabla \mathbf{f}$, while $\nabla_i f^j$ is a mixed component of $\nabla \mathbf{f}$.

4.7   Covariant Derivative of a Second-Order Tensor

Let us find a partial derivative of a tensor $\mathbf{A}$ of order two:

$$\begin{aligned}
\frac{\partial}{\partial q^k} \mathbf{A} &= \frac{\partial}{\partial q^k}(a^{ij} \mathbf{r}_i \mathbf{r}_j) \\
&= \frac{\partial a^{ij}}{\partial q^k} \mathbf{r}_i \mathbf{r}_j + a^{ij}(\mathbf{r}_{ik} \mathbf{r}_j + \mathbf{r}_i \mathbf{r}_{jk}) \\
&= \frac{\partial a^{ij}}{\partial q^k} \mathbf{r}_i \mathbf{r}_j + a^{ij}(\Gamma_{ik}^t \mathbf{r}_t \mathbf{r}_j + \mathbf{r}_i \Gamma_{jk}^t \mathbf{r}_t).
\end{aligned}$$

Changing dummy indices we get

$$\frac{\partial}{\partial q^k} \mathbf{A} = \left(\frac{\partial a^{ij}}{\partial q^k} + \Gamma_{ks}^i a^{sj} + \Gamma_{ks}^j a^{is} \right) \mathbf{r}_i \mathbf{r}_j.$$

The parenthetical expression is designated as a covariant derivative:

$$\nabla_k a^{ij} = \frac{\partial a^{ij}}{\partial q^k} + \Gamma_{ks}^i a^{sj} + \Gamma_{ks}^j a^{is}.$$

Similarly

$$\frac{\partial}{\partial q^k}\mathbf{A} = \frac{\partial}{\partial q^k}(a_{ij}\mathbf{r}^i\mathbf{r}^j)$$

$$= \frac{\partial a_{ij}}{\partial q^k}\mathbf{r}^i\mathbf{r}^j + a_{ij}\left(\frac{\partial \mathbf{r}^i}{\partial q^k}\mathbf{r}^j + \mathbf{r}^i\frac{\partial \mathbf{r}^j}{\partial q^k}\right)$$

$$= \frac{\partial a_{ij}}{\partial q^k}\mathbf{r}^i\mathbf{r}^j - a_{ij}(\Gamma^i_{kt}\mathbf{r}^t\mathbf{r}^j + \mathbf{r}^i\Gamma^j_{kt}\mathbf{r}^t)$$

$$= \left(\frac{\partial a_{ij}}{\partial q^k} - \Gamma^s_{ki}a_{sj} - \Gamma^s_{kj}a_{is}\right)\mathbf{r}^i\mathbf{r}^j.$$

As before we denote

$$\nabla_k a_{ij} = \frac{\partial a_{ij}}{\partial q^k} - \Gamma^s_{ki}a_{sj} - \Gamma^s_{kj}a_{is}.$$

Exercise 4.16. Show that

$$\frac{\partial}{\partial q^k}\mathbf{A} = \left(\frac{\partial a_i^{\cdot j}}{\partial q^k} - \Gamma^s_{ki}a_s^{\cdot j} + \Gamma^j_{ks}a_i^{\cdot s}\right)\mathbf{r}^i\mathbf{r}_j,$$

$$\frac{\partial}{\partial q^k}\mathbf{A} = \left(\frac{\partial a^i_{\cdot j}}{\partial q^k} + \Gamma^i_{ks}a^s_{\cdot j} - \Gamma^s_{kj}a^i_{\cdot s}\right)\mathbf{r}_i\mathbf{r}^j.$$

The expressions in parentheses are all denoted the same way:

$$\nabla_k a_i^{\cdot j} = \frac{\partial a_i^{\cdot j}}{\partial q^k} - \Gamma^s_{ki}a_s^{\cdot j} + \Gamma^j_{ks}a_i^{\cdot s},$$

$$\nabla_k a^i_{\cdot j} = \frac{\partial a^i_{\cdot j}}{\partial q^k} + \Gamma^i_{ks}a^s_{\cdot j} - \Gamma^s_{kj}a^i_{\cdot s}.$$

Quick summary

For a tensor **A** of order two, we have

$$\frac{\partial}{\partial q^k}\mathbf{A} = \nabla_k a^{ij}\mathbf{r}_i\mathbf{r}_j = \nabla_k a_{ij}\mathbf{r}^i\mathbf{r}^j = \nabla_k a_i^{\cdot j}\mathbf{r}^i\mathbf{r}_j = \nabla_k a^i_{\cdot j}\mathbf{r}_i\mathbf{r}^j$$

where

$$\nabla_k a^{ij} = \frac{\partial a^{ij}}{\partial q^k} + \Gamma^i_{ks}a^{sj} + \Gamma^j_{ks}a^{is}, \quad \nabla_k a_{ij} = \frac{\partial a_{ij}}{\partial q^k} - \Gamma^s_{ki}a_{sj} - \Gamma^s_{kj}a_{is},$$

$$\nabla_k a_i^{\cdot j} = \frac{\partial a_i^{\cdot j}}{\partial q^k} - \Gamma^s_{ki}a_s^{\cdot j} + \Gamma^j_{ks}a_i^{\cdot s}, \quad \nabla_k a^i_{\cdot j} = \frac{\partial a^i_{\cdot j}}{\partial q^k} + \Gamma^i_{ks}a^s_{\cdot j} - \Gamma^s_{kj}a^i_{\cdot s}.$$

Exercise 4.17. Show that any of the covariant derivatives of any component of the metric tensor is equal to zero.

Exercise 4.18. Demonstrate that the components of the metric tensor behave as constants under covariant differentiation of components:

$$\nabla_k g^{st} a_t = g^{st} \nabla_k a_t, \qquad \nabla_k g_{st} a^t = g_{st} \nabla_k a^t.$$

4.8 Differential Operations

Here we look further at various differential operations that may be performed on vector and tensor fields. We begin with the rotation of a vector. By (4.7) and (4.29) we have

$$\nabla \times \mathbf{f} = \mathbf{r}^i \times \frac{\partial \mathbf{f}}{\partial q^i} = \mathbf{r}^i \times \mathbf{r}^j \nabla_i f_j = \epsilon^{ijk} \mathbf{r}_k \nabla_i f_j.$$

The use of (4.28) allows us to write this as

$$\nabla \times \mathbf{f} = \epsilon^{ijk} \mathbf{r}_k \left(\frac{\partial f_j}{\partial q^i} - \Gamma_{ij}^n f_n \right). \tag{4.30}$$

Considering the second term in parentheses we note that

$$\epsilon^{ijk} \Gamma_{ij}^n = \epsilon^{ijk} \Gamma_{ji}^n = \epsilon^{jik} \Gamma_{ij}^n$$

by (4.12) and a subsequent renaming of dummy indices. But $\epsilon^{jik} = -\epsilon^{ijk}$, hence $\epsilon^{ijk} \Gamma_{ij}^n = -\epsilon^{ijk} \Gamma_{ij}^n$ so that $\epsilon^{ijk} \Gamma_{ij}^n = 0$. Equation (4.30) is thereby reduced to

$$\nabla \times \mathbf{f} = \mathbf{r}_k \epsilon^{ijk} \frac{\partial f_j}{\partial q^i}.$$

As an example, we recall that the vector field $\mathbf{E}$ of electrostatics satisfies $\nabla \times \mathbf{E} = 0$. Such a field is said to be *irrotational*. So the condition for any vector field $\mathbf{f}$ to be irrotational can be written in generalized coordinates as

$$\epsilon^{ijk} \frac{\partial f_j}{\partial q^i} = 0.$$

Of course, the operation $\nabla \times \mathbf{f}$ is given in a Cartesian frame by the familiar formula

$$\nabla \times \mathbf{f} = \begin{vmatrix} \hat{\mathbf{x}} & \hat{\mathbf{y}} & \hat{\mathbf{z}} \\ \dfrac{\partial}{\partial x} & \dfrac{\partial}{\partial y} & \dfrac{\partial}{\partial z} \\ f_x & f_y & f_z \end{vmatrix}.$$

Let us turn to the divergence of $\mathbf{f}$. We start by writing

$$\nabla \cdot \mathbf{f} = \mathbf{r}^i \cdot \frac{\partial \mathbf{f}}{\partial q^i} = \mathbf{r}^i \cdot \mathbf{r}^j \nabla_i f_j = g^{ij} \nabla_i f_j.$$

By the result of Exercise 4.18 and equation (4.27) this can be written as

$$\nabla \cdot \mathbf{f} = \nabla_i g^{ij} f_j = \nabla_i f^i = \frac{\partial f^i}{\partial q^i} + \Gamma^i_{in} f^n. \tag{4.31}$$

As was the case with (4.30) this can be simplified; we must first develop a useful identity for Γ^i_{in}. This is done as follows. We begin by writing

$$\frac{\partial \sqrt{g}}{\partial q^n} = \frac{\partial}{\partial q^n} [\mathbf{r}_1 \cdot (\mathbf{r}_2 \times \mathbf{r}_3)] = \frac{\partial \mathbf{r}_1}{\partial q^n} \cdot (\mathbf{r}_2 \times \mathbf{r}_3) + \mathbf{r}_1 \cdot \frac{\partial}{\partial q^n} (\mathbf{r}_2 \times \mathbf{r}_3)$$

where

$$\mathbf{r}_1 \cdot \frac{\partial}{\partial q^n} (\mathbf{r}_2 \times \mathbf{r}_3) = \mathbf{r}_1 \cdot \mathbf{r}_2 \times \frac{\partial \mathbf{r}_3}{\partial q^n} + \mathbf{r}_1 \cdot \frac{\partial \mathbf{r}_2}{\partial q^n} \times \mathbf{r}_3$$

$$= \frac{\partial \mathbf{r}_3}{\partial q^n} \cdot (\mathbf{r}_1 \times \mathbf{r}_2) + \frac{\partial \mathbf{r}_2}{\partial q^n} \cdot (\mathbf{r}_3 \times \mathbf{r}_1)$$

so that

$$\frac{\partial \sqrt{g}}{\partial q^n} = \frac{\partial \mathbf{r}_1}{\partial q^n} \cdot (\mathbf{r}_2 \times \mathbf{r}_3) + \frac{\partial \mathbf{r}_2}{\partial q^n} \cdot (\mathbf{r}_3 \times \mathbf{r}_1) + \frac{\partial \mathbf{r}_3}{\partial q^n} \cdot (\mathbf{r}_1 \times \mathbf{r}_2).$$

Continuing to rewrite this we have

$$\frac{\partial \sqrt{g}}{\partial q^n} = \mathbf{r}_{1n} \cdot (\mathbf{r}_2 \times \mathbf{r}_3) + \mathbf{r}_{2n} \cdot (\mathbf{r}_3 \times \mathbf{r}_1) + \mathbf{r}_{3n} \cdot (\mathbf{r}_1 \times \mathbf{r}_2)$$

$$= \Gamma^i_{1n} \mathbf{r}_i \cdot (\mathbf{r}_2 \times \mathbf{r}_3) + \Gamma^i_{2n} \mathbf{r}_i \cdot (\mathbf{r}_3 \times \mathbf{r}_1) + \Gamma^i_{3n} \mathbf{r}_i \cdot (\mathbf{r}_1 \times \mathbf{r}_2)$$

$$= \Gamma^1_{1n} \sqrt{g} + \Gamma^2_{2n} \sqrt{g} + \Gamma^3_{3n} \sqrt{g}$$

$$= \sqrt{g} \, \Gamma^i_{in}.$$

Hence

$$\Gamma^i_{in} = \frac{1}{\sqrt{g}} \frac{\partial \sqrt{g}}{\partial q^n}.$$

This is the needed identity, and with it (4.31) may be written as

$$\nabla \cdot \mathbf{f} = \frac{1}{\sqrt{g}} \frac{\partial}{\partial q^i} \left(\sqrt{g} f^i \right). \tag{4.32}$$

We know that the condition of incompressibility of a liquid in hydromechanics is expressed as

$$\nabla \cdot \mathbf{v} = 0$$

where $\mathbf{v}$ is the velocity of a material point, so in general coordinates we can write it out as

$$\frac{1}{\sqrt{g}} \frac{\partial}{\partial q^i} \left(\sqrt{g} v^i \right) = 0.$$

In electromagnetic theory the magnetic source law states that $\nabla \cdot \mathbf{B} = 0$ where $\mathbf{B}$ is the vector field known as magnetic flux density. In general, a vector field $\mathbf{f}$ is called *solenoidal* if $\nabla \cdot \mathbf{f} = 0$. Of course, $\nabla \cdot \mathbf{f}$ is given in Cartesian frames by the familiar expression

$$\nabla \cdot \mathbf{f} = \frac{\partial f_x}{\partial x} + \frac{\partial f_y}{\partial y} + \frac{\partial f_z}{\partial z}.$$

Exercise 4.19. Use (4.32) to express $\nabla \cdot \mathbf{f}$ in the cylindrical and spherical coordinate systems.

The curl of a tensor field $\mathbf{A}$ may be computed as follows. We start with

$$\nabla \times \mathbf{A} = \mathbf{r}^k \times \frac{\partial}{\partial q^k} \mathbf{A}$$
$$= \mathbf{r}^k \times \mathbf{r}^i \mathbf{r}^j \nabla_k a_{ij}$$
$$= \epsilon^{kin} \mathbf{r}_n \mathbf{r}^j \nabla_k a_{ij}$$
$$= \epsilon^{kin} \mathbf{r}_n \mathbf{r}^j \left(\frac{\partial a_{ij}}{\partial q^k} - \Gamma^s_{ki} a_{sj} - \Gamma^s_{kj} a_{is} \right).$$

We then use

$$\epsilon^{kin} \Gamma^s_{ki} = 0, \qquad -\mathbf{r}^j \Gamma^s_{kj} = \frac{\partial \mathbf{r}^s}{\partial q^k},$$

to get

$$\nabla \times \mathbf{A} = \epsilon^{kin} \mathbf{r}_n \left(\mathbf{r}^j \frac{\partial a_{ij}}{\partial q^k} + \frac{\partial \mathbf{r}^s}{\partial q^k} a_{is} \right)$$

and hence

$$\nabla \times \mathbf{A} = \epsilon^{kin} \mathbf{r}_n \frac{\partial}{\partial q^k} \left(\mathbf{r}^j a_{ij} \right).$$

For the divergence of a tensor field $\mathbf{A}$, we write

$$\nabla \cdot \mathbf{A} = \mathbf{r}^k \cdot \frac{\partial}{\partial q^k} \mathbf{A}$$
$$= \mathbf{r}^k \cdot \mathbf{r}_i \mathbf{r}_j \nabla_k a^{ij}$$
$$= \mathbf{r}_j \nabla_i a^{ij}$$
$$= \mathbf{r}_j \left(\frac{\partial a^{ij}}{\partial q^i} + \Gamma^i_{is} a^{sj} + \Gamma^j_{is} a^{is} \right).$$

But

$$\mathbf{r}_j \Gamma^j_{is} = \frac{\partial \mathbf{r}_s}{\partial q^i}$$

so we have

$$\nabla \cdot \mathbf{A} = \mathbf{r}_j \frac{\partial a^{ij}}{\partial q^i} + \mathbf{r}_j \frac{1}{\sqrt{g}} \frac{\partial \sqrt{g}}{\partial q^s} a^{sj} + \frac{\partial \mathbf{r}_s}{\partial q^i} a^{is}$$

$$= \frac{\partial}{\partial q^i} \left(\mathbf{r}_j a^{ij} \right) + \mathbf{r}_j \frac{1}{\sqrt{g}} \frac{\partial \sqrt{g}}{\partial q^i} a^{ij}.$$

Finally then,

$$\nabla \cdot \mathbf{A} = \frac{1}{\sqrt{g}} \frac{\partial}{\partial q^i} \left(\sqrt{g} a^{ij} \mathbf{r}_j \right). \tag{4.33}$$

Many problems of mathematical physics reduce to Poisson's equation or Laplace's equation. The unknown function in these equations can be a scalar function or (as in electrodynamics) a vector function. The equations of the linear theory of elasticity in displacements also contain the Laplacian operator and another type of operation involving the nabla operator. In Cartesian frames it is simple to write out corresponding expressions. Solving corresponding problems with the use of curvilinear coordinates, we need to find the representation of these formulas. We begin with the formulas that relate to the second-order tensor $\nabla \nabla f$. We have

$$\nabla \nabla f = \mathbf{r}^i \frac{\partial}{\partial q^i} \left(\mathbf{r}^j \frac{\partial}{\partial q^j} f \right) = \mathbf{r}^i \left(\frac{\partial}{\partial q^i} \frac{\partial f}{\partial q^j} - \Gamma_{ij}^k \frac{\partial f}{\partial q^k} \right) \mathbf{r}^j,$$

hence

$$\nabla \nabla f = \mathbf{r}^i \mathbf{r}^j \left(\frac{\partial^2 f}{\partial q^i \partial q^j} - \Gamma_{ij}^k \frac{\partial f}{\partial q^k} \right). \tag{4.34}$$

From this we see that

$$(\nabla \nabla f)^T = \nabla \nabla f \tag{4.35}$$

which is obvious by symmetry (in i and j) of the Christoffel coefficient and the rest of the expression on the right side of (4.34). A formal insertion[2] of the dot product operation between the vectors of (4.34) allows us to generate an expression for the Laplacian $\nabla^2 f \equiv \nabla \cdot \nabla f$:

$$\nabla^2 f = \mathbf{r}^i \cdot \mathbf{r}^j \left(\frac{\partial^2 f}{\partial q^i \partial q^j} - \Gamma_{ij}^k \frac{\partial f}{\partial q^k} \right) = g^{ij} \left(\frac{\partial^2 f}{\partial q^i \partial q^j} - \Gamma_{ij}^k \frac{\partial f}{\partial q^k} \right). \tag{4.36}$$

---

[2] This sort of operation can be done with any dyad of vectors; we can generate a scalar $\mathbf{a} \cdot \mathbf{b}$ from $\mathbf{ab}$ by inserting a dot. This operation can have additional meaning when done with a tensor $\mathbf{A}$: if we write the tensor in mixed form $a_i^{\cdot j} \mathbf{r}^i \mathbf{r}_j$ we obtain $a_i^{\cdot j} \mathbf{r}^i \cdot \mathbf{r}_j = a_i^{\cdot i}$. This is the first invariant of $\mathbf{A}$, known as the trace of $\mathbf{A}$.

Hence Laplace's equation $\nabla^2 f = 0$ appears in generalized coordinates as

$$g^{ij}\left(\frac{\partial^2 f}{\partial q^i \partial q^j} - \Gamma^k_{ij}\frac{\partial f}{\partial q^k}\right) = 0.$$

In a Cartesian frame (4.36) gives us

$$\nabla^2 f = \frac{\partial^2 f}{\partial x^2} + \frac{\partial^2 f}{\partial y^2} + \frac{\partial^2 f}{\partial z^2},$$

while in cylindrical and spherical frames

$$\nabla^2 f = \frac{1}{\rho}\frac{\partial}{\partial \rho}\left(\rho\frac{\partial f}{\partial \rho}\right) + \frac{1}{\rho^2}\frac{\partial^2 f}{\partial \phi^2} + \frac{\partial^2 f}{\partial z^2},$$

$$\nabla^2 f = \frac{1}{r^2}\frac{\partial}{\partial r}\left(r^2\frac{\partial f}{\partial r}\right) + \frac{1}{r^2 \sin\theta}\frac{\partial}{\partial \theta}\left(\sin\theta\frac{\partial f}{\partial \theta}\right) + \frac{1}{r^2 \sin^2\theta}\frac{\partial^2 f}{\partial \phi^2},$$

respectively.

Exercise 4.20. Show that a formal insertion of the cross product operation between the vectors of (4.34) leads to the useful identity

$$\nabla \times \nabla f = 0,$$

holding for any scalar field f.

Since usual and covariant differentiation amount to the same thing for a scalar f, we can write

$$\nabla\nabla f = \mathbf{r}^i\frac{\partial}{\partial q^i}\mathbf{r}^j\frac{\partial}{\partial q^j}f = \mathbf{r}^i\frac{\partial}{\partial q^i}\left(\mathbf{r}^j\nabla_j f\right) = \mathbf{r}^i\mathbf{r}^j\nabla_i\nabla_j f = \mathbf{r}^j\mathbf{r}^i\nabla_i\nabla_j f$$

where in the last step we used (4.35). The corresponding result for the Laplacian is

$$\nabla^2 f = \mathbf{r}^i \cdot \mathbf{r}^j\nabla_i\nabla_j f = g^{ij}\nabla_i\nabla_j f = \nabla^j\nabla_j f$$

where

$$\nabla^j \equiv g^{ij}\nabla_i.$$

The Laplacian of a vector arises in physical applications such as electromagnetic field theory. For this we have

$$\nabla^2\mathbf{f} = \nabla \cdot \nabla\mathbf{f} = \mathbf{r}^k \cdot \frac{\partial}{\partial q^k}\mathbf{r}^i\mathbf{r}_j\nabla_i f^j = \mathbf{r}^k \cdot \mathbf{r}^i\mathbf{r}_j\nabla_k\nabla_i f^j = g^{ki}\mathbf{r}_j\nabla_k\nabla_i f^j$$

hence

$$\nabla^2\mathbf{f} = \mathbf{r}_j\nabla^i\nabla_i f^j.$$

Also

$$\nabla\nabla \cdot \mathbf{f} = \mathbf{r}^i \frac{\partial}{\partial q^i} \left[\frac{1}{\sqrt{g}} \frac{\partial}{\partial q^j} \left(\sqrt{g} f^j \right) \right] = \mathbf{r}^i \nabla_i \nabla_j f^j.$$

It is possible to demonstrate that

$$\nabla \times \nabla \times \mathbf{f} = \nabla\nabla \cdot \mathbf{f} - \nabla^2 \mathbf{f}.$$

4.9 Orthogonal Coordinate Systems

The most frequently used coordinate frames are Cartesian, cylindrical, and spherical. All of these are orthogonal. There are many other orthogonal coordinate frames in use as well. It is sensible to give a general treatment of these systems because mutual orthogonality of the frame vectors leads to simplification in many formulae. Additional motivation is provided by the fact that in applications it is important to know the magnitudes of field components. The general formulations given above are inconvenient for this; the general frame vectors are not of unit length, hence the magnitude of the projection of a vector onto an orthogonal frame direction is not the corresponding component of the vector. The physical components of a vector are conveniently displayed in orthogonal frames with use of the Lamé coefficients.

In this section we consider frames where the coordinate vectors are mutually orthogonal:

$$\mathbf{r}_i \cdot \mathbf{r}_j = 0, \qquad i \neq j.$$

The Lamé coefficients H_i are

$$(H_i)^2 = g_{ii} = \mathbf{r}_i \cdot \mathbf{r}_i \qquad (i = 1, 2, 3).$$

Note that H_i is the length of the frame vector $\mathbf{r}_i$. At each (q^1, q^2, q^3) the coordinate frame is orthogonal. In the orthogonal frame, by the construction of the reciprocal basis, the vectors $\mathbf{r}^i$ are co-directed with $\mathbf{r}_i$ and the product of their lengths is 1. Hence

$$\mathbf{r}^i = \mathbf{r}_i / (H_i)^2 \qquad (i = 1, 2, 3).$$

Let us introduce the frame whose vectors are co-directed with the basis vectors but have unit length:

$$\hat{\mathbf{r}}_i = \mathbf{r}_i / H_i \qquad (i = 1, 2, 3).$$

Thus at each point the vectors $\hat{\mathbf{r}}_i$ form a Cartesian basis, a frame that rotates when the origin of the frame moves from point to point. We shall present all the main formulas of differentiation when vectors are given in this basis. They are not convenient in theory since many useful properties of symmetry are lost, but they are necessary when doing calculations in corresponding coordinates. We begin by noting that

$$\mathbf{r}_i = H_i\hat{\mathbf{r}}_i, \qquad \mathbf{r}^i = \hat{\mathbf{r}}_i/H_i \qquad (i = 1, 2, 3).$$

Let us compute the $\hat{\mathbf{r}}_i$ in the cylindrical and spherical systems. In a cylindrical frame where

$$H_1 = 1, \qquad H_2 = \rho, \qquad H_3 = 1,$$

we have

$$\hat{\mathbf{r}}_1 = \hat{\mathbf{x}}\cos\phi + \hat{\mathbf{y}}\sin\phi,$$
$$\hat{\mathbf{r}}_2 = -\hat{\mathbf{x}}\sin\phi + \hat{\mathbf{y}}\cos\phi,$$
$$\hat{\mathbf{r}}_3 = \hat{\mathbf{z}}.$$

In a spherical frame where

$$H_1 = 1, \qquad H_2 = r, \qquad H_3 = r\sin\theta,$$

we have

$$\hat{\mathbf{r}}_1 = \hat{\mathbf{x}}\sin\theta\cos\phi + \hat{\mathbf{y}}\sin\theta\sin\phi + \hat{\mathbf{z}}\cos\theta,$$
$$\hat{\mathbf{r}}_2 = \hat{\mathbf{x}}\cos\theta\cos\phi + \hat{\mathbf{y}}\cos\theta\sin\phi - \hat{\mathbf{z}}\sin\theta,$$
$$\hat{\mathbf{r}}_3 = -\hat{\mathbf{x}}\sin\phi + \hat{\mathbf{y}}\cos\phi.$$

Differentiation in the orthogonal basis

Using the definition we can represent the ∇-operator in new terms as

$$\nabla = \frac{\hat{\mathbf{r}}_i}{H_i}\frac{\partial}{\partial q^i}.$$

(In this formula there *is* summation over i, and this continues to hold in all formulas of this section below. The convention on summation over sub- and super-indices is modified here since in a Cartesian system, which the frame $\hat{\mathbf{r}}_i$ locally constitutes, the reciprocal and main bases coincide.) Now let us find the formulas of differentiation of the new frame vectors. We begin with the formula (4.18):

$$\mathbf{r}_{ij} \cdot \mathbf{r}_t = \frac{1}{2} \left(\frac{\partial g_{it}}{\partial q^j} + \frac{\partial g_{tj}}{\partial q^i} - \frac{\partial g_{ji}}{\partial q^t} \right)$$

$$= H_t \frac{\partial H_i}{\partial q^j} \delta_{it} + H_t \frac{\partial H_j}{\partial q^i} \delta_{jt} - H_i \frac{\partial H_j}{\partial q^t} \delta_{ij}.$$

Here we used the fact that $g_{ij} = 0$ if $i \neq j$. On the other hand

$$\mathbf{r}_{ij} \cdot \mathbf{r}_t = \frac{\partial}{\partial q^j} (H_i \hat{\mathbf{r}}_i) \cdot H_t \hat{\mathbf{r}}_t$$

$$= \frac{\partial H_i}{\partial q^j} \hat{\mathbf{r}}_i \cdot H_t \hat{\mathbf{r}}_t + H_i \frac{\partial \hat{\mathbf{r}}_i}{\partial q^j} \cdot H_t \hat{\mathbf{r}}_t$$

$$= \frac{\partial H_i}{\partial q^j} H_t \delta_{it} + H_i H_t \frac{\partial \hat{\mathbf{r}}_i}{\partial q^j} \cdot \hat{\mathbf{r}}_t.$$

Thus

$$\frac{\partial \hat{\mathbf{r}}_i}{\partial q^j} \cdot \hat{\mathbf{r}}_t = \frac{1}{H_i} \frac{\partial H_t}{\partial q^i} \delta_{jt} - \frac{1}{H_t} \frac{\partial H_i}{\partial q^t} \delta_{ij}.$$

Therefore

$$\frac{\partial \hat{\mathbf{r}}_i}{\partial q^j} = \sum_{t=1}^{3} \left(\frac{1}{H_i} \frac{\partial H_t}{\partial q^i} \delta_{jt} - \frac{1}{H_t} \frac{\partial H_i}{\partial q^t} \delta_{ij} \right) \hat{\mathbf{r}}_t.$$

(Note that the components of vectors are given in the frame of the same $\hat{\mathbf{r}}_i$.) Using this we derive

$$\nabla f = \sum_{i,j=1}^{3} \frac{\hat{\mathbf{r}}_j}{H_j} \frac{\partial}{\partial q^j} (f_i \hat{\mathbf{r}}_i).$$

The gradient, divergence, and rotation of a vector field $\mathbf{f}$ are given by

$$\nabla \mathbf{f} = \hat{\mathbf{r}}_i \hat{\mathbf{r}}_j \left(\frac{1}{H_i} \frac{\partial f_j}{\partial q^i} - \frac{f_i}{H_i H_j} \frac{\partial H_i}{\partial q^j} + \delta_{ij} \frac{f_k}{H_k} \frac{1}{H_i} \frac{\partial H_i}{\partial q^k} \right)$$

$$\nabla \cdot \mathbf{f} = \frac{1}{H_1 H_2 H_3} \left(\frac{\partial}{\partial q^1} (H_2 H_3 f_1) + \frac{\partial}{\partial q^2} (H_3 H_1 f_2) + \frac{\partial}{\partial q^3} (H_1 H_2 f_3) \right) \tag{4.37}$$

and

$$\nabla \times \mathbf{f} = \frac{1}{2} \frac{\hat{\mathbf{r}}_i \times \hat{\mathbf{r}}_j}{H_i H_j} \left(\frac{\partial}{\partial q^i} (H_j f_j) - \frac{\partial}{\partial q^j} (H_i f_i) \right). \tag{4.38}$$

Using the gradient we can write out the tensor of small strains for a displacement vector $\mathbf{u} = u_i \hat{\mathbf{r}}_i$, which is the main object of linear elasticity:

$$
\begin{aligned}
\varepsilon &= \frac{1}{2} \left(\nabla \mathbf{u} + \nabla \mathbf{u}^T \right) \\
&= \frac{1}{2} \hat{\mathbf{r}}_i \hat{\mathbf{r}}_j \left(\frac{1}{H_i} \frac{\partial u_i}{\partial q^i} + \frac{1}{H_j} \frac{\partial u_j}{\partial q^j} - \frac{u_i}{H_i H_j} \frac{\partial H_i}{\partial q^j} \right. \\
&\left. \quad - \frac{u_j}{H_i H_j} \frac{\partial H_j}{\partial q^i} + 2\delta_{ij} \frac{u_t}{H_i H_j} \frac{\partial H_j}{\partial q^t} \right).
\end{aligned}
\tag{4.39}
$$

The Laplacian of a scalar field f is

$$
\begin{aligned}
\nabla^2 f &= \frac{1}{H_1 H_2 H_3} \left[\frac{\partial}{\partial q^1} \left(\frac{H_2 H_3}{H_1} \frac{\partial f}{\partial q^1} \right) \right. \\
&\left. + \frac{\partial}{\partial q^2} \left(\frac{H_3 H_1}{H_2} \frac{\partial f}{\partial q^2} \right) + \frac{\partial}{\partial q^3} \left(\frac{H_1 H_2}{H_3} \frac{\partial f}{\partial q^3} \right) \right].
\end{aligned}
\tag{4.40}
$$

In the cylindrical and spherical coordinate systems, for example, (4.37) yields

$$
\nabla \cdot \mathbf{f} = \frac{1}{\rho} \frac{\partial}{\partial \rho} (\rho f_\rho) + \frac{1}{\rho} \frac{\partial f_\phi}{\partial \phi} + \frac{\partial f_z}{\partial z}
$$

and

$$
\nabla \cdot \mathbf{f} = \frac{1}{r^2} \frac{\partial}{\partial r} (r^2 f_r) + \frac{1}{r \sin \theta} \frac{\partial}{\partial \theta} (\sin \theta f_\theta) + \frac{1}{r \sin \theta} \frac{\partial f_\phi}{\partial \phi},
$$

while (4.38) yields

$$
\nabla \times \mathbf{f} = \hat{\boldsymbol{\rho}} \left(\frac{1}{\rho} \frac{\partial f_z}{\partial \phi} - \frac{\partial f_\phi}{\partial z} \right) + \hat{\boldsymbol{\phi}} \left(\frac{\partial f_\rho}{\partial z} - \frac{\partial f_z}{\partial \rho} \right) + \hat{\mathbf{z}} \frac{1}{\rho} \left(\frac{\partial (\rho f_\phi)}{\partial \rho} - \frac{\partial f_\rho}{\partial \phi} \right)
$$

and

$$
\begin{aligned}
\nabla \times \mathbf{f} &= \hat{\mathbf{r}} \frac{1}{r \sin \theta} \left(\frac{\partial}{\partial \theta} (\sin \theta f_\phi) - \frac{\partial f_\theta}{\partial \phi} \right) \\
&+ \hat{\boldsymbol{\theta}} \frac{1}{r \sin \theta} \left(\frac{\partial f_r}{\partial \phi} - \sin \theta \frac{\partial (r f_\phi)}{\partial r} \right) \\
&+ \hat{\boldsymbol{\phi}} \frac{1}{r} \left(\frac{\partial (r f_\theta)}{\partial r} - \frac{\partial f_r}{\partial \theta} \right).
\end{aligned}
$$

The specializations of (4.40) to cylindrical and spherical coordinates were given in § 4.8.

4.10   Some Formulas of Integration

Let $f(x_1, x_2, x_3)$ be a continuous function of the Cartesian coordinates x_1, x_2, x_3 in a compact volume V. Let q^1, q^2, q^3 be curvilinear coordinates in the same volume, in one-to-one continuously differentiable correspondence with the Cartesian coordinates, so after transformation of the coordinates we shall write out the same function as $f(q^1, q^2, q^3)$. The transformation is

$$\int_V f(x_1, x_2, x_3)\, dx_1\, dx_2\, dx_3 = \int_V f(q^1, q^2, q^3) J\, dq^1\, dq^2\, dq^3$$

where

$$J = \sqrt{g} = \left| \frac{\partial x_i}{\partial q^j} \right|$$

is the Jacobian.

Exercise 4.21. Show that the Jacobian determinants of the transformations from Cartesian coordinates to cylindrical and spherical coordinates are, respectively,

$$\left| \frac{\partial(x, y, z)}{\partial(\rho, \phi, z)} \right| = \rho, \qquad \left| \frac{\partial(x, y, z)}{\partial(r, \theta, \phi)} \right| = r^2 \sin\theta.$$

Exercise 4.22. Two successive coordinate transformations are given by $x^i = x^i(q^j)$ and $q^i = q^i(\tilde{q}^j)$. Show that the Jacobian determinant of the composite transformation $x^i = x^i(\tilde{q}^j)$ is given by

$$\left| \frac{\partial(x^1, x^2, x^3)}{\partial(\tilde{q}^1, \tilde{q}^2, \tilde{q}^3)} \right| = \left| \frac{\partial(x^1, x^2, x^3)}{\partial(q^1, q^2, q^3)} \right| \left| \frac{\partial(q^1, q^2, q^3)}{\partial(\tilde{q}^1, \tilde{q}^2, \tilde{q}^3)} \right|.$$

For functions $f(x_1, x_2, x_3)$ and $g(x_1, x_2, x_3)$ that are continuously differentiable on a compact volume V with piecewise smooth boundary S, there is the well known Gauss–Ostrogradsky formula for integration by parts:

$$\int_V \frac{\partial f}{\partial x_k} g\, dx_1\, dx_2\, dx_3 = -\int_V \frac{\partial g}{\partial x_k} f\, dx_1\, dx_2\, dx_3 + \int_S f g n_k\, dS,$$

where dS is the differential element of area on S and n_k is the projection of the outward unit normal $\mathbf{n}$ from S onto the axis $\mathbf{i}_k$. In the particular case $g = 1$ we have

$$\int_V \frac{\partial f}{\partial x_k}\, dx_1\, dx_2\, dx_3 = \int_S f n_k\, dS. \tag{4.41}$$

Let us use (4.41) to derive some formulas involving the nabla operator that are frequently used in applications. These formulas will be valid in curvilinear coordinates, despite the fact that the intermediate transformations will be done in Cartesian coordinates, because the final results will be written in non-coordinate form. We begin with the integral of ∇f:

$$\int_V \nabla f \, dV = \int_V \mathbf{i}_k \frac{\partial f}{\partial x_k} \, dx_1 \, dx_2 \, dx_3 = \mathbf{i}_k \int_S f n_k \, dS = \int_S f \mathbf{n} \, dS.$$

Now consider an analogous formula for a vector function $\mathbf{f}$:

$$\int_V \nabla \mathbf{f} \, dV = \int_V \mathbf{i}_k \frac{\partial f_t}{\partial x_k} \mathbf{i}_t \, dx_1 \, dx_2 \, dx_3$$

$$= \mathbf{i}_k \mathbf{i}_t \int_S n_k f_t \, dS$$

$$= \int_S n_k \mathbf{i}_k f_t \mathbf{i}_t \, dS$$

$$= \int_S \mathbf{n} \mathbf{f} \, dS.$$

Since the left- and right-hand sides are written in non-coordinate form we can use this formula with any coordinate frame with $dV = \sqrt{g} \, dq^1 \, dq^2 \, dq^3$. This is the formula for a tensor $\nabla \mathbf{f}$; for its trace we have

$$\int_V \nabla \cdot \mathbf{f} \, dV = \int_V \mathbf{i}_k \frac{\partial f_t}{\partial x_k} \cdot \mathbf{i}_t \, dx_1 \, dx_2 \, dx_3$$

$$= \mathbf{i}_k \cdot \mathbf{i}_t \int_S n_k f_t \, dS$$

$$= \int_S n_k \mathbf{i}_k \cdot f_t \mathbf{i}_t \, dS$$

$$= \int_S \mathbf{n} \cdot \mathbf{f} \, dS.$$

In a similar fashion we can derive

$$\int_V \nabla \times \mathbf{f} \, dV = \int_S \mathbf{n} \times \mathbf{f} \, dS.$$

It is easily seen that

$$\int_V \nabla \mathbf{f}^T \, dV = \int_S (\mathbf{n} \mathbf{f})^T \, dS = \int_S \mathbf{f} \mathbf{n} \, dS.$$

In a similar fashion the reader can use (4.41) to derive the formulas

$$\int_V \nabla \mathbf{A} \, dV = \int_S \mathbf{n} \mathbf{A} \, dS,$$

$$\int_V \nabla \cdot \mathbf{A}\, dV = \int_S \mathbf{n} \cdot \mathbf{A}\, dS,$$

and

$$\int_V \nabla \times \mathbf{A}\, dV = \int_S \mathbf{n} \times \mathbf{A}\, dS$$

for a tensor field **A**. Finally let us write out Stokes's formula in non-coordinate form. Recall that this relates a vector function **f** given on a simply-connected surface S to its circulation over the piecewise smooth boundary contour Γ:

$$\oint_\Gamma \mathbf{f} \cdot d\mathbf{r} = \int_S (\mathbf{n} \times \nabla) \cdot \mathbf{f}\, dS.$$

For a tensor **A** of second-order this formula extends to the two formulas

$$\oint_\Gamma d\mathbf{r} \cdot \mathbf{A} = \int_S (\mathbf{n} \times \nabla) \cdot \mathbf{A}\, dS$$

and

$$\oint_\Gamma \mathbf{A} \cdot d\mathbf{r} = \int_S (\mathbf{n} \times \nabla) \cdot \mathbf{A}^T\, dS.$$

We leave these for the reader to prove. We should note that Stokes's formulas hold only when S is simply-connected; for a doubly- or multiply-connected surface, the formulas must be amended by some cyclic constants.

4.11 Problems

In this problem set, we let **u** denote a vector field; f, g, h smooth functions; a, b, c arbitrary constants; **A** a second-order tensor; **r** the position vector of a body point; **n** the unit external normal to the body boundary.

4.1 Let $f = f(r)$, where $r^2 = \mathbf{r} \cdot \mathbf{r}$. Find ∇f and $\nabla^2 f$.

4.2 Let a, b, c be arbitrary constants, f, g arbitrary functions, and **u** a given vector field. Find $\nabla \mathbf{u}$ and $\nabla \cdot \mathbf{u}$ for the following **u**.

    (a) $\mathbf{u} = ax_1\mathbf{i}_1 + bx_2\mathbf{i}_2 + cx_3\mathbf{i}_3$;

    (b) $\mathbf{u} = ax_2\mathbf{i}_1$;

    (c) $\mathbf{u} = a\mathbf{r}$;

    (d) $\mathbf{u} = f(r)\mathbf{e}_r$ (assume polar coordinates);

    (e) $\mathbf{u} = f(r)\mathbf{e}_\phi$ (assume polar coordinates);

(f) $\mathbf{u} = f(r)\mathbf{e}_z$ (assume cylindrical coordinates);

(g) $\mathbf{u} = f(r)\mathbf{e}_r$ (assume spherical coordinates);

(h) $\mathbf{u} = \boldsymbol{\omega} \times \mathbf{r}$, $\boldsymbol{\omega} = \text{const}$;

(i) $\mathbf{u} = f(\phi)\mathbf{e}_z + g(\phi)\mathbf{e}_\phi$ (assume cylindrical coordinates);

(j) $\mathbf{u} = f(z)\mathbf{e}_z + g(\phi)\mathbf{e}_\phi$ (assume cylindrical coordinates);

(k) $\mathbf{u} = \mathbf{A} \cdot \mathbf{r}$, $\mathbf{A} = \text{const}$.

4.3 Demonstrate:

(a) $\nabla \cdot (\mathbf{A} \cdot \mathbf{f}) = (\nabla \cdot \mathbf{A}) \cdot \mathbf{f} + \mathbf{A}^T \cdot\cdot \nabla \mathbf{f}$;

(b) $\nabla \cdot (\mathbf{A} \cdot \mathbf{B}) = (\nabla \cdot \mathbf{A}) \cdot \mathbf{B} + \mathbf{A}^T \cdot\cdot \nabla \mathbf{B}$;

(c) $\nabla \cdot (\mathbf{A} \times \mathbf{r}) = (\nabla \cdot \mathbf{A}) \times \mathbf{r}$, if $\mathbf{A}$ is a symmetric tensor: $\mathbf{A} = \mathbf{A}^T$;

(d) $\nabla \cdot [(\mathbf{E} \times \boldsymbol{\omega}) \times \mathbf{r}] = 2\boldsymbol{\omega} + (\nabla \times \boldsymbol{\omega}) \times \mathbf{r}$;

(e) $\nabla \times (\nabla \times \mathbf{A}) = \nabla(\nabla \cdot \mathbf{A}) - \nabla \cdot (\nabla \mathbf{A})$;

(f) $\nabla \times (\mathbf{f} \times \mathbf{r}) = \mathbf{r} \cdot \nabla \mathbf{f} - \mathbf{r}(\nabla \cdot \mathbf{f}) + 2\mathbf{f}$;

(g) $\operatorname{tr}[\nabla \times (\mathbf{E} \times \boldsymbol{\omega})] = -2\nabla \cdot \boldsymbol{\omega}$;

(h) $\nabla \cdot [(\nabla \mathbf{f})^T - (\nabla \cdot \mathbf{f})\mathbf{E}] = \mathbf{0}$;

(i) $\nabla \cdot [\nabla f \times \nabla g] = 0$;

(j) $\nabla \times (\nabla \times \mathbf{A})^T$ is a symmetric tensor if $\mathbf{A}$ is symmetric;

(k) $\nabla \cdot (\mathbf{f}\,\mathbf{f}) = \mathbf{f}\nabla \cdot \mathbf{f} + (\nabla \mathbf{f}) \cdot \mathbf{f} - \mathbf{f} \times (\nabla \times \mathbf{f})$;

(l) $\operatorname{tr}[\nabla \times (\mathbf{A} \cdot \mathbf{B})] = \mathbf{B} \cdot\cdot (\nabla \times \mathbf{A}) - \mathbf{A}^T \cdot\cdot (\nabla \times \mathbf{B}^T)$;

(m) $(\nabla \mathbf{f})_\times = \nabla \times \mathbf{f}$.

4.4 Find

(a) $\nabla \cdot (\mathbf{E}r)$;

(b) $\nabla \times [\nabla \times (\mathbf{r} \times \mathbf{A} \times \mathbf{r})]^T$, if $\mathbf{A} = \mathbf{A}^T = \text{const}$;

(c) $\nabla \cdot (\mathbf{A} \cdot \mathbf{r})$ if $\mathbf{A} = \text{const}$;

(d) $\nabla \cdot (f\mathbf{E})$;

(e) $\nabla \cdot (\mathbf{r}\mathbf{E})$;

(f) $\nabla \cdot (\mathbf{r}\mathbf{r})$.

4.5 Let f, g, h be arbitrary smooth functions and $r^2 = \mathbf{r} \cdot \mathbf{r}$. Find the divergence of the tensors $\mathbf{A}$ given by the following formulas.

(a) $\mathbf{A} = f(r)\mathbf{e}_r\mathbf{e}_r + g(r)\mathbf{e}_\phi\mathbf{e}_\phi + h(z)\mathbf{e}_z\mathbf{e}_z$;

(b) $\mathbf{A} = f(r)\mathbf{e}_r\mathbf{e}_r + g(r)\mathbf{e}_\phi\mathbf{e}_\phi + g(r)\mathbf{e}_\theta\mathbf{e}_\theta$;

(c) $\mathbf{A} = f(r)\mathbf{e}_r\mathbf{e}_r + g(r)\mathbf{e}_\phi\mathbf{e}_\phi + h(r)\mathbf{e}_z\mathbf{e}_z$;

(d) $\mathbf{A} = f(x_1)\mathbf{i}_1\mathbf{i}_1 + g(x_2)\mathbf{i}_2\mathbf{i}_2 + h(x_3)\mathbf{i}_3\mathbf{i}_3;$

(e) $\mathbf{A} = f(x_2)\mathbf{i}_1\mathbf{i}_1 + g(x_3)\mathbf{i}_2\mathbf{i}_2 + h(x_1)\mathbf{i}_3\mathbf{i}_3;$

(f) $\mathbf{A} = f(r)\mathbf{e}_r\mathbf{e}_\phi + g(r)\mathbf{e}_\phi\mathbf{e}_r + h(r)\mathbf{e}_r\mathbf{e}_z;$

(g) $\mathbf{A} = f(z)\mathbf{e}_r\mathbf{e}_z + g(z)\mathbf{e}_\phi\mathbf{e}_r + h(z)\mathbf{e}_z\mathbf{e}_r.$

4.6 Let $\mathbf{A}$ be a symmetric second-order tensor depending on the coordinates. Denote $\mathbf{f} = \nabla \cdot \mathbf{A}$. Find $\nabla \cdot (\mathbf{A} \times \mathbf{r})$ in terms of $\mathbf{f}$.

4.7 The elliptic cylindrical coordinates are related to Cartesian coordinates by the formulas

$$x_1 = a\sigma\tau,$$
$$x_2 = \pm a\sqrt{(\sigma^2 - 1)(1 - \tau^2)},$$
$$x_3 = z,$$

where $\sigma \geq 1$, $|\tau| \leq 1$, and a is a positive parameter. So σ, τ, z are internal coordinates for the cylinder. Show that the internal coordinates are orthogonal and find their Lamé coefficients.

4.8 Parabolic coordinates σ, τ, ϕ in space are related to the Cartesian coordinates x_1, x_2, x_3 by the formulas

$$x_1 = \sigma\tau \cos\phi,$$
$$x_2 = \sigma\tau \sin\phi,$$
$$x_3 = \frac{1}{2}(\tau^2 - \sigma^2),$$

where a is a positive parameter. Show that the parabolic system of coordinates is orthogonal. Find its Lamé coefficients.

4.9 The bipolar cylindrical coordinates σ, τ, z are related to the Cartesian coordinates x_1, x_2, x_3 by the formulas

$$x_1 = \frac{a \sinh\tau}{\cosh\tau - \cos\sigma},$$
$$x_2 = \frac{a \sin\sigma}{\cosh\tau - \cos\sigma},$$
$$x_3 = z,$$

where a is a positive parameter. Show that the bicylindrical coordinate system is orthogonal. Find its Lamé coefficients.

4.10 The bipolar coordinates σ, τ, ϕ are related to the Cartesian coordinates x_1, x_2, x_3 by the formulas

$$x_1 = \frac{a \sin \sigma}{\cosh \tau - \cos \sigma} \cos \phi,$$

$$x_2 = \frac{a \sin \sigma}{\cosh \tau - \cos \sigma} \sin \phi,$$

$$x_3 = \frac{a \sinh \tau}{\cosh \tau - \cos \sigma},$$

where $0 \le \sigma < \pi$, $0 \le \phi < 2\pi$, and a is a positive parameter. Show that the bipolar coordinate system is orthogonal. Find its Lamé coefficients.

4.11 The toroidal coordinates σ, τ, ϕ are related to Cartesian coordinates by the formulas

$$x_1 = \frac{a \sinh \tau}{\cosh \tau - \cos \sigma} \cos \phi,$$

$$x_2 = \frac{a \sinh \tau}{\cosh \tau - \cos \sigma} \sin \phi,$$

$$x_3 = \frac{a \sin \sigma}{\cosh \tau - \cos \sigma},$$

where $-\pi \le \sigma \le \pi$, $0 \le \tau$, $0 \le \phi < 2\pi$, and a is a positive parameter. Show that the toroidal coordinate system is orthogonal. Find its Lamé coefficients.

4.12 Use the Gauss–Ostrogradsky theorem to show that

$$\int_S \mathbf{n} \mathbf{r} \, dS = V \mathbf{E},$$

where V is the volume of the domain bounded by surface S.

4.13 Show that the volume V of the body bounded by surface S is given by the following formulas.

(a)

$$V = \frac{1}{6} \int_S \mathbf{n} \cdot \nabla r^2 \, dS, \quad r^2 = \mathbf{r} \cdot \mathbf{r},$$

(b)

$$V = \frac{1}{3} \int_S \mathbf{n} \cdot \mathbf{r} \, dS.$$

4.14 Let **a** be a vector field satisfying the condition $\nabla \cdot \mathbf{a} = 0$. Using the Gauss–Ostrogradsky formula, demonstrate that

$$\int_S f\mathbf{n}\mathbf{a}\, dS = \int_V (\nabla f) \cdot \mathbf{a}\, dV.$$

4.15 Let S be a closed surface. Demonstrate that

$$\int_S \mathbf{n}\, dS = \mathbf{0}.$$

4.16 Let the second-order tensor **A** be symmetric. Prove the following identity.

$$\int_S \mathbf{r} \times (\mathbf{n} \cdot \mathbf{A})\, dS = \int_V \mathbf{r} \times (\nabla \cdot \mathbf{A})\, dV.$$

4.17 Prove the identity

$$\int_V \nabla^2 \mathbf{A}\, dV = \int_S \mathbf{n} \cdot \nabla \mathbf{A}\, dS.$$

4.18 Let **A** be a given second-order tensor. Denote $\mathbf{f} = \nabla \cdot \mathbf{A}$ in volume V and $\mathbf{n} \cdot \mathbf{A}|_S = \mathbf{g}$ over S, the boundary surface of V. Find

$$\int_V \mathbf{A}\, dV.$$

4.19 Prove the identity $(\mathbf{n} \times \nabla) \cdot \mathbf{A} = \mathbf{n} \cdot \operatorname{rot} \mathbf{A}$.

Chapter 5

Elements of Differential Geometry

The standard fare of high school geometry consists mostly of material collected two millennia ago when geometry stood at the center of natural philosophy. The ancient Greeks, however, did not limit their investigations to the circles and straight lines of Euclid's *Elements*. Archimedes, using methods and ideas that were later to underpin the analysis of infinitesimal quantities, could calculate the length of a spiral and the areas and volumes of other complex figures. In elementary algebra we learn to graph simple quadratic functions such as the parabola and hyperbola, and then in analytic geometry we learn to handle space figures such as the ellipsoid. The methods involved are essentially due to Descartes, who connected the ideas of geometry with those of algebra, and their application is largely limited to objects whose describing equations are of the second order. Finally, in elementary calculus we study formulas that permit us to calculate the length of a curve given in Cartesian coordinates, etc. These more powerful methods are now incorporated into a branch of mathematics known as differential geometry.

Differential geometry allows us to characterize curves and figures of a very general nature. The practical importance of this is well illustrated by the problem of optimal pursuit, wherein one object tries to catch another moving object in the shortest possible time.[1] Of course, we shall often make use of standard figures such as circles, parabolas, etc., as specific examples since we are fully familiar with their properties; in this way the objects of both elementary and analytic geometry enter into the more general subject of differential geometry.

---

[1] A rather humorous statement of one such problem has two old ladies traveling in opposite directions around the base of a hemispherical mountain while a mathematically-minded fly travels back and forth between their noses in the least possible time.

5.1   Elementary Facts from the Theory of Curves

In Chapter 4 we introduced the idea of a coordinate curve in space, which is described by the tip of the radius vector as it moves in such a way that one of the coordinates q^1, q^2, q^3 changes while the other two remain fixed. Now we consider a general curve described in a similar manner by a radius vector whose initial point is the origin and whose terminal point moves through space along the curve. We can describe the position of the radius vector using some parameter t (for a coordinate curve this was the coordinate value q^i). Thus a curve is described as

$$\mathbf{r} = \mathbf{r}(t) \tag{5.1}$$

where t runs through some set along the real axis. Each value of t corresponds to a point of the curve. Unless otherwise stated we shall suppose that the dependence of $\mathbf{r}$ on t is smooth enough that $\mathbf{r}'(t)$ is continuous in t at each point and, where necessary, that the same holds for $\mathbf{r}''(t)$. The notion of vector norm is required for this ($\S 2.7$). Here we denote this norm using the ordinary notation for the magnitude of a vector, e.g., $|\mathbf{r}|$. Recall that in Chapter 4 we introduced frame vectors $\mathbf{r}_i$ tangential to the coordinate lines. We now introduce the tangential vector to an arbitrary curve at a point t, which is $\mathbf{r}'(t)$. The differential of the radius vector corresponding to the curve (5.1) is

$$d\mathbf{r}(t) = \mathbf{r}'(t)\, dt.$$

The length of an elementary section of the curve is

$$ds = |\mathbf{r}'(t)|\ dt,$$

and the length of the portion of the curve corresponding to $t \in [a, b]$ is

$$s = \int_a^b |\mathbf{r}'(t)|\ dt.$$

Exercise 5.1. (a) Find the length of one turn of the helix

$$\mathbf{r}(t) = \mathbf{i}_1 \cos t + \mathbf{i}_2 \sin t + \mathbf{i}_3 t.$$

(b) Calculate the perimeter of the ellipse

$$\frac{x^2}{A^2} + \frac{y^2}{B^2} = 1$$

that lies in the $z = 0$ plane.

Exercise 5.2. Show that the general formulas for arc length in the rectangular, cylindrical, and spherical coordinate systems are

$$s = \int_a^b \left[\left(\frac{dx}{dt} \right)^2 + \left(\frac{dy}{dt} \right)^2 + \left(\frac{dz}{dt} \right)^2 \right]^{1/2} dt,$$

$$s = \int_a^b \left[\left(\frac{d\rho}{dt} \right)^2 + \rho^2 \left(\frac{d\phi}{dt} \right)^2 + \left(\frac{dz}{dt} \right)^2 \right]^{1/2} dt,$$

$$s = \int_a^b \left[\left(\frac{dr}{dt} \right)^2 + r^2 \left(\frac{d\theta}{dt} \right)^2 + r^2 \sin^2 \theta \left(\frac{d\phi}{dt} \right)^2 \right]^{1/2} dt.$$

Most convenient theoretically is the *natural parametrization* of a curve where the parameter represents the length of the curve calculated from the endpoint:

$$\mathbf{r} = \mathbf{r}(s).$$

With this parametrization

$$ds = |\mathbf{r}'(s)| \, ds \qquad \text{so} \qquad |\mathbf{r}'(s)| = 1.$$

Hence when a curve is parametrized naturally $\boldsymbol{\tau}(s) = \mathbf{r}'(s)$ is the unit tangential vector at point s. In this section s shall denote the length parameter of a curve.

Exercise 5.3. Re-parametrize the helix of Exercise 5.1 in terms of its natural length parameter s. Then calculate the unit tangent to the helix as a function of s.

If there is another parametrization of the curve that relates with the natural parametrization $s = s(t)$, then the vector $d\mathbf{r}(s(t))/dt$ also is tangent to the curve at the point $s = s(t)$. For the derivative of a vector function the chain rule

$$\frac{d\mathbf{r}(s(t))}{dt} = \frac{d\mathbf{r}(s)}{ds} \bigg|_{s=s(t)} \frac{ds(t)}{dt}$$

holds.

As is known from analytic geometry, when we know the position of a point $\mathbf{a}$ of a straight line and its directional vector $\mathbf{b}$, then the line can be represented in the parametric form

$$\mathbf{r} = \mathbf{a} + \lambda \mathbf{b}.$$

This gives the vector equation of the line tangent to a curve at point $\mathbf{r}(t_0)$:

$$\mathbf{r} = \mathbf{r}(t_0) + \lambda \mathbf{r}'(t_0).$$

In Cartesian coordinates (x, y, z) this equation takes the form

$$x = x(t_0) + \lambda x'(t_0),$$
$$y = y(t_0) + \lambda y'(t_0),$$
$$z = z(t_0) + \lambda z'(t_0), \tag{5.2}$$

which in nonparametric form is

$$\frac{x - x(t_0)}{x'(t_0)} = \frac{y - y(t_0)}{y'(t_0)} = \frac{z - z(t_0)}{z'(t_0)}. \tag{5.3}$$

Exercise 5.4. Describe the plane curve

$$\mathbf{r}(t) = e^t(\mathbf{i}_1 \cos t + \mathbf{i}_2 \sin t),$$

and find the line tangent to this curve at the point $t = \pi/4$.

Exercise 5.5. (a) Write out the equations for the tangent line to a plane curve corresponding to (5.2) and (5.3) in polar coordinates. (b) Find the angle between the tangent at a point of the curve and the radius vector (from the origin) at this point.

Exercise 5.6. Write out the equations for the tangent line to a curve corresponding to (5.2) and (5.3) in cylindrical and spherical coordinates.

Note that construction of a tangent vector is possible if $\mathbf{r}'(t_0) \neq \mathbf{0}$. A point t_0 where $\mathbf{r}'(t_0) = \mathbf{0}$ is called *singular*.

Exercise 5.7. For a sufficiently smooth curve $\mathbf{r} = \mathbf{r}(t)$ find a tangent at a singular point.

Exercise 5.8. Under the conditions of the previous exercise write out the representation of the type (5.2).

Curvature

For an element of circumference of a circle corresponding to the length Δs, the radius R relates to the central angle $\Delta \varphi$ according to $\Delta s = R \Delta \varphi$. Thus the *curvature* $k = 1/R$ is

$$k = \frac{\Delta \varphi}{\Delta s}.$$

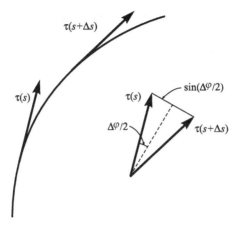

Fig. 5.1 Calculation of curvature.

The curvature of an arbitrary plane curve at a point s is defined as the limit

$$k = \lim_{\Delta s \to 0} \frac{\Delta \varphi}{\Delta s}$$

where $\Delta \varphi$ is the change in angle of the tangent to the element of the curve. For a spatial curve the role of the tangent is played by the tangent unit vector $\boldsymbol{\tau}(s)$. When a point moves through the element of the curve corresponding to Δs, the tangent turns through an angle $\Delta \varphi$. Fig. 5.1 demonstrates that $|\boldsymbol{\tau}(s + \Delta s) - \boldsymbol{\tau}(s)|$ is equal to $2\sin(\Delta \varphi / 2)$. Since

$$\lim_{\Delta \varphi \to 0} \frac{\sin(\Delta \varphi / 2)}{(\Delta \varphi / 2)} = 1,$$

the curvature of a spatial curve is given by

$$\begin{aligned}
k &= \lim_{\Delta s \to 0} \frac{|\Delta \varphi|}{\Delta s} \\
&= \lim_{\Delta s \to 0} \frac{|\boldsymbol{\tau}(s + \Delta s) - \boldsymbol{\tau}(s)|}{\Delta s} \\
&= \lim_{\Delta s \to 0} \frac{|\mathbf{r}'(s + \Delta s) - \mathbf{r}'(s)|}{\Delta s} \\
&= |\mathbf{r}''(s)| .
\end{aligned} \tag{5.4}$$

We intentionally introduced the definition of curvature in such a way that it remains nonnegative. For a plane curve the definition is normally given so that k can be positive or negative depending on the sense of convexity ("concave up" or "concave down").

Exercise 5.9. A helix is described by

$$\mathbf{r}(t) = \mathbf{i}_1\alpha\cos t + \mathbf{i}_2\alpha\sin t + \mathbf{i}_3\beta t.$$

Study the curvature as a function of the parameters α and β, showing that $k \to 0$ as $\alpha \to 0$ and $k \to 1/\alpha$ as $\beta \to 0$. Explain.

Moving trihedron

Let us note that $\tau^2 = \boldsymbol{\tau} \cdot \boldsymbol{\tau} = 1$ for all s. This means that

$$\frac{d}{ds}\tau^2(s) = 2\boldsymbol{\tau}(s) \cdot \boldsymbol{\tau}'(s) = 0, \tag{5.5}$$

hence $\boldsymbol{\tau}'(s)$ is orthogonal to $\boldsymbol{\tau}(s)$.

Remark 5.1. It is clear that $\boldsymbol{\tau}(s)$ in (5.5) need not be a tangent vector. Thus we have a general statement: any unit vector $\mathbf{e}(s)$ is orthogonal to its derivative $\mathbf{e}'(s)$. That is, $\mathbf{e}(s) \cdot \mathbf{e}'(s) = 0$.

We define the *principal normal* $\boldsymbol{\nu}$ at point s by the equation

$$\boldsymbol{\nu} = \frac{\mathbf{r}''(s)}{k}. \tag{5.6}$$

If the curve $\mathbf{r} = \mathbf{r}(s)$ lies in a plane, it is clear that $\boldsymbol{\nu}$ lies in the same plane. Any plane through the point s that contains a tangent to the curve at this point is called a tangent plane. Among all the tangent planes at the same point of a non-planar curve, there is a unique one that plays the role of the plane containing a plane curve. It is the plane that contains $\boldsymbol{\tau}$ and $\boldsymbol{\nu}$ simultaneously. This plane is said to be *osculating*, and can be thought of as "the most locally tangent" plane to the curve at the point s. A unit normal to the osculating plane at the same point s is introduced using the relation

$$\boldsymbol{\beta} = \boldsymbol{\tau} \times \boldsymbol{\nu}.$$

These vectors $\boldsymbol{\tau}$, $\boldsymbol{\nu}$, and $\boldsymbol{\beta}$ constitute what is called the *moving trihedron* of the curve. Associated with this frame at each point along the curve is a set of three mutually perpendicular planes. As we stated above, the plane of $\boldsymbol{\tau}$ and $\boldsymbol{\nu}$ is called the osculating plane. The plane of $\boldsymbol{\nu}$ and $\boldsymbol{\beta}$ is called the *normal plane*, and the plane of $\boldsymbol{\beta}$ and $\boldsymbol{\tau}$ is called the *rectifying plane*.

The equation of the osculating plane at point s_0 is

$$(\mathbf{r} - \mathbf{r}(s_0)) \cdot \boldsymbol{\beta}(s_0) = 0$$

where $\mathbf{r}$ is the radius vector of a point of the osculating plane. For general parametrization of the curve $\mathbf{r} = \mathbf{r}(t)$ the equation of the osculating plane in Cartesian coordinates is

$$\begin{vmatrix} x - x(t_0) & y - y(t_0) & z - z(t_0) \\ x'(t_0) & y'(t_0) & z'(t_0) \\ x''(t_0) & y''(t_0) & z''(t_0) \end{vmatrix} = 0.$$

Exercise 5.10. Calculate $\boldsymbol{\nu}$ and $\boldsymbol{\beta}$ for the helix of Exercise 5.9. Then find the equation of the rectifying plane at $t = t_0$.

Using the chain rule we can present the expression for k for an arbitrary parametrization of a curve:

$$k^2 = \frac{(\mathbf{r}'(t) \times \mathbf{r}''(t))^2}{(\mathbf{r}'^2(t))^3}. \tag{5.7}$$

We shall now denote the curvature by k_1, because there is another quantity that characterizes how a curve differs from a straight line. In terms of k_1 we may define the *radius of curvature* as the number $R = 1/k_1$.

Exercise 5.11. Derive expression (5.7).

The principal normal and the binormal to a space curve are not defined uniquely when $\boldsymbol{\tau}'(s) = \mathbf{0}$. Any point at which this condition holds is called a *point of inflection* of the curve. We shall assume that our curves satisfy $\boldsymbol{\tau}'(s) \neq \mathbf{0}$ for all s.

Curves in the plane

The equations of this section take special forms when the curve under consideration lies in a plane. In such a case it is expedient to work in a concrete coordinate system such as rectangular coordinates or plane polar coordinates. In a rectangular coordinate frame $(\hat{\mathbf{x}}, \hat{\mathbf{y}})$ where

$$\mathbf{r} = \hat{\mathbf{x}}x(t) + \hat{\mathbf{y}}y(t),$$

it is easily seen that the length of the part of the curve corresponding to the interval $[a, b]$ of the parameter t is

$$s = \int_a^b \sqrt{[x'(t)]^2 + [y'(t)]^2}\, dt.$$

The curvature is given by

$$k_1 = \frac{x'(t)y''(t) - x''(t)y'(t)}{\{[x'(t)]^2 + [y'(t)]^2\}^{3/2}}.$$

These formulas correspond to the familiar formulas

$$s = \int_a^b \sqrt{1 + [f'(x)]^2}\, dx, \qquad k_1 = \frac{f''(x)}{\{1 + [f'(x)]^2\}^{3/2}}$$

from elementary calculus, which are written for a curve expressed in the non-parametric form $y = f(x)$. In polar coordinates where the curve is expressed in the form $r = r(\theta)$, we have

$$s = \int_a^b \sqrt{[r(\theta)]^2 + [r'(\theta)]^2}\, d\theta, \qquad k_1 = \frac{[r(\theta)]^2 + 2[r'(\theta)]^2 - r(\theta)r''(\theta)}{\{[r(\theta)]^2 + [r'(\theta)]^2\}^{3/2}}.$$

Note that for a plane curve the curvature k_1 possesses an algebraic sign.

Exercise 5.12. (a) Find the radius of curvature of the curve $y = x^3$ at the point $(1, 1)$. Repeat for the curve $y = x^4$ at the point $(0, 0)$. (b) Locate the point of maximum curvature of the parabola $y = ax^2 + bx + c$.

Exercise 5.13. Suppose that all the tangents to a smooth curve pass through the same point. Demonstrate that the curve is a part of a straight line or the whole line.

Exercise 5.14. Suppose that all the tangents to a smooth curve are parallel to a plane. Show that the curve lies in a plane.

Exercise 5.15. Suppose that all the principal normals of a smooth curve are parallel to a plane. Does this curve lie in a plane? Repeat when all the binormals are parallel to a plane.

5.2   The Torsion of a Curve

When a curve lies in a plane, the binormal $\boldsymbol{\beta}$ is normal to this plane. Moreover, $\boldsymbol{\beta}$ is normal to the osculating plane to the curve at a point, so the rate of rotation of this plane, which is measured by the rate of turn of the binormal, characterizes how the curve is "non-planar." By analogy to the curvature of a curve we introduce this characteristic of "non-planeness" called the *torsion* or *second curvature*, and define it as the limit of the ratio $\Delta\vartheta/\Delta s$ where $\Delta\vartheta$ is the angle of turn of the binormal $\boldsymbol{\beta}$. Since $\boldsymbol{\beta}$ is a unit vector we can use the same reasoning as we used to derive the expression for the curvature (5.4). Let us denote the torsion by k_2 and write

$$k_2 = \lim_{\Delta s \to 0} \frac{\Delta \vartheta}{\Delta s}.$$

This quantity has an algebraic sign as we explain further below.[2]

Theorem 5.1. *Let* $\mathbf{r} = \mathbf{r}(s)$ *be a three times continuously differentiable vector function of* s. *At any point where* $k_1 \neq 0$, *the torsion of the curve is*

$$k_2 = -\frac{(\mathbf{r}'(s) \times \mathbf{r}''(s)) \cdot \mathbf{r}'''(s)}{k_1^2}.$$

Proof. At a point s where $k_1 \neq 0$ the binormal is defined uniquely as is $\boldsymbol{\nu}$. Since $\boldsymbol{\beta}$ is a unit vector we can define the absolute value of the turn of the binormal, when it moves along the element corresponding to Δs, in the same manner as we used to introduce the curvature of a curve. Now $|\boldsymbol{\beta}(s + \Delta s) - \boldsymbol{\beta}(s)| = 2 |\sin \Delta \vartheta / 2|$. Thus we have

$$\begin{aligned}
|k_2| &= \lim_{\Delta s \to 0} \left| \frac{\Delta \vartheta}{\Delta s} \right| \\
&= \lim_{\Delta s \to 0} \left| \frac{\Delta \vartheta}{2 \sin(\Delta \vartheta / 2)} \right| \left| \frac{2 \sin(\Delta \vartheta / 2)}{\Delta s} \right| \\
&= \lim_{\Delta s \to 0} \left| \frac{\boldsymbol{\beta}(s + \Delta s) - \boldsymbol{\beta}(s)}{\Delta s} \right| \\
&= |\boldsymbol{\beta}'(s)|.
\end{aligned}$$

Let us demonstrate that $\boldsymbol{\beta}'(s)$ and $\boldsymbol{\nu}$ are parallel. The derivative of any unit vector $\mathbf{x}(s)$ is normal to the vector:

$$0 = \frac{d}{ds}(\mathbf{x}(s) \cdot \mathbf{x}(s)) = 2\mathbf{x}(s) \cdot \mathbf{x}'(s).$$

So $\boldsymbol{\beta}'(s)$ is orthogonal to $\boldsymbol{\beta}(s)$ and hence parallel to the osculating plane. Next

$$\boldsymbol{\beta}'(s) = (\boldsymbol{\tau}(s) \times \boldsymbol{\nu}(s))' = \boldsymbol{\tau}'(s) \times \boldsymbol{\nu}(s) + \boldsymbol{\tau}(s) \times \boldsymbol{\nu}'(s) = \boldsymbol{\tau}(s) \times \boldsymbol{\nu}'(s) \quad (5.8)$$

where we used the fact that $\boldsymbol{\tau}'(s)$ is parallel to $\boldsymbol{\nu}(s)$. By (5.8) it follows that $\boldsymbol{\beta}'(s)$ is orthogonal to $\boldsymbol{\tau}(s)$, so we have established the needed property. Since $|\boldsymbol{\nu}(s)| = 1$ it follows that $|\boldsymbol{\beta}'(s) \cdot \boldsymbol{\nu}(s)| = |\boldsymbol{\beta}'(s)|$ and thus

$$|k_2| = |\boldsymbol{\beta}'(s) \cdot \boldsymbol{\nu}(s)|.$$

---

[2]Basically, a curve having positive torsion will twist in the manner of a right-hand screw thread as s increases.

Let us use the fact that $\boldsymbol{\nu} = \mathbf{r}''/k_1$, and thus

$$\boldsymbol{\nu}' = \mathbf{r}'''/k_1 - \mathbf{r}''k_1'/k_1^2, \qquad \boldsymbol{\beta} = (\mathbf{r}' \times \mathbf{r}'')/k_1.$$

We get

$$|k_2| = |(\boldsymbol{\tau} \times \boldsymbol{\nu}') \cdot \boldsymbol{\nu}| = \frac{|(\mathbf{r}' \times \mathbf{r}'') \cdot \mathbf{r}'''|}{k_1^2}.$$

Here we used the properties of the mixed product. We now define

$$k_2 = -\frac{(\mathbf{r}' \times \mathbf{r}'') \cdot \mathbf{r}'''}{k_1^2}.$$

The rule for the sign is introduced as follows: if the binormal $\boldsymbol{\beta}$ turns in the direction from $\boldsymbol{\beta}$ to $\boldsymbol{\nu}$, then the sign is positive; otherwise, it is negative. The sign of k_2 is taken in such a way that

$$\boldsymbol{\beta}' = k_2\boldsymbol{\nu}. \tag{5.9}$$

This completes the proof. $\qquad\square$

Let us also mention that if we consider another parametrization of the curve, then a simple calculation using the chain rule yields

$$k_2 = -\frac{(\mathbf{r}'(t) \times \mathbf{r}''(t)) \cdot \mathbf{r}'''(t)}{(\mathbf{r}'(t) \times \mathbf{r}''(t))^2}. \tag{5.10}$$

We have said that the value of the torsion indicates the rate at which the curve distinguishes itself from a plane curve. Let us demonstrate this more clearly. Let $k_2 = 0$ for all s. We show that the curve lies in a plane. Indeed $0 = |k_2| = |\boldsymbol{\beta}'(s) \cdot \boldsymbol{\nu}(s)|$, thus $|\boldsymbol{\beta}'(s)| = 0$ (since $\boldsymbol{\nu}$ is a unit vector) and so $\boldsymbol{\beta}(s) = \boldsymbol{\beta}^0 = \text{const}$. The tangent vector $\boldsymbol{\tau}$ is orthogonal to $\boldsymbol{\beta}^0$, so $0 = \boldsymbol{\tau} \cdot \boldsymbol{\beta}^0 = \mathbf{r}'(s) \cdot \boldsymbol{\beta}^0$ and thus, integrating, we get $(\mathbf{r}(s) - \mathbf{r}(s_0)) \cdot \boldsymbol{\beta}^0 = 0$. This means that the curve lies in a plane.

By formula (5.9), $k_2 = |d\boldsymbol{\beta}/ds|$. So $k_2 = \lim_{\Delta s \to 0} |\Delta\boldsymbol{\beta}/\Delta s|$. But up to small quantities of the second order of Δs, the change $|\Delta\mathbf{e}|$ of a unit vector $\mathbf{e}$ is equal to the angle of rotation of $\mathbf{e}$ when moved through a distance Δs. Thus $|\Delta\boldsymbol{\beta}|$ is approximately equal to the angle of rotation of the binormal during the shift of the point through Δs, and so k_2 measures the rate of rotation of the binormal when a point moves along the curve.

Exercise 5.16. Demonstrate that a smooth curve lies in a plane only if $k_2 = 0$.

Exercise 5.17. Demonstrate that in Cartesian coordinates

$$k_1 = \frac{\left[(y'z'' - z'y'')^2 + (z'x'' - x'z'')^2 + (x'y'' - y'x'')^2\right]^{1/2}}{\left[(x')^2 + (y')^2 + (z')^2\right]^{3/2}}$$

and

$$k_2 = \frac{\begin{vmatrix} x' & y' & z' \\ x'' & y'' & z'' \\ x''' & y''' & z''' \end{vmatrix}}{(y'z'' - z'y'')^2 + (z'x'' - x'z'')^2 + (x'y'' - y'x'')^2}$$

where the prime denotes d/dt.

Exercise 5.18. What happens to k_1 and k_2 (and the moving trihedron of a curve) if the direction of change of the parameter is reversed $(t \mapsto (-t))$?

Exercise 5.19. Calculate k_2 for the helix of Exercise 5.9.

5.3 Frenet–Serret Equations

The natural triad τ, ν, β can serve as coordinate axes of the space; this is used when studying local properties of a curve. Frenet established a system of ordinary differential equations which governs the triad along the curve. We have already derived two of these three equations: (5.6) and (5.9). The former will be written as $\tau' = k_1\nu$. Let us derive the third formula of the Frenet–Serret system. We have $\nu = \beta \times \tau$. By this,

$$\begin{aligned}
\nu' &= (\beta \times \tau)' \\
&= \beta' \times \tau + \beta \times \tau' \\
&= k_2\nu \times \tau + \beta \times (k_1\nu) \\
&= -k_1\tau - k_2\beta.
\end{aligned}$$

Let us collect the Frenet–Serret equations together:

$$\begin{aligned}
\tau' &= k_1\nu, \\
\nu' &= -k_1\tau - k_2\beta, \\
\beta' &= k_2\nu.
\end{aligned} \tag{5.11}$$

We recall that these equations are written out when the curve has the natural parametrization with the length parameter s.

Note that if the curvatures $k_1(s)$ and $k_2(s)$ are given functions of s, the system (5.11) becomes a linear system of ordinary differential equations; in component form, it becomes a system of nine equations in nine unknowns. Fixing some point of the curve in space and an orthonormal triad $\boldsymbol{\tau}, \boldsymbol{\nu}, \boldsymbol{\beta}$ at the point, by this system we can define $\boldsymbol{\tau}(s)$, $\boldsymbol{\nu}(s)$, and $\boldsymbol{\beta}(s)$ uniquely; then, by the equation $\mathbf{r}'(s) = \boldsymbol{\tau}(s)$, we define the curve $\mathbf{r} = \mathbf{r}(s)$ uniquely as well. Thus $k_1(s)$ and $k_2(s)$ define the curve up to a motion in space. That is why the pair of equations for $k_1(s), k_2(s)$ is called the set of natural equations of the curve.

Let us demonstrate how to use the Frenet–Serret equations to characterize the curve locally. We use the Taylor expansion of the radius vector of a curve at point s:

$$\mathbf{r}(s + \Delta s) = \mathbf{r}(s) + \Delta s \mathbf{r}'(s) + \frac{(\Delta s)^2}{2}\mathbf{r}''(s) + \frac{(\Delta s)^3}{6}\mathbf{r}'''(s) + o(|\Delta s|^3).$$

By the definition $\boldsymbol{\nu} = \mathbf{r}''/k_1$ and by the second of equations (5.11) we get

$$\mathbf{r}''' = (k_1\boldsymbol{\nu})' = k_1'\boldsymbol{\nu} + k_1\boldsymbol{\nu}' = k_1'\boldsymbol{\nu} + k_1(-k_1\boldsymbol{\tau} - k_2\boldsymbol{\beta}).$$

Substituting these we get

$$\mathbf{r}(s + \Delta s) = \mathbf{r}(s) + \Delta s \boldsymbol{\tau} + \frac{(\Delta s)^2}{2}k_1\boldsymbol{\nu} + \frac{(\Delta s)^3}{6}(k_1'\boldsymbol{\nu} - k_1^2\boldsymbol{\tau} - k_1 k_2\boldsymbol{\beta}) + o(|\Delta s|^3).$$

This shows that, to the order of $(\Delta s)^2$, the curve lies in the osculating plane at point s. At a point s the triad $\boldsymbol{\tau}, \boldsymbol{\nu}, \boldsymbol{\beta}$ is Cartesian. Let us fix this frame and place its origin at $\mathbf{r}(s) = 0$. Defining x, y, z as the components of $\mathbf{r}(s + \Delta s)$ in this frame we obtain the approximate representation for the curve

$$x = \Delta s - \frac{k_1^2(\Delta s)^3}{6} + o(|\Delta s|^3),$$

$$y = \frac{k_1(\Delta s)^2}{2} + \frac{k_1'(\Delta s)^3}{6} + o(|\Delta s|^3),$$

$$z = -\frac{k_1 k_2(\Delta s)^3}{6} + o(|\Delta s|^3).$$

As another application of the Frenet–Serret equations, we find the velocity and acceleration of a point moving in space. Let s be the length parameter of the trajectory of the point, and let t denote time. We shall use the triad $(\boldsymbol{\tau}, \boldsymbol{\nu}, \boldsymbol{\beta})$ of the trajectory $\mathbf{r} = \mathbf{r}(s)$. The position of the point is given by the equation

$$\mathbf{r} = \mathbf{r}(s(t)).$$

The velocity of the point is

$$\mathbf{v} = \frac{d}{dt}\mathbf{r}(s(t)) = \frac{d\mathbf{r}}{ds}\frac{ds}{dt} = \frac{ds}{dt}\boldsymbol{\tau}.$$

Denoting $v = ds/dt$, the particle speed at each point along its path, we can write $\mathbf{v} = v\boldsymbol{\tau}$. Now let us find the acceleration of the point in the same frame, which is

$$\mathbf{a} = \frac{d^2}{dt^2}\mathbf{r}(s(t)) = \frac{d}{dt}\mathbf{v}(s(t)) = \frac{d}{dt}\left(\frac{ds}{dt}\boldsymbol{\tau}\right)$$

$$= \frac{d^2s}{dt^2}\boldsymbol{\tau} + \frac{ds}{dt}\frac{d\boldsymbol{\tau}}{ds}\frac{ds}{dt} = s''(t)\boldsymbol{\tau} + v^2\frac{d\boldsymbol{\tau}}{ds}.$$

Using the Frenet–Serret equations we get

$$\mathbf{a} = s''(t)\boldsymbol{\tau} + k_1 v^2 \boldsymbol{\nu}$$

where k_1 is the principal curvature of the trajectory. In mechanics this is commonly written as

$$\mathbf{a} = s''(t)\boldsymbol{\tau} + (v^2/\rho)\boldsymbol{\nu}$$

where $\rho = 1/k_1$ is the radius of curvature of the curve. Thus the acceleration vector lies in the osculating plane at each point along the trajectory. On this fact several practical methods of finding the acceleration are based.

Exercise 5.20. Express $d^3\mathbf{r}/ds^3$ in terms of the moving trihedron.

Exercise 5.21. By defining a vector $\boldsymbol{\delta} = k_1\boldsymbol{\beta} - k_2\boldsymbol{\tau}$, show that the Frenet–Serret equations can be written in the form

$$\boldsymbol{\tau}' = \boldsymbol{\delta} \times \boldsymbol{\tau}, \qquad \boldsymbol{\nu}' = \boldsymbol{\delta} \times \boldsymbol{\nu}, \qquad \boldsymbol{\beta}' = \boldsymbol{\delta} \times \boldsymbol{\beta}.$$

The vector $\boldsymbol{\delta}$ is known as the *Darboux vector*.

Exercise 5.22. A particle moves through space in such a way that its position vector is given by

$$\mathbf{r}(t) = \mathbf{i}_1(1 - \cos t) + \mathbf{i}_2 t + \mathbf{i}_3 \sin t.$$

Find the tangential and normal components of the acceleration.

5.4 Elements of the Theory of Surfaces

The reader is familiar with many standard surfaces: the sphere, the cone, the cylinder, etc. These are easy to visualize, and in Cartesian coordinates are described by simple equations. For example, the equation of a sphere reflects the definition of a sphere: all points (x, y, z) of a sphere have the same distance R from the center (x_0, y_0, z_0):

$$(x - x_0)^2 + (y - y_0)^2 + (z - z_0)^2 = R^2. \tag{5.12}$$

An infinite circular cylinder with generator parallel to the z-axis is given by the equation

$$(x - x_0)^2 + (y - y_0)^2 = R^2,$$

in which the variable z does not appear. We can obtain other types of cylindrical surfaces by considering a set of spatial points satisfying the equation

$$f(x, y) = 0.$$

This equation does not depend on z, hence by drawing a generator parallel to the z-axis through the point $(x, y, 0)$ we get a more general cylindrical surface. A paraboloid is an example of a more complex surface:

$$z = ax^2 + 2bxy + cy^2 + dx + ey + f, \tag{5.13}$$

where a, b, c, d, e, f are numerical coefficients. Depending on the values of a, b, c, the paraboloid can be elliptic, hyperbolic or parabolic. These terms from analytic geometry will be used to characterize the shape of a surface at a point. The areas and volumes associated with such standard surfaces (or portions thereof) can be calculated by integration or, sometimes, by the use of elementary formulas.

From a naive viewpoint a surface is something that fully or partially bounds a spatial body. However, a precise definition of the term "surface" is not easy to give. An attempt could be based on a local description, regarding an elementary portion of a surface as a continuous image in space of a small disk in the plane. Unfortunately such a definition would place under consideration surfaces of extremely complex structure. To use the ordinary tools of calculus we must invoke the idea of smoothness of a surface, even if we do not define this term precisely (it is common in the natural sciences to use notions and study objects that are not explicitly introduced, and about which we know only some things).

As a first step we could view a surface as something in space that resembles the figures mentioned above in the sense that it has zero thickness. To use calculus we must represent the coordinates of the points of a surface using functions. Note that the paraboloid (5.13) can be described using a single function of the two variables x and y, whereas this is not possible with the sphere (5.12). But we can divide a sphere into hemispheres in such a way that each can be described by a separate function of x and y. This can be done with more general surfaces, although it may not be possible to have z as the dependent variable for all portions of a surface. In general, the position of any point of a surface is determined by the values of a pair (u, v) of independent parameters which may or may not be Cartesian coordinates. It is convenient to use the position vector

$$\mathbf{r} = \mathbf{r}(u, v)$$

to locate this point with respect to the coordinate origin. For a Cartesian frame this vector equation is equivalent to the three scalar equations

$$x = x(u, v), \qquad y = y(u, v), \qquad z = z(u, v),$$

but the use of other coordinate frames is common. The pair (u, v) can be regarded as the coordinates of a point in the uv-plane, and a portion of the surface may be regarded as the image of a domain in this plane; in this sense we can say that a surface is two-dimensional. As was the case in earlier sections, we shall consider only sufficiently smooth functions. We have seen that the use of tensor symbolism simplifies many formulas and renders them clear. So instead of the notation (u, v) we shall use (u^1, u^2) to denote the *intrinsic coordinates* of a surface.[3]

One goal of this book is to collect major results and explain how these can be used by practitioners of many sciences. In particular, we now present formulas used in the mechanics of elastic shells, which relies heavily on the theory of surfaces. First we show how to calculate distances and angles on a general surface. This is done by introducing the "first fundamental form" of the surface. In this way we introduce a metric form of the surface analogous to the one used for three-dimensional space. Other surface properties can be described once we possess the notion of the surface normal. The structure of a surface at a point may cause us to experience differences as we depart from the point in various directions. To study this we introduce the "second fundamental form" of the surface. The two fundamental forms provide a full local description of the surface. We shall study other notions as well, concluding our treatment by adapting tensor analysis to two dimensions and applying the resulting tools to the study of surfaces.

First fundamental form

We have said that a coordinate line in space is a set of points generated by fixing two of the coordinates q^1, q^2, q^3 and changing the remaining one.

---

[3]We use this term to indicate that the coordinates refer to the surface; they are taken in the surface, not from the surrounding space somehow.

A *coordinate level surface* is generated by fixing one of the coordinates q^1, q^2, q^3 and changing the remaining two. A similar idea is used to introduce an arbitrary surface in space. The position of a point on a surface can be characterized by a radius vector initiated from some origin of the space and with terminus that moves along the surface. A point of a surface can be characterized by two parameters that we shall denote by u^1, u^2 or sometimes, when convenient, by u, v:

$$\mathbf{r} = \mathbf{r}(u^1, u^2). \tag{5.14}$$

Thus we consider a surface as the spatial image of some domain in the (u^1, u^2) coordinate plane. For simplicity we shall restrict ourselves to those surfaces that are smooth everywhere except possibly for some simple curves or points which lie thereon. So we shall consider the case when (5.14) is as smooth as we like (it is normally necessary for (5.14) to have all of its second (and sometimes third) partial derivatives continuous except for some lines or poles, as is the case with the vertex of a cone). Similar to the case of a three-dimensional space, on a surface there are coordinate lines that arise when in (5.14) we fix one of coordinates. In similar fashion we can introduce the tangent vectors

$$\mathbf{r}_i = \frac{\partial \mathbf{r}(u^1, u^2)}{\partial u^i} \qquad (i = 1, 2).$$

We suppose that, except at some singular lines or poles in the (u^1, u^2) plane, these vectors are continuously differentiable in the coordinates. We also suppose that, except at the same points, these two vectors are not collinear so the vector

$$\mathbf{N} = \mathbf{r}_1 \times \mathbf{r}_2 \neq \mathbf{0}.$$

By this we can introduce the unit normal to the surface

$$\mathbf{n} = \frac{\mathbf{r}_1 \times \mathbf{r}_2}{|\mathbf{r}_1 \times \mathbf{r}_2|},$$

which is simultaneously normal to the plane that is osculating to the surface at point (u^1, u^2). Let us demonstrate this. Let the plane through point (u^1, u^2) have normal $\mathbf{n}$. The osculating plane (Fig. 5.2) is the plane for which

$$\lim_{\Delta s \to 0} \frac{h}{\Delta s} = 0$$

where

$$\Delta s = \left| \mathbf{r}(u^1 + \Delta u^1, u^2 + \Delta u^2) - \mathbf{r}(u^1, u^2) \right|$$

and h is the distance from the point with coordinates $(u^1 + \Delta u^1, u^2 + \Delta u^2)$ to the plane. Note that this must hold for any curve on the surface through the point. Let the plane be orthogonal to $\mathbf{n}$. Then

$$h = \left| \left(\mathbf{r}(u^1 + \Delta u^1, u^2 + \Delta u^2) - \mathbf{r}(u^1, u^2) \right) \cdot \mathbf{n} \right|.$$

Using the definition of the differential we have

$$\frac{h}{\Delta s} = \frac{\left| \left(\mathbf{r}(u^1 + \Delta u^1, u^2 + \Delta u^2) - \mathbf{r}(u^1, u^2) \right) \cdot \mathbf{n} \right|}{\left| \mathbf{r}(u^1 + \Delta u^1, u^2 + \Delta u^2) - \mathbf{r}(u^1, u^2) \right|}$$

$$= \frac{\left| \left(\mathbf{r}_i \Delta u^i + o(\left| \Delta u^1 \right| + \left| \Delta u^2 \right|) \right) \cdot \mathbf{n} \right|}{\left| \mathbf{r}_i \Delta u^i + o(\left| \Delta u^1 \right| + \left| \Delta u^2 \right|) \right|}$$

$$= \frac{o(\left| \Delta u^1 \right| + \left| \Delta u^2 \right|)}{\left| \mathbf{r}_i \Delta u^i + o(\left| \Delta u^1 \right| + \left| \Delta u^2 \right|) \right|}.$$

This means that the plane we chose is osculating.

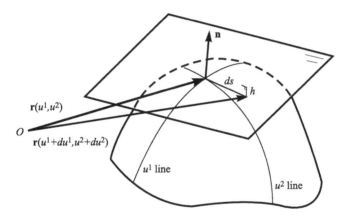

Fig. 5.2 Osculating plane to a surface.

It is clear that much of what was said about $\mathbf{r}_i$ in Chapter 4 should be simply reformulated for the present case if we use the vectors $\mathbf{r}_1, \mathbf{r}_2, \mathbf{r}^3 = \mathbf{n}$ as a basis in $\mathbb{R}^3$. First, we can introduce the reciprocal basis in space whose third vector is $\mathbf{n}$ again and whose two others are defined by the equations

$$\mathbf{r}^i \cdot \mathbf{r}_j = \delta^i_j \quad (i = 1, 2), \qquad \mathbf{r}^i \cdot \mathbf{n} = 0.$$

The vectors $\mathbf{r}^i$ ($i = 1, 2$) constitute the reciprocal frame to $\mathbf{r}_i$ ($i = 1, 2$) in the plane osculating to the surface at (u^1, u^2).

Let us consider the differential

$$d\mathbf{r} = \mathbf{r}_i \, du^i. \tag{5.15}$$

In this chapter, when summing, the indices take the values 1 and 2. The differential is the main linear part in du^i of the difference

$$\Delta\mathbf{r} = \mathbf{r}(u^1 + du^1, u^2 + du^2) - \mathbf{r}(u^1, u^2).$$

The main part of the distance between nearby points $\mathbf{r}(u^1 + du^1, u^2 + du^2)$ and $\mathbf{r}(u^1, u^2)$ is given by

$$(ds)^2 = d\mathbf{r} \cdot d\mathbf{r} = \mathbf{r}_i \, du^i \cdot \mathbf{r}_j \, du^j = g_{ij} \, du^i \, du^j$$

where $g_{ij} = \mathbf{r}_i \cdot \mathbf{r}_j$ is the metric tensor of the surface. The form

$$(ds)^2 = g_{ij} \, du^i \, du^j$$

is called the *first fundamental form* of the surface. Beginning with Gauss the metric coefficients were denoted by

$$E = g_{11}, \qquad F = g_{12} = g_{21}, \qquad G = g_{22},$$

so the first fundamental form is

$$(ds)^2 = E(du^1)^2 + 2F \, du^1 \, du^2 + G(du^2)^2.$$

Exercise 5.23. Find the first fundamental form of a sphere of radius a.

Besides the length of an elementary curve on the surface, formula (5.15) allows us to find an angle between two elementary curves $d\mathbf{r} = \mathbf{r}_i \, du^i$ and $d\tilde{\mathbf{r}} = \mathbf{r}_i \, d\tilde{u}^i$. Indeed, by the definition of the dot product we have

$$\cos\varphi = \frac{d\mathbf{r} \cdot d\tilde{\mathbf{r}}}{|d\mathbf{r}| \, |d\tilde{\mathbf{r}}|}$$

$$= \frac{E \, du^1 \, d\tilde{u}^1 + F(du^1 \, d\tilde{u}^2 + du^2 \, d\tilde{u}^1) + G \, du^2 \, d\tilde{u}^2}{\sqrt{E(du^1)^2 + 2F \, du^1 \, du^2 + G(du^2)^2}\sqrt{E(d\tilde{u}^1)^2 + 2F \, d\tilde{u}^1 \, d\tilde{u}^2 + G(d\tilde{u}^2)^2}}.$$

$$\tag{5.16}$$

Exercise 5.24. (a) Find the expression for the angle between coordinate lines of a surface. (b) Calculate the angle at which the curve $\theta = \phi$ crosses the equator of a sphere.

The area of the parallelogram that is based on the elementary vectors $\mathbf{r}_1 \, du^1$ and $\mathbf{r}_2 \, du^2$ which are tangent to the coordinate lines is

$$dS = \left| \mathbf{r}_1 \, du^1 \times \mathbf{r}_2 \, du^2 \right|.$$

This approximates up to higher order terms the area of the corresponding curvilinear figure bordered by the coordinate lines

$$u^1 = \text{const}, \quad u^2 = \text{const}, \quad u^1 + du^1 = \text{const}, \quad u^2 + du^2 = \text{const}.$$

Let us find dS in terms of the first fundamental form. We need to calculate $|\mathbf{r}_1 \times \mathbf{r}_2|$. Let us demonstrate that

$$|\mathbf{r}_1 \times \mathbf{r}_2| = \sqrt{EG - F^2}. \tag{5.17}$$

Indeed, by definition of the dot and cross product we have

$$|\mathbf{r}_1 \times \mathbf{r}_2|^2 + (\mathbf{r}_1 \cdot \mathbf{r}_2)^2 = |\mathbf{r}_1|^2 |\mathbf{r}_2|^2 \sin^2 \varphi + |\mathbf{r}_1|^2 |\mathbf{r}_2|^2 \cos^2 \varphi = |\mathbf{r}_1|^2 |\mathbf{r}_2|^2.$$

Thus

$$|\mathbf{r}_1 \times \mathbf{r}_2|^2 = |\mathbf{r}_1|^2 |\mathbf{r}_2|^2 - (\mathbf{r}_1 \cdot \mathbf{r}_2)^2 = EG - F^2$$

and (5.17) follows.

Summing the elementary areas corresponding to a domain A in the (u^1, u^2) plane and doing the limit passage we get the value

$$S = \int_A \sqrt{EG - F^2} \, du^1 \, du^2.$$

It can be shown that for a smooth surface this limit S does not depend on the parametrization and hence is the area of the portion A of the surface.

Exercise 5.25. A cone is described by the position vector

$$\mathbf{r}(u^1, u^2) = \mathbf{i}_1 u^1 \cos u^2 + \mathbf{i}_2 u^1 \sin u^2 + \mathbf{i}_3 u^1,$$

where $0 \le u^1 \le a$ and $0 \le u^2 < 2\pi$. Find the surface area of the cone.

Exercise 5.26. Find the above formulas for a figure of revolution described by the Cartesian coordinate position vector

$$\mathbf{r} = \hat{\mathbf{x}} \rho \cos \phi + \hat{\mathbf{y}} \rho \sin \phi + \hat{\mathbf{z}} f(\rho)$$

where $f(\rho)$ is a suitable profile function.

Exercise 5.27. Let two smooth surfaces be parametrized in such a way that the coefficients of their first fundamental forms are proportional at any point with the same coordinates (so this defines a map of one surface onto the other). Show that the map preserves the angles between corresponding directions on the surfaces.

Geodesics

A fundamental problem in surface theory is to find the curve of minimum length between two given points on a surface S. To treat such a problem it is natural to begin with the first fundamental form

$$(ds)^2 = g_{ij} \, du^i \, du^j.$$

We know that parametric equations

$$u^1 = u^1(t), \qquad u^2 = u^2(t),$$

can specify a curve C on S. Writing $(ds)^2$ as

$$(ds)^2 = g_{ij} \frac{du^i}{dt} \frac{du^j}{dt} (dt)^2,$$

the expression for arc length along C from $t = a$ to $t = b$ becomes

$$s = \int_a^b \left(g_{ij} \frac{du^i}{dt} \frac{du^j}{dt} \right)^{1/2} dt. \tag{5.18}$$

We seek a curve of minimum length between a given pair of endpoints: i.e., we seek functions $u^1(t)$ and $u^2(t)$ that minimize s.

Since

$$g_{ij} = g_{ij}(u^1(t), u^2(t)),$$

we see that (5.18) has the form

$$s(\mathbf{u}) = \int_a^b f(t, \mathbf{u}, \dot{\mathbf{u}}) \, dt \tag{5.19}$$

where $\mathbf{u} = \mathbf{u}(t) = (u^1(t), u^2(t))$ and the overdot denotes differentiation with respect to t. Because $s(\mathbf{u})$ is a correspondence that assigns a real number s to each function $\mathbf{u}$ in some class of functions, we call it a *functional* in $\mathbf{u}$. Here the class of functions includes all admissible routes $\mathbf{u}$ between the given endpoints on the surface.

The minimization of functionals is treated in the *calculus of variations*. Fortunately the main ideas of this extensive subject lend themselves to a brief discussion aimed at the present task. (Additional coverage appears in Chapter 6, where we consider the variational principles of elasticity.)

The approach to finding a *minimizer* $\mathbf{u}(t)$ for the functional (5.19) hinges on replacing $\mathbf{u}(t)$ by a new function $\mathbf{u}(t) + \varepsilon\boldsymbol{\varphi}(t)$, where $\boldsymbol{\varphi}(t)$ is an *admissible variation* of the function $\mathbf{u}(t)$ and ε is a small real parameter. In this discussion we shall consider a variation $\boldsymbol{\varphi}(t)$ to be admissible if $\varphi(a) = \varphi(b) = 0$; that is, if each curve of the form $\mathbf{u}(t) + \varepsilon\varphi(t)$ connects the same endpoints as the curve $\mathbf{u}(t)$. Such a replacement gives

$$s(\mathbf{u} + \varepsilon\varphi) = \int_a^b f(t, \mathbf{u} + \varepsilon\varphi, \dot{\mathbf{u}} + \varepsilon\dot{\varphi})\, dt. \qquad (5.20)$$

For fixed $\mathbf{u}$ and φ this is a function of the real variable ε, and takes its minimum at $\varepsilon = 0$ for any φ. Let us vary the components of $\mathbf{u}(t)$ one at a time. If we take φ of the special form $\varphi_1(t) = (\varphi(t), 0)$, then (5.20) becomes

$$s(\mathbf{u} + \varepsilon\varphi_1) = \int_a^b f(t, u^1(t) + \varepsilon\varphi(t), u^2(t), \dot{u}^1(t) + \varepsilon\dot{\varphi}(t), \dot{u}^2(t))\, dt.$$

We now set

$$\left.\frac{ds(\mathbf{u} + \varepsilon\varphi_1)}{d\varepsilon}\right|_{\varepsilon=0} = 0 \qquad (5.21)$$

and handle the left-hand side using the chain rule:

$$\left.\frac{d}{d\varepsilon} \int_a^b f(t, u^1 + \varepsilon\varphi, u^2, \dot{u}^1 + \varepsilon\dot{\varphi}, \dot{u}^2)\, dt\right|_{\varepsilon=0} =$$

$$= \int_a^b \left[\frac{\partial}{\partial u^1} f(t, u^1, u^2, \dot{u}^1, \dot{u}^2)\varphi + \frac{\partial}{\partial \dot{u}^1} f(t, u^1, u^2, \dot{u}^1, \dot{u}^2)\dot{\varphi}\right] dt.$$

This is the *first variation* of the functional $s(\mathbf{u})$ with respect to variations of the form $\mathbf{u} + \varepsilon\varphi_1$; we equate it to zero as in (5.21), and employ integration by parts in the second term of the integrand to get

$$\int_a^b \left[\frac{\partial}{\partial u^1} f(t, u^1, u^2, \dot{u}^1, \dot{u}^2) - \frac{d}{dt}\frac{\partial}{\partial \dot{u}^1} f(t, u^1, u^2, \dot{u}^1, \dot{u}^2)\right] \varphi\, dt = 0.$$

According to the *fundamental lemma* of the calculus of variations (cf., [Lebedev and Cloud (2003)]; one version of this lemma is presented in Theorem 6.8), this equation can hold for *all* admissible variations φ only if the bracketed quantity vanishes:

$$\frac{\partial}{\partial u^1} f(t, u^1, u^2, \dot{u}^1, \dot{u}^2) - \frac{d}{dt}\frac{\partial}{\partial \dot{u}^1} f(t, u^1, u^2, \dot{u}^1, \dot{u}^2) = 0.$$

Repeating this process for variations of the special form $\varphi_2(t) = (0, \varphi(t))$ we obtain a similar result

$$\frac{\partial}{\partial u^2} f(t, u^1, u^2, \dot{u}^1, \dot{u}^2) - \frac{d}{dt}\frac{\partial}{\partial \dot{u}^2} f(t, u^1, u^2, \dot{u}^1, \dot{u}^2) = 0.$$

This system of two equations, which we can write in more compact notation as

$$\frac{\partial f}{\partial u^i} - \frac{d}{dt}\frac{\partial f}{\partial \dot{u}^i} = 0 \qquad (i = 1, 2), \tag{5.22}$$

is known as the system of *Euler equations* for the functional (5.19). Satisfaction of this system is a necessary condition for $\mathbf{u} = (u^1(t), u^2(t))$ to be a minimizer of s. In the calculus of variations a solution to an Euler equation (or, in this case, a system of such equations) is called an *extremal*. This is analogous to a stationary point of an ordinary function: further testing is required to ascertain whether the extremal actually yields a minimum of the functional (it could yield a maximum, say). We refer to solutions of the system (5.22) as *geodesics* on the surface S.

We now use the fact that

$$f^2 = g_{jk}\dot{u}^j\dot{u}^k \tag{5.23}$$

to compute the derivatives in (5.22). First we have

$$2f\frac{\partial f}{\partial u^i} = \frac{\partial g_{jk}}{\partial u^i}\dot{u}^j\dot{u}^k$$

so that

$$\frac{\partial f}{\partial u^i} = \frac{1}{2f}\frac{\partial g_{jk}}{\partial u^i}\dot{u}^j\dot{u}^k. \tag{5.24}$$

Expanding the notation in (5.23) we see that

$$f^2 = g_{11}\dot{u}^1\dot{u}^1 + g_{12}\dot{u}^1\dot{u}^2 + g_{21}\dot{u}^2\dot{u}^1 + g_{22}\dot{u}^2\dot{u}^2,$$

hence

$$2f\frac{\partial f}{\partial \dot{u}^1} = 2g_{11}\dot{u}^1 + g_{12}\dot{u}^2 + g_{21}\dot{u}^2 = 2(g_{11}\dot{u}^1 + g_{12}\dot{u}^2) = 2g_{1j}\dot{u}^j$$

and $\partial/\partial\dot{u}^2$ is taken similarly. So

$$f\frac{\partial f}{\partial \dot{u}^i} = g_{ij}\dot{u}^j,$$

which gives

$$\frac{d}{dt}\frac{\partial f}{\partial \dot{u}^i} = \frac{d}{dt}\left(\frac{g_{ij}\dot{u}^j}{f}\right)$$

$$= \frac{1}{f}\frac{d}{dt}(g_{ij}\dot{u}^j) + g_{ij}\dot{u}^j\frac{d}{dt}\left(\frac{1}{f}\right)$$

$$= \frac{1}{f}\left(g_{ij}\ddot{u}^j + \dot{u}^j\frac{d}{dt}g_{ij}\right) + g_{ij}\dot{u}^j\left(-\frac{1}{f^2}\frac{df}{dt}\right).$$

Here

$$\frac{d}{dt} g_{ij} = \frac{\partial g_{ij}}{\partial u^k} \dot{u}^k$$

by the chain rule, so

$$\frac{d}{dt} \frac{\partial f}{\partial \dot{u}^i} = \frac{1}{f} \left(g_{ij} \ddot{u}^j + \frac{\partial g_{ij}}{\partial u^k} \dot{u}^j \dot{u}^k \right) - \frac{g_{ij} \dot{u}^j}{f^2} \frac{df}{dt}.$$

It simplifies things if we take t to be the arc length parameter; then $f \equiv 1$ and $df/dt \equiv 0$ so that

$$\frac{d}{dt} \frac{\partial f}{\partial \dot{u}^i} = g_{ij} \ddot{u}^j + \frac{\partial g_{ij}}{\partial u^k} \dot{u}^j \dot{u}^k. \tag{5.25}$$

Putting (5.24) and (5.25) into (5.22) we have

$$g_{ij} \ddot{u}^j + \left[\frac{\partial g_{ij}}{\partial u^k} - \frac{1}{2} \frac{\partial g_{jk}}{\partial u^i} \right] \dot{u}^j \dot{u}^k = 0. \tag{5.26}$$

The second term on the left can be rewritten:

$$\left[\frac{\partial g_{ij}}{\partial u^k} - \frac{1}{2} \frac{\partial g_{jk}}{\partial u^i} \right] = \left[\frac{1}{2} \left(\frac{\partial g_{ij}}{\partial u^k} + \frac{\partial g_{ik}}{\partial u^j} \right) - \frac{1}{2} \frac{\partial g_{jk}}{\partial u^i} \right] \dot{u}^j \dot{u}^k$$

$$= \frac{1}{2} \left[\frac{\partial g_{ij}}{\partial u^k} + \frac{\partial g_{ik}}{\partial u^j} - \frac{\partial g_{jk}}{\partial u^i} \right] \dot{u}^j \dot{u}^k$$

$$= \Gamma_{jki} \dot{u}^j \dot{u}^k.$$

Putting this into (5.26) and raising the index i, we have finally

$$\ddot{u}^n + \Gamma^n_{jk} \dot{u}^j \dot{u}^k = 0 \qquad (n = 1, 2). \tag{5.27}$$

This is a system of two nonlinear differential equations for the unknown $u^n(t)$. Again, t is assumed to be the natural length parameter; this means that the constraint $g_{ij} \dot{u}^j \dot{u}^k = 1$ must be enforced at each value of t along the curve.

As a simple example we may treat a cylinder $\rho = a$. It is clear in advance that if we cut the cylinder along a generator and "unroll" it onto a plane, then the shortest route between two points is the direct segment connecting them. Rolling this plane back into a cylinder we get a curve of the form

$$\phi(t) = c_1 t + c_2, \qquad z(t) = c_3 t + c_4, \tag{5.28}$$

i.e., a helix. Now let us obtain the same result using (5.27). For this particular surface $(u^1, u^2) = (\phi, z)$ and

$$\mathbf{r} = \hat{\mathbf{x}} a \cos \phi + \hat{\mathbf{y}} a \sin \phi + \hat{\mathbf{z}} z.$$

We find

$$\frac{\partial \mathbf{r}}{\partial \phi} = -\hat{\mathbf{x}} a \sin \phi + \hat{\mathbf{y}} a \cos \phi, \qquad \frac{\partial \mathbf{r}}{\partial z} = \hat{\mathbf{z}},$$

giving $g_{11} = a^2$, $g_{12} = g_{21} = 0$, $g_{22} = 1$. Since these are all constants the Christoffel coefficients are all zero, and (5.27) reduces to the system

$$\ddot{\phi} = 0, \qquad \ddot{z} = 0.$$

Integration replicates the result (5.28).

Exercise 5.28. Find the geodesics of a sphere. Show that they are great circles of the sphere.

This exercise shows that a geodesic is not always a shortest route, since we can get either part of a great circle on the sphere connecting two given points, and both of them are geodesics.

5.5　The Second Fundamental Form of a Surface

Using the Taylor expansion we can find the change of $\mathbf{r}(u^1, u^2)$ to a higher degree of approximation:

$$\mathbf{r}(u^1 + \Delta u^1, u^2 + \Delta u^2) = \mathbf{r}(u^1, u^2) + \frac{\partial \mathbf{r}(u^1, u^2)}{\partial u^i} \Delta u^i +$$
$$+ \frac{1}{2} \frac{\partial^2 \mathbf{r}(u^1, u^2)}{\partial u^i \partial u^j} \Delta u^i \Delta u^j + o\left((\Delta u^1)^2 + (\Delta u^2)^2\right).$$

Here we apply the o-notation to vector quantities, indicating that the norm of the remainder is of a higher order of smallness. We are interested in the deviation of the surface from the osculating plane in a neighborhood of a point (u^1, u^2). This value is given through the dot product by the normal vector $\mathbf{n}$:

$$(\mathbf{r}(u^1 + \Delta u^1, u^2 + \Delta u^2) - \mathbf{r}(u^1, u^2)) \cdot \mathbf{n}$$
$$= \frac{\partial \mathbf{r}(u^1, u^2)}{\partial u^i} \cdot \mathbf{n} \Delta u^i + \frac{1}{2} \frac{\partial^2 \mathbf{r}(u^1, u^2)}{\partial u^i \partial u^j} \cdot \mathbf{n} \Delta u^i \Delta u^j + o\left((\Delta u^1)^2 + (\Delta u^2)^2\right)$$
$$= \frac{1}{2} \frac{\partial^2 \mathbf{r}(u^1, u^2)}{\partial u^i \partial u^j} \cdot \mathbf{n} \Delta u^i \Delta u^j + o\left((\Delta u^1)^2 + (\Delta u^2)^2\right).$$

We have used the fact that $\mathbf{r}_i \cdot \mathbf{n} = 0$. The terms of the second order of smallness in Δu^i constitute the *second fundamental form*. Denoting

$$\frac{\partial^2 \mathbf{r}(u^1, u^2)}{(\partial u^1)^2} \cdot \mathbf{n} = L(u^1, u^2) = L,$$

$$\frac{\partial^2 \mathbf{r}(u^1, u^2)}{\partial u^1 \partial u^2} \cdot \mathbf{n} = M(u^1, u^2) = M,$$

$$\frac{\partial^2 \mathbf{r}(u^1, u^2)}{(\partial u^2)^2} \cdot \mathbf{n} = N(u^1, u^2) = N,$$

we introduce the second fundamental form by

$$d^2 \mathbf{r} \cdot \mathbf{n} = L(du^1)^2 + 2M du^1 du^2 + N(du^2)^2. \tag{5.29}$$

The deviation of the surface from the osculating plane in a small neighborhood of (u^1, u^2) is given by the following approximation:

$$z = \frac{1}{2} \left(L(du^1)^2 + 2M du^1 du^2 + N(du^2)^2 \right), \tag{5.30}$$

and so (5.30) defines the nature of the behavior of the surface in this neighborhood. This formula defines a paraboloid

$$z = \frac{1}{2} \left(L(v^1)^2 + 2M v^1 v^2 + N(v^2)^2 \right) \tag{5.31}$$

written in the triad $\left(\mathbf{r}_1(u^1, u^2), \mathbf{r}_2(u^1, u^2), \mathbf{n}(u^1, u^2) \right)$ when the origin is the point of the surface with coordinates (u^1, u^2). Depending on the coefficients L, M, N, this paraboloid can be elliptic, hyperbolic or parabolic. If $L = M = N = 0$ then the corresponding point is called a planar point.

For the second fundamental form of the surface there is a representation different from (5.29). Differentiating the identity $\mathbf{r}_i \cdot \mathbf{n} = 0$ we get

$$\frac{\partial \mathbf{r}_i}{\partial u^j} \cdot \mathbf{n} = -\mathbf{r}_i \cdot \frac{\partial \mathbf{n}}{\partial u^j}. \tag{5.32}$$

Dot multiplying $d\mathbf{r} \cdot d\mathbf{n}$ we get

$$\mathbf{r}_i \, du^i \cdot \frac{\partial \mathbf{n}}{\partial u^j} \, du^j = - \left(\frac{\partial \mathbf{r}_i}{\partial u^j} \, du^i \, du^j \right) \cdot \mathbf{n} = -d^2 \mathbf{r} \cdot \mathbf{n},$$

which means that

$$-d\mathbf{r} \cdot d\mathbf{n} = L(du^1)^2 + 2M \, du^1 \, du^2 + N(du^2)^2.$$

Exercise 5.29. Let a surface be given in Cartesian components as

$$\mathbf{r} = x(u, v)\hat{\mathbf{x}} + y(u, v)\hat{\mathbf{y}} + z(u, v)\hat{\mathbf{z}}.$$

Let the subscripts u and v denote corresponding partial derivatives with respect to u and v. Demonstrate that the coefficients of the first fundamental form are

$$E = \mathbf{r}_u^2 = x_u^2 + y_u^2 + z_u^2,$$
$$F = \mathbf{r}_u \cdot \mathbf{r}_v = x_u x_v + y_u y_v + z_u z_v,$$
$$G = \mathbf{r}_v^2 = x_v^2 + y_v^2 + z_v^2.$$

Then show that the coefficients of the second fundamental form are

$$L = \frac{\begin{vmatrix} x_{uu} & y_{uu} & z_{uu} \\ x_u & y_u & z_u \\ x_v & y_v & z_v \end{vmatrix}}{\sqrt{EG - F^2}}, \quad M = \frac{\begin{vmatrix} x_{uv} & y_{uv} & z_{uv} \\ x_u & y_u & z_u \\ x_v & y_v & z_v \end{vmatrix}}{\sqrt{EG - F^2}}, \quad N = \frac{\begin{vmatrix} x_{vv} & y_{vv} & z_{vv} \\ x_u & y_u & z_u \\ x_v & y_v & z_v \end{vmatrix}}{\sqrt{EG - F^2}}.$$

Exercise 5.30. Find L, M, N for a surface of revolution.

Normal curvature of the surface

Let us consider a curve lying on the surface. The curve can be uniquely defined by a parameter s in such a way that $u^1 = u^1(s)$, $u^2 = u^2(s)$; for simplicity we assume s to be the length parameter. Then the equation of the curve is

$$\mathbf{r} = \mathbf{r}(u^1(s), u^2(s)) = \mathbf{r}(s)$$

and a tangential vector to the curve at (u^1, u^2) is $(\mathbf{r}_i du^i/ds)\, ds$. Consider the curvature of this curve, which can be found using $\mathbf{r}''(s)$: the normal $\boldsymbol{\nu}$ to the curve has the same direction and k_1 equals $|\mathbf{r}''(s)|$. It follows that

$$\mathbf{r}''(s) \cdot \mathbf{n} = k_1 \cos \vartheta \tag{5.33}$$

where ϑ is the angle between $\boldsymbol{\nu}$ and $\mathbf{n}$. Let us rewrite (5.33) in the form

$$k_1 \cos \vartheta =$$

$$= \frac{\frac{\partial^2 \mathbf{r}}{(\partial u^1)^2} \left(\frac{du^1}{ds}\right)^2 (ds)^2 + 2\frac{\partial^2 \mathbf{r}}{\partial u^1 \partial u^2} \frac{du^1}{ds} \frac{du^2}{ds}(ds)^2 + \frac{\partial^2 \mathbf{r}}{(\partial u^2)^2} \left(\frac{du^2}{ds}\right)^2 (ds)^2}{(ds)^2} \cdot \mathbf{n}$$

$$= \frac{L(du^1(s))^2 + 2M\, du^1(s)\, du^2(s) + N(du^2(s))^2}{E(du^1(s))^2 + 2F\, du^1(s)\, du^2(s) + G(du^2(s))^2}. \tag{5.34}$$

For all the curves with the same direction defined by the constant ratio $du^1(s) : du^2(s)$, the right-hand side of (5.34) is the same as the ratio of the second to the first fundamental forms. We define this value k_0 as the *normal curvature* of the surface in the direction $du^1(s) : du^2(s)$. Geometrically, this is the curvature of the curve formed by intersecting the surface with the plane through $(u^1(s), u^2(s))$ that is parallel to $\mathbf{n}$ and $\mathbf{r}_i\, du^i(s)$. This normal curvature satisfies the equality

$$k_0 = k_1 \cos \vartheta$$

which is the main part of *Meusnier's theorem*. From this formula for curvatures it follows that the normal curvature of the surface in the direction $du^1 : du^2$ has minimal absolute value among all the curvatures of the curves on the surface through $(u^1(s), u^2(s))$ having the same direction $du^1 : du^2$. Let us note that the curvature of the osculating paraboloid is characterized by the same equation. The normal curvature depends on the direction $du^1 : du^2 = x : y$ only. Let us rewrite its expression as

$$k_0 = \frac{Lx^2 + 2Mxy + Ny^2}{Ex^2 + 2Fxy + Gy^2}. \tag{5.35}$$

The right-hand side of (5.35) is a homogeneous form with respect to x, y of zero order; it is easy to see that the minimum and maximum of k_0 can be found as the minimum and maximum of the quadratic form

$$Lx^2 + 2Mxy + Ny^2$$

when

$$Ex^2 + 2Fxy + Gy^2 = 1. \tag{5.36}$$

This problem can be solved with use of the Lagrange multiplier. On the curve (5.36) denote

$$\min(Lx^2 + 2Mxy + Ny^2) = k_{\min}, \quad \max(Lx^2 + 2Mxy + Ny^2) = k_{\max}.$$

Let the corresponding values for x, y be (x_1, y_1) and (x_2, y_2), respectively. If $k_{\min} \neq k_{\max}$ these values define only two directions $x_j : y_j$ $(j = 1, 2)$ and it can be shown that

$$Ex_1x_2 + F(x_1y_2 + x_2y_1) + Gy_1y_2 = 0,$$

which, by (5.16), means that the directions corresponding to the extreme values of the normal curvature are mutually orthogonal. In the case when $k_{\min} = k_{\max}$ the normal curvature is the same in all directions (the osculating paraboloid corresponds to a paraboloid of revolution). These values $k_{\min}, k_{\max}$ are the extreme normal curvatures of the surface at a point; they characterize the surface at a point and hence are invariant with respect to the change of coordinates of the surface. So are two other characteristics of the surface at a point: the *mean curvature*

$$H = \frac{1}{2}(k_{\min} + k_{\max}),$$

and the *Gaussian* or *complete curvature*

$$K = k_{\min}k_{\max}.$$

We present without proof the equations in terms of the fundamental forms of the surface:

$$H = \frac{1}{2}\frac{LG - 2MF + NE}{EG - F^2}, \qquad K = \frac{LN - M^2}{EG - F^2}.$$

Since $EG - F^2 > 0$, the sign of K is the sign of $LN - M^2$. When $K = 0$ everywhere the surface is developable.[4] These and many other facts on the properties of surfaces with given H and K can be found in any textbook on differential geometry (e.g., [Pogorelov (1957)]).

Exercise 5.31. Find the mean and Gaussian curvatures for a sphere of radius a.

Exercise 5.32. Find the mean and Gaussian curvatures at $x = y = 0$ of the following paraboloids: (a) $z = axy$; (b) $z = a(x^2+y^2)$; (c) $z = ax^2+by^2$.

Exercise 5.33. Suppose a surface is given in Cartesian coordinates:

$$\mathbf{r} = \hat{\mathbf{x}}x + \hat{\mathbf{y}}y + \hat{\mathbf{z}}f(x, y).$$

Demonstrate that (a)

$$E = \mathbf{r}_x^2 = 1 + f_x^2, \qquad F = \mathbf{r}_x \cdot \mathbf{r}_y = f_xf_y, \qquad G = \mathbf{r}_y^2 = 1 + f_y^2;$$

(b)

$$EG - F^2 = 1 + f_x^2 + f_y^2;$$

(c)

$$L = \frac{f_{xx}}{\sqrt{1 + f_x^2 + f_y^2}}, \qquad M = \frac{f_{xy}}{\sqrt{1 + f_x^2 + f_y^2}}, \qquad N = \frac{f_{yy}}{\sqrt{1 + f_x^2 + f_y^2}};$$

---

[4]Roughly speaking, a *developable surface* is one that can be flattened into a portion of a plane without compressing or stretching any part of it. Examples include cones and cylinders, but not spheres. A developable surface can be generated by sweeping a straight line (*generator*) along a curve through space. See page 168 for a more precise definition.

(d) the area of a portion of the surface is

$$S = \int_D \sqrt{1 + f_x^2 + f_y^2}\, dx\, dy;$$

(e) the Gaussian curvature is

$$K = \frac{LN - M^2}{EG - F^2} = \frac{f_{xx}f_{yy} - f_{xy}^2}{\left(1 + f_x^2 + f_y^2\right)^2}.$$

5.6   Derivation Formulas

At each point of a surface there is defined the frame triad

$$\mathbf{r}_1(u^1, u^2), \quad \mathbf{r}_2(u^1, u^2), \quad \mathbf{n}(u^1, u^2).$$

In the theory of shells they introduce the curvilinear coordinates in a neighborhood of the surface in such a way that on the surface the triad $(\mathbf{r}_1, \mathbf{r}_2, \mathbf{n})$ is preserved. Since we seek different characteristics of fields given on the surface and outside of it, we must find the derivatives of the triad with respect to the coordinates. The goal of this section is to present these in terms of the surface we have introduced: i.e., in terms of the first and second fundamental forms of the surface.

We start with the representation for the derivatives through the Christoffel notation. This is valid because, by assumption, $(\mathbf{r}_1, \mathbf{r}_2, \mathbf{n})$ is a basis of the space of vectors. So

$$\mathbf{r}_{ij} = \frac{\partial^2 \mathbf{r}}{\partial u^i \partial u^j} = \Gamma_{ij}^t \mathbf{r}_t + \lambda_{ij} \mathbf{n}. \tag{5.37}$$

We recall that we use indices i, j, t taking values from the set $\{1, 2\}$, which is why we introduce the notation λ_{ij} for the coefficients of $\mathbf{n}$. Similarly let us introduce the expansion for the derivatives of $\mathbf{n}$. To derive the coefficients λ_{ij} we dot multiply (5.37) by $\mathbf{n}$. Using the expressions for the second fundamental form we get

$$\lambda_{11} = L, \qquad \lambda_{12} = \lambda_{21} = M, \qquad \lambda_{22} = N.$$

Next, dot multiplying (5.37) first by $\mathbf{r}_1$ and then by $\mathbf{r}_2$, we get six equations in the six unknown Christoffel symbols:

$$\frac{1}{2}\frac{\partial E}{\partial u^1} = \Gamma_{11}^1 E + \Gamma_{11}^2 F,$$

$$\frac{\partial F}{\partial u^1} - \frac{1}{2}\frac{\partial E}{\partial u^2} = \Gamma_{11}^1 F + \Gamma_{11}^2 G,$$

$$\frac{1}{2}\frac{\partial E}{\partial u^2} = \Gamma_{12}^1 E + \Gamma_{12}^2 F,$$

$$\frac{1}{2}\frac{\partial G}{\partial u^1} = \Gamma_{12}^1 F + \Gamma_{12}^2 G,$$

$$\frac{\partial F}{\partial u^2} - \frac{1}{2}\frac{\partial G}{\partial u^1} = \Gamma_{22}^1 E + \Gamma_{22}^2 F,$$

$$\frac{1}{2}\frac{\partial G}{\partial u^2} = \Gamma_{22}^1 F + \Gamma_{22}^2 G.$$

This system can be easily solved for the Christoffel coefficients since the system splits into pairs of equations with respect to the pairs of Christoffel symbols. The first pair yields, for instance,

$$\Gamma_{11}^1 = \frac{1}{EG - F^2}\left[\frac{G}{2}\frac{\partial E}{\partial u^1} - F\frac{\partial F}{\partial u^1} + \frac{F}{2}\frac{\partial E}{\partial u^2}\right],$$

and a similar expression for Γ_{11}^2. We leave the rest of this work to the reader, mentioning that the expressions for the Christoffel symbols are composed only of the coefficients of the first fundamental form of the surface. Next, we consider the expressions for the derivatives of the normal $\mathbf{n}$:

$$\mathbf{n}_i \equiv \frac{\partial \mathbf{n}}{\partial u^i} = \mu_i^{\cdot t}\mathbf{r}_t + \mu_i^{\cdot 3}\mathbf{n}. \tag{5.38}$$

Let us find the coefficients of these expansions through the coefficients of the fundamental forms of the surface. Because of the equality

$$0 = \frac{\partial}{\partial u^i}1 = \frac{\partial}{\partial u^i}(\mathbf{n}\cdot\mathbf{n}) = 2\mathbf{n}_i\cdot\mathbf{n}$$

we have

$$\mu_i^{\cdot 3} = 0.$$

Let us dot multiply (5.38) when $i = 1$ by $\mathbf{r}_1$ and $\mathbf{r}_2$ successively. Using the definitions for the coefficients of the first and second fundamental forms we have the equations

$$-L = \mu_1^{\cdot 1}E + \mu_1^{\cdot 2}F, \qquad -M = \mu_1^{\cdot 1}F + \mu_1^{\cdot 2}G,$$

from which we get

$$\mu_1^{\cdot 1} = \frac{-LG + MF}{EG - F^2}, \qquad \mu_1^{\cdot 2} = \frac{LF - ME}{EG - F^2}.$$

Repeating this procedure for (5.38) when $i = 2$ we similarly obtain

$$\mu_2^{\cdot 1} = \frac{NF - MG}{EG - F^2}, \qquad \mu_2^{\cdot 2} = \frac{-NE + MF}{EG - F^2}.$$

Exercise 5.34. Derive the Christoffel symbols for the case when the first fundamental form is $(du^1)^2 + G(du^2)^2$.

Exercise 5.35. For a surface of revolution with coordinate lines being parallels and meridians, derive all coefficients of the expansions (5.37) and (5.38).

Some useful formulas

Certain identities are frequently employed. Their derivation is lengthy so we omit it here. Let us redenote the coefficients of the second fundamental form using index notations:

$$b_{11} = L, \qquad b_{12} = b_{21} = M, \qquad b_{22} = N.$$

These are the covariant components of a corresponding symmetric tensor of order two. The first formula is due to Gauss:

$$b_{11}b_{22} - b_{12}^2 = \frac{\partial^2 g_{12}}{\partial u^1 \partial u^2} - \frac{1}{2}\left(\frac{\partial^2 g_{11}}{(\partial u^2)^2} + \frac{\partial^2 g_{22}}{(\partial u^1)^2}\right) + \left(\Gamma_{12}^i\Gamma_{12}^j - \Gamma_{11}^i\Gamma_{22}^j\right)g_{ij},$$

from which (and the form of K, see page 152) it follows that the Gaussian curvature of the surface can be expressed purely in terms of the first fundamental form. The next formulas are due to Peterson and Codazzi:

$$\frac{\partial b_{i1}}{\partial u^2} - \frac{\partial b_{i2}}{\partial u^1} = \Gamma_{i2}^t b_{t1} - \Gamma_{i1}^t b_{t2} \qquad (i = 1, 2).$$

Using the formulas of this paragraph, we can derive the formulas for differentiating a spatial vector

$$\mathbf{f} = f^i\mathbf{r}_i + f_3\mathbf{n} = f_i\mathbf{r}^i + f_3\mathbf{n}$$

given on the surface. Denoting $\mathbf{r}_3 = \mathbf{r}^3 = \mathbf{n}$, we can derive the following formulas that mimic the formulas of Chapter 4:

$$\frac{\partial \mathbf{f}}{\partial u^i} = \nabla_i f^t \mathbf{r}_t = \nabla_i f_t \mathbf{r}^t.$$

In these, the covariant derivatives are

$$\nabla_i f^j = \frac{\partial f^j}{\partial u^i} + \Gamma^j_{it} f^t - b^j_i f_3,$$

$$\nabla_i f_j = \frac{\partial f_j}{\partial u^i} + \Gamma^t_{ij} f_t - b_{ij} f_3,$$

$$\nabla_i f_3 = \frac{\partial f_3}{\partial u^i} + b_{it} f^t = \frac{\partial f_3}{\partial u^i} + b^t_i f_t.$$

Here i, j take the values $1, 2$, while t takes the values $1, 2, 3$. Summation is carried out over i, j, and t in the respective ranges. When $\mathbf{f} = \mathbf{n}$, i.e., when $f^1 = f^2 = 0$ and $f^3 = 1$, we get the following formula for the derivatives of $\mathbf{n}$:

$$\frac{\partial \mathbf{n}}{\partial u^i} = -b^t_i \mathbf{r}_t = -b_{it} \mathbf{r}^t. \tag{5.39}$$

Introducing the *surface gradient* or *surface nabla operator* by the formula

$$\widetilde{\nabla} = \mathbf{r}^i \frac{\partial}{\partial u^i} \qquad (i = 1, 2),$$

we find that

$$\widetilde{\nabla} \mathbf{f} = \nabla_i f^t \mathbf{r}^i \mathbf{r}_t = \nabla_i f_t \mathbf{r}^i \mathbf{r}^t, \quad \widetilde{\nabla} \mathbf{n} = -\mathbf{B},$$

where $\mathbf{B}$ is the *surface curvature tensor* given by

$$\mathbf{B} = b^j_i \mathbf{r}^i \mathbf{r}_j = b_{ij} \mathbf{r}^i \mathbf{r}^j = b^{ij} \mathbf{r}_i \mathbf{r}_j.$$

One may introduce the surface divergence and rotation operations

$$\operatorname{div} \mathbf{f} = \widetilde{\nabla} \cdot \mathbf{f} = \mathbf{r}^i \cdot \frac{\partial \mathbf{f}}{\partial u^i}, \quad \operatorname{rot} \mathbf{f} = \widetilde{\nabla} \times \mathbf{f} = \mathbf{r}^i \times \frac{\partial \mathbf{f}}{\partial u^i}.$$

The applications of the differential calculus on a surface are presented in Chapter 7, where shell theories are considered.

5.7 Implicit Representation of a Curve; Contact of Curves

Our work in Chapter 5 has stressed technical topics needed to derive equations describing certain natural objects. We now turn to some less technical but quite important questions that can further our understanding of differential geometry.

We have described a curve by the representation $\mathbf{r} = \mathbf{r}(t)$. From this point of view, a curve is the image of some one-dimensional domain of the parameter t. Similarly, a surface was an image of a two-dimensional domain, given by a formula

$$\mathbf{r} = \mathbf{r}(u^1, u^2). \tag{5.40}$$

However, we know that in Cartesian coordinates a sphere is described by the equation

$$(x - x_0)^2 + (y - y_0)^2 + (z - z_0)^2 = R^2,$$

and this cannot be represented in the form (5.40) for all points simultaneously. This is an example of the implicit form of description of a surface. We can describe any small part of a sphere in an explicit form of the type (5.40). This is a point from which the theory of manifolds arose: a surface such as a sphere can be divided into small portions, each of which can be described in the needed way.

However, implicit form descriptions of surfaces and curves are common in practice. Let us suppose that the equation of a surface is

$$F(\mathbf{r}) = 0.$$

In Cartesian coordinates this would look like

$$F(x, y, z) = 0.$$

Earlier we described coordinate curves in space as sets given by the equation $\mathbf{r} = \mathbf{r}(q^1, q^2, q^3)$ when two of the three coordinate parameters q^1, q^2, q^3 are fixed. Each such curve can be described alternatively as the intersection of two coordinate surfaces. Consider for instance a q^1 coordinate curve — the curve corresponding to $q^2 = q_0^2$, $q^3 = q_0^3$, is the intersection of the surfaces $\mathbf{r} = \mathbf{r}(q^1, q^2, q_0^3)$ and $\mathbf{r} = \mathbf{r}(q^1, q_0^2, q^3)$. Similarly a curve in space can be represented as the set of points described by a vector $\mathbf{r}$ that satisfies two simultaneous equations

$$F_1(\mathbf{r}) = 0, \qquad F_2(\mathbf{r}) = 0,$$

or, in coordinate form,

$$F_1(q^1, q^2, q^3) = 0, \qquad F_2(q^1, q^2, q^3) = 0.$$

A tangent $\mathbf{t}$ to the curve at a point must be orthogonal to the two surface normals at the point; if $\mathbf{n}_1$ and $\mathbf{n}_2$ are these normals, we can write

$$\mathbf{t} = \mathbf{n}_1 \times \mathbf{n}_2.$$

Contact of curves

A curve in the plane can be also shown implicitly, but using just one equation:

$$F(\mathbf{r}) = 0 \qquad \text{or} \qquad F(q^1, q^2) = 0.$$

We would like to study the problem of approximating a given plane curve, at a given point, by another curve taken from a parametric family of curves. Let us begin with the following problem:

Given a smooth curve A and point C on it, select from all straight lines in the plane that which best approximates the behavior of A at C.

Of course, one solution is the line tangent to A at C. We see this by considering a line through C that intersects A at a point D close to C, and then producing a limit passage under which this line rotates about C in such a way that D tends to C. The line approached in the limit is the tangent line.

Let us extend this idea somewhat and seek an approximation that can reflect both the direction of the curve A at C and its curvature at that point. Since straight lines cannot reflect curvature behavior, we shall have to employ some other family of approximating curves. We could use the family of all circles in the plane, since any three points in the plane determine a unique circle. Taking two points D_1 and D_2 near C and drawing the circle through these three points, we could attempt a limit passage under which both D_1 and D_2 tend to C. This should yield a circle capable of reflecting both desired properties (tangent and curvature) of A at C. The family of all circles in the plane is described in Cartesian coordinates by the equation

$$(x - x_0)^2 + (y - y_0)^2 = R^2, \tag{5.41}$$

where the three values x_0, y_0 and R are free parameters that should be properly chosen to approximate a given curve at a point. So we see that a three-parameter family of plane curves will be needed to get the better approximation we seek. We could employ a parabola from the family $y = ax^2 + bx + c$, but this manner of approximation would be less informative.

In general we can seek to approximate a plane curve

$$\mathbf{r} = \mathbf{r}(t) \tag{5.42}$$

at a point $t = t_0$ by introducing a parametric family Φ_n of curves

$$F(\mathbf{r}, a_1, \ldots, a_n) = 0,$$

where $a_1, \ldots, a_n$ are free parameters.[5] Suppose we take n points

$$\mathbf{r}(t_k) \qquad (k = 0, \ldots, n - 1)$$

of the curve (5.42), where all the t_k are close to t_0. With n free parameters at our disposal we should be able to find a curve from Φ_n that goes through these points. This means that the system

$$F(\mathbf{r}(t_0), a_1, \ldots, a_n) = 0,$$
$$F(\mathbf{r}(t_1), a_1, \ldots, a_n) = 0,$$
$$\vdots$$
$$F(\mathbf{r}(t_{n-1}), a_1, \ldots, a_n) = 0, \qquad (5.43)$$

is satisfied by some set of parameters $a_1, \ldots, a_n$. We assume that the a_i depend on the t_k in such a way that when all the t_k tend to t_0 the a_i tend continuously to some respective values b_i. Thus we find the curve $F(\mathbf{r}, b_1, \ldots, b_n) = 0$ that best approximates the behavior of (5.42) at t_0, and we have

$$F(\mathbf{r}(t_0), b_1, \ldots, b_n) = 0. \qquad (5.44)$$

As a practical matter it is not convenient to use such a limit passage to obtain the needed curve. It would be better to have conditions in which only those characteristics of the curves at t_0 are involved. Since we suppose the above limit passage is well-defined, we can take the ordered set of points t_k:

$$t_0 < t_1 < \cdots < t_{n-1}.$$

Let $a_1, \ldots, a_n$ be a solution to (5.43) in this case, and consider the function

$$f(t) = F(\mathbf{r}(t), a_1, \ldots, a_n)$$

of the single variable t. This function vanishes at $t = t_k$ for $k = 0, \ldots, n-1$. By Rolle's theorem there exists $t'_k \in [t_{k-1}, t_k]$ $(k = 1, \ldots, n - 1)$ such that $f'(t'_k) = 0$. During the limit passage all the t'_k also tend to t_0, so we have

$$\frac{d}{dt} F(\mathbf{r}(t), b_1, \ldots, b_n) \bigg|_{t=t_0} = 0.$$

Similarly, the function

$$f'(t) = F_t(\mathbf{r}(t), a_1, \ldots, a_n)$$

---

[5] We suppose that the curve (5.42) is sufficiently smooth at t_0, as is each curve taken from the family Φ_n.

takes $n - 1$ zeroes at the points $t'_1 < \cdots < t'_{n-1}$. So by Rolle's theorem there is a point t''_k on each segment $[t'_{k-1}, t'_k]$ such that $f''(t''_k) = 0$. After the main limit passage we will have

$$\left. \frac{d^2}{dt^2} F(\mathbf{r}(t), b_1, \ldots, b_n) \right|_{t=t_0} = 0.$$

Repeating this procedure for each derivative up to order $n - 1$, we obtain the conditions

$$\left. \frac{d^k}{dt^k} F(\mathbf{r}(t), b_1, \ldots, b_n) \right|_{t=t_0} = 0 \qquad (k \le n - 1).$$

These, taken together with (5.44), constitute n conditions that should be fulfilled by the needed curve from the family Φ_n. We define the *order of contact* between this curve and (5.42) as the number $n - 1$.

Contact of a curve with a circle; evolutes

Let us return to our previous problem and apply these conditions to approximate a curve

$$\mathbf{r} = \mathbf{i}_1 x(t) + \mathbf{i}_2 y(t)$$

at a point $t = t_0$ by a circle of the family (5.41). Our three free parameters x_0, y_0, R should satisfy the system

$$(x(t_0) - x_0)^2 + (y(t_0) - y_0)^2 = R^2,$$
$$2(x(t_0) - x_0) x'(t_0) + 2(y(t_0) - y_0) y'(t_0) = 0,$$
$$2x'(t_0)^2 + 2(x(t_0) - x_0) x''(t_0) + 2y'(t_0)^2 + 2(y(t_0) - y_0) y''(t_0) = 0.$$

In particular, solution of these yields

$$R^2 = \frac{\{[x'(t_0)]^2 + [y'(t_0)]^2\}^3}{[x'(t_0)y''(t_0) - x''(t_0)y'(t_0)]^2}.$$

We see that the curvature $1/R$ of the circle coincides with the curvature k_1 of the curve. As the formulas for x_0 and y_0 are cumbersome, it is worthwhile to recast the problem in vector notation.

Let $\mathbf{r}_0$ locate the center of the contact circle

$$(\mathbf{r} - \mathbf{r}_0)^2 = R^2.$$

Let the curve be given in natural parametrization as $\mathbf{r} = \mathbf{r}(s)$. To solve this problem of second-order contact, we define

$$F(s) = (\mathbf{r}(s) - \mathbf{r}_0)^2 - R^2$$

and have

$$F(s_0) = 0, \qquad F'(s_0) = 0, \qquad F''(s_0) = 0,$$

or

$$(\mathbf{r}(s_0) - \mathbf{r}_0)^2 - R^2 = 0,$$
$$2(\mathbf{r}(s_0) - \mathbf{r}_0) \cdot \boldsymbol{\tau}(s_0) = 0,$$
$$2 + 2(\mathbf{r}(s_0) - \mathbf{r}_0) \cdot \boldsymbol{\nu}(s_0)k_1 = 0.$$

By the second equation the vector $\mathbf{r}(s_0) - \mathbf{r}_0$ is orthogonal to the tangent $\boldsymbol{\tau}(s_0)$, hence is directed along the normal $\boldsymbol{\nu}(s_0)$ and we have

$$(\mathbf{r}(s_0) - \mathbf{r}_0) \cdot \boldsymbol{\nu}(s_0) = -|\mathbf{r}(s_0) - \mathbf{r}_0| = -R.$$

This and the third equation yield $1 - k_1 R = 0$, hence $R = 1/k_1$ as stated above.

The locus of centers of all contact circles for a given curve is called the *evolute* of the curve. The equation of the evolute is

$$\boldsymbol{\rho}(t) = \mathbf{r}(t) + R\boldsymbol{\nu}(t), \qquad R = 1/k_1$$

or in parametric Cartesian form ($\boldsymbol{\rho} = (\xi, \eta)$)

$$\xi = x - y' \frac{x'^2 + y'^2}{x'y'' - x''y'}, \qquad \eta = y + x' \frac{x'^2 + y'^2}{x'y'' - x''y'}.$$

Contact of nth order between a curve and a surface

Quite similarly we can solve a problem of nth-order contact between a space curve $\mathbf{r} = \mathbf{r}(t)$ at $t = t_0$ and a surface from an $n + 1$ parameter family of surfaces given implicitly by

$$F(\mathbf{r}, a_1, \ldots, a_{n+1}) = 0.$$

Everything from the previous pages should be repeated word for word. First we choose a surface from the family in such a way that it coincides with the curve in $n + 1$ points close to t_0. This gives $n + 1$ equations:

$$F(\mathbf{r}(t_0), a_1, \ldots, a_{n+1}) = 0,$$
$$F(\mathbf{r}(t_1), a_1, \ldots, a_{n+1}) = 0,$$
$$\vdots$$
$$F(\mathbf{r}(t_{n-1}), a_1, \ldots, a_{n+1}) = 0.$$

Our previous reasoning carries through and we obtain

$$\left. \frac{d^k}{dt^k} F(\mathbf{r}(t), b_1, b_2, \ldots, b_{n+1}) \right|_{t=t_0} = 0 \qquad (k = 0, 1, \ldots, n),$$

as the equations that should hold at the point of nth-order contact.

We have introduced the osculating plane to a curve at a point A as the plane through A that contains $\boldsymbol{\tau}$ and $\boldsymbol{\nu}$. Note that it could also be defined as a surface of second-order contact. The result is the same.

The reader should consider how to apply these considerations to the problem of nth-order contact between a given surface and a surface from a many-parameter family of surfaces.

Exercise 5.36. Treat the problem of third-order contact between a space curve and a sphere. Show that denoting $R = 1/k_1$ and ρ the radius of the sphere we get (in natural parametrization)

$$\rho^2 = R^2 + R'^2/k_2^2.$$

Also show that the center of the contact sphere lies on the straight line through the center of principal curvature that is parallel to the binormal at the point of the curve.

5.8   Osculating Paraboloid

When considering the structure of a surface at a point, it is often helpful to approximate the surface using another surface whose behavior is more easily visualized. A spherical surface would be insufficient for this purpose because it has the same normal curvature in all directions. We can, however, use the osculating paraboloid introduced in (5.31). Let us reconsider this paraboloid from the point of view of local approximation.

Let O be a fixed point of a surface, and assume the surface is sufficiently smooth at O. At O we determine the osculating plane and introduce a Cartesian frame $(\mathbf{i}_1, \mathbf{i}_2, \mathbf{n})$, where $(\mathbf{i}_1, \mathbf{i}_2)$ is a Cartesian frame with origin O on the osculating plane and $\mathbf{n}$ is normal to both the surface and the osculating plane. For a smooth surface, Cartesian coordinates (x, y) can play the role of surface coordinates since they uniquely define any point of the surface at O. The surface in the vicinity of O can be described by the equation $z = z(x, y)$. We suppose that $z(x, y)$ is twice continuously differentiable near O. In the neighborhood of $(0, 0)$ we can use the Taylor expansion of $z(x, y)$:

$$z(x, y) = z(0, 0) + z_x(0, 0)x + z_y(0, 0)y$$
$$+ \frac{1}{2} \left(z_{xx}x^2 + 2z_{xy}xy + z_{yy}y^2 \right) + o(x^2 + y^2),$$

where the indices x, y indicate that we take partial derivatives with respect to x, y, respectively (and in this section, evaluate at the point $(0, 0)$). Since the coordinates (x, y) are in the osculating plane, the partial derivatives

$$z_x(0, 0) = 0, \qquad z_y(0, 0) = 0.$$

By choice of the origin, $z(0, 0) = 0$. Thus the Taylor expansion is

$$z(x, y) = \frac{1}{2} \left(z_{xx}x^2 + 2z_{xy}xy + z_{yy}y^2 \right) + o(x^2 + y^2). \tag{5.45}$$

Let us consider the paraboloid

$$z(x, y) = \frac{1}{2} \left(z_{xx}x^2 + 2z_{xy}xy + z_{yy}y^2 \right). \tag{5.46}$$

The difference between the surfaces described by (5.45) and (5.46) is small (as indicated by the o term). This allows us to show that it is an osculating paraboloid that approximates the behavior of the surface at point O. A change of coordinate frames can show that the paraboloid (5.46) coincides with (5.31). Its coefficients are L, M, N at the point O.

Let us note that the second fundamental form of the initial surface at a point coincides with that of the osculating paraboloid, and the situation is the same with the normal curvatures. We denote

$$r = z_{xx}(0, 0), \qquad s = z_{xy}(0, 0), \qquad t = z_{yy}(0, 0),$$

so that the equation of the osculating paraboloid is

$$z = \frac{1}{2} \left(rx^2 + 2sxy + ty^2 \right).$$

Let us draw the projection onto the xy-coordinate plane of the cross-sections of the osculating paraboloid by the two planes $z = \pm h$, $h > 0$. This is the curve defined by

$$\frac{1}{2} \left| rx^2 + 2sxy + ty^2 \right| = h. \tag{5.47}$$

The curve is an ellipse when the paraboloid is elliptical ($rt - s^2 > 0$) or a hyperbola when it is hyperbolic ($rt - s^2 < 0$), or a family of straight lines when it is parabolic ($rt - s^2 = 0$). It can be shown that the radius of normal curvature of the surface in the direction $x : y$ is proportional to the squared distance from the origin to the point of the curve (5.47) taken in the same direction. This curve is called the *Dupin indicatrix*.

The Dupin indicatrix is a curve of second order on the plane, and thus it has some special directions (axes) and special values characterizing the curve. The special directions are known as *principal directions*. In the next section we consider this from another vantage point.

5.9 The Principal Curvatures of a Surface

Let us discuss in more detail the properties of a surface connected with its second fundamental form. We now denote the coordinates without indices, using $u = u^1, v = u^2$. We also use subscripts u, v to indicate partial differentiation:

$$\mathbf{r}_1 = \mathbf{r}_u = \frac{\partial \mathbf{r}}{\partial u}, \qquad \mathbf{r}_2 = \mathbf{r}_v = \frac{\partial \mathbf{r}}{\partial v}, \qquad \mathbf{n}_u = \frac{\partial \mathbf{n}}{\partial u}, \qquad \mathbf{n}_v = \frac{\partial \mathbf{n}}{\partial v}.$$

The second fundamental form is

$$L \, du^2 + 2M \, du \, dv + N \, dv^2 = -d\mathbf{r} \cdot d\mathbf{n},$$

where

$$L = \mathbf{r}_{uu} \cdot \mathbf{n}, \qquad M = \mathbf{r}_{uv} \cdot \mathbf{n}, \qquad N = \mathbf{r}_{vv} \cdot \mathbf{n}.$$

Consider the first differential of $\mathbf{n}$ at a point P:

$$d\mathbf{n} = \mathbf{n}_u \, du + \mathbf{n}_v \, dv.$$

Since $\mathbf{n}$ is a unit vector, its differential $d\mathbf{n}$ is orthogonal to $\mathbf{n}$ and thus lies in the osculating plane to the surface at P. We know that the differential $d\mathbf{r} = \mathbf{r}_u \, du + \mathbf{r}_v \, dv$ also lies in the plane osculating at P. Let us consider the relation between the differentials $d\mathbf{r}$ and $d\mathbf{n}$ with respect to the variables du, dv now considered as independent variables. It is clearly a linear correspondence $d\mathbf{r} \mapsto d\mathbf{n}$. Thus it defines a tensor $\mathbf{A}$ in two-dimensional space such that

$$d\mathbf{n} = \mathbf{A} \cdot d\mathbf{r}.$$

This tensor is completely defined by its values $\mathbf{n}_u = \mathbf{A} \cdot \mathbf{r}_u$ and $\mathbf{n}_v = \mathbf{A} \cdot \mathbf{r}_v$.

Lemma 5.1. *The tensor $\mathbf{A}$ is symmetric.*

Proof. It is enough to establish the equality

$$\mathbf{x}_1 \cdot (\mathbf{A} \cdot \mathbf{x}_2) = \mathbf{x}_2 \cdot (\mathbf{A} \cdot \mathbf{x}_1)$$

for a pair of linearly independent vectors $(\mathbf{x}_1, \mathbf{x}_2)$. To show symmetry of $\mathbf{A}$ consider

$$\mathbf{r}_u \cdot (\mathbf{A} \cdot \mathbf{r}_v) = \mathbf{r}_u \cdot \mathbf{n}_v.$$

Similarly

$$\mathbf{r}_v \cdot (\mathbf{A} \cdot \mathbf{r}_u) = \mathbf{r}_v \cdot \mathbf{n}_u.$$

The symmetry of $\mathbf{A}$ follows from the identity

$$\mathbf{r}_u \cdot \mathbf{n}_v = \mathbf{r}_v \cdot \mathbf{n}_u;$$

this is derived by differentiating the identity $\mathbf{n} \cdot \mathbf{r}_u = 0$ with respect to v, then differentiating $\mathbf{n} \cdot \mathbf{r}_v = 0$ with respect to u and eliminating the term containing $\mathbf{r}_{uv}$. □

We know that a symmetric tensor has real eigenvalues; in this case there are no more than two eigenvalues λ_1 and λ_2 to which there correspond mutually orthogonal eigenvectors $\mathbf{x}_1, \mathbf{x}_2$, respectively (if $\lambda_1 \neq \lambda_2$).

Let us find the equation for the eigenvalues. A vector in the osculating plane can be represented as $x\mathbf{r}_u + y\mathbf{r}_v$ since $(\mathbf{r}_u, \mathbf{r}_v)$ is a basis in it. By definition of $\mathbf{A}$ we have

$$\mathbf{A} \cdot (x\mathbf{r}_u + y\mathbf{r}_v) = x\mathbf{n}_u + y\mathbf{n}_v.$$

On the other hand, $x\mathbf{r}_u + y\mathbf{r}_v \neq 0$ is an eigenvector if there exists λ such that

$$\mathbf{A} \cdot (x\mathbf{r}_u + y\mathbf{r}_v) = \lambda(x\mathbf{r}_u + y\mathbf{r}_v).$$

Thus, for the same vector we get

$$x\mathbf{n}_u + y\mathbf{n}_v = \lambda(x\mathbf{r}_u + y\mathbf{r}_v). \tag{5.48}$$

We see that an eigenvector is defined by the same condition as the principal directions of the previous section, since this equation means that we find the direction in which $d\mathbf{r}$ and $d\mathbf{n}$ are parallel.

From the last vector equation, let us derive scalar equations. For this, dot multiply (5.48) first by $\mathbf{r}_u$, then by $\mathbf{r}_v$. By (5.32) we have

$$L = -\mathbf{n}_u \cdot \mathbf{r}_u, \qquad M = -\mathbf{n}_u \cdot \mathbf{r}_v, \qquad N = -\mathbf{n}_v \cdot \mathbf{r}_v,$$

hence

$$-Lx - My = \lambda(Ex + Fy),$$
$$-Mx - Ny = \lambda(Fx + Gy).$$

Rewriting this as

$$(L + \lambda E)x + (M + \lambda F)y = 0,$$
$$(M + \lambda F)x + (N + \lambda G)y = 0, \tag{5.49}$$

we have a homogeneous linear system of algebraic equations. This system has nontrivial solutions when its determinant vanishes:

$$(EG - F^2)\lambda^2 - (2MF - EN - LG)\lambda + (LN - M^2) = 0. \tag{5.50}$$

In the general case, there are two roots λ_1 and λ_2, to which there correspond two directions previously called the principal directions. They can be found by elimination of λ from (5.49):

$$\begin{vmatrix} -x^2 & xy & -y^2 \\ E & F & G \\ L & M & N \end{vmatrix} = 0.$$

By this we define two mutually orthogonal directions $x : y$. If $\lambda_1 = \lambda_2$, then all the directions are principal so we can choose any two mutually orthogonal directions and regard them as principal.

Now we would like to consider another question that will bring us to the same equations. It is the question of finding the extreme normal curvatures at a point of a surface. We have established that in the direction $x : y$ the normal curvature of a surface is given by the formula

$$k = \frac{Lx^2 + 2Mxy + Ny^2}{Ex^2 + 2Fxy + Gy^2}.$$

Because of the second-order homogeneity in x, y of the numerator and denominator we can reformulate the problem of finding extreme curvatures as the problem of finding extremal points of the function

$$Lx^2 + 2Mxy + Ny^2$$

under the restriction

$$Ex^2 + 2Fxy + Gy^2 = 1.$$

Applying the theory of Lagrange multipliers to this, we should find the extreme points of the function

$$Lx^2 + 2Mxy + Ny^2 - k(Ex^2 + 2Fxy + Gy^2),$$

which leads to the equations

$$(L - kE)x + (M - kF)y = 0,$$
$$(M - kF)x + (N - kG)y = 0. \tag{5.51}$$

The systems (5.49) and (5.51) coincide if we put $k = -\lambda$, which means that the principal directions and the directions found here are the same. Moreover, it is seen that λ_1 and λ_2 found as the eigenvalues of the tensor $\mathbf{A}$ give the extreme curvatures of the surface at a point, which are $k_1 = -\lambda_1$ and $k_2 = -\lambda_2$. Remembering that they are the roots of the polynomial (5.50) and using the Viète theorem, we get

$$k_1 + k_2 = \frac{EN - 2FM + GL}{EG - F^2}, \qquad k_1 k_2 = \frac{LN - M^2}{EG - F^2}.$$

We have met these expressions in the Gaussian curvature $K = k_1 k_2$ and the mean curvature $H = (k_1 + k_2)/2$.

We see that both are expressed through the coefficients of the first and second fundamental forms of the surface. There is a famous theorem due to Gauss that K can be expressed only in terms of the coefficients of the first fundamental form.

When the principal curvatures and their directions are known, the Euler formula gives the normal curvature at any direction of the curve which composes the angle ϕ with the first principal direction at the same point, corresponding to k_1:

$$k_\phi = k_1 \cos \phi + k_2 \sin \phi.$$

The principal directions of a surface define the *lines of curvature* of the surface. A line in the surface is called a line of curvature if at each point its tangent is directed along one of the principal directions at this point of the surface.

Two lines of curvature pass through each point. Denoting their directions by $du : dv$ and $\delta u : \delta v$ we have two equations, the first of which means orthogonality of the directions and the second their conjugation:

$$E\, du\, \delta u + F(du\, \delta v + dv\, \delta u) + G\, dv\, \delta v = 0,$$
$$L\, du\, \delta u + M(du\, \delta v + dv\, \delta u) + N\, dv\, \delta v = 0.$$

It is convenient to take the family of the lines of curvature of the surface as the coordinate lines of the surface. For these curvilinear coordinates $F = M = 0$.

Two surfaces are called *isometric* if there is a one-to-one correspondence between the points of the surfaces such that corresponding curves in the surfaces have the same length.

A planar surface and a cylindrical surface provide an example of isometric surfaces since we can develop the cylindrical surface over the plane.

The correspondence between points is defined by coincidence of the points in this developing.

We can refer to local isometry of a surface at point A_1 to another surface at point B_2 if there are neighborhoods of points on the surfaces that are isometric.

Two smooth surfaces are locally isometric if and only if there are parametrizations of the surfaces at corresponding points such that the coefficients of the first principal forms of the surfaces coincide:

$$E_1 = E_2, \qquad F_1 = F_2, \qquad G_1 = G_2.$$

A surface is *developable* if it is locally isometric to the plane at each point. It turns out that a surface is developable if and only if its Gaussian curvature is zero at each point.

Surfaces with zero Gaussian curvature appear once more in the problem of finding spatial surfaces of minimal area that have given boundaries. This popular physics problem describes the shape assumed by a soap film on a wire frame. The energy of such a film is proportional to its area, and thus the form actually taken by such a film is the surface of minimum area.

5.10 Surfaces of Revolution

Surfaces of revolution are quite frequent in practice. Suppose that the surface is formed by rotation of the profile curve

$$x = \phi(u), \qquad z = \psi(u), \qquad (5.52)$$

in the xz-plane about the z-axis (Fig. 5.3). When we fix u we get a circle having center on the z axis; it is called a *parallel*. To define a point on the parallel we introduce the angle of rotation v from the xz-plane. When we fix v we get a *meridian*, a curve congruent to the initial curve (5.52).

It is easy to see that the equations of the surface of revolution corresponding to (5.52) are

$$x = \phi(u) \cos v, \qquad y = \phi(u) \sin v, \qquad z = \psi(u).$$

Let us find the coefficients of the first fundamental form of the surface:

$$E = x_u^2 + y_u^2 + z_u^2 = (\phi' \cos v)^2 + (\phi' \sin v)^2 + \psi'^2 = \phi'^2 + \psi'^2,$$
$$F = x_u x_v + y_u y_v + z_u z_v = (\phi' \cos v)(-\phi \sin v) + (\phi' \sin v)(\phi \cos v) = 0,$$
$$G = x_v^2 + y_v^2 + z_v^2 = (-\phi \sin v)^2 + (\phi \cos v)^2 = \phi^2.$$

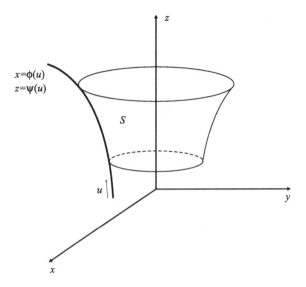

Fig. 5.3   Surface of revolution about the z-axis.

Note that $F = 0$ means the orthogonality of the parametrization net. Thus the first fundamental form is

$$(ds)^2 = \left(\phi'^2 + \psi'^2\right) du^2 + \phi^2 \, dv^2.$$

For the components of the second fundamental form we have

$$L = \frac{\mathbf{r}_{uu} \cdot (\mathbf{r}_u \times \mathbf{r}_v)}{|\mathbf{r}_u \times \mathbf{r}_v|} = \frac{\begin{vmatrix} x_{uu} & y_{uu} & z_{uu} \\ x_u & y_u & z_u \\ x_v & y_v & z_v \end{vmatrix}}{\sqrt{EG - F^2}} = \frac{\psi''\phi' - \phi''\psi'}{\sqrt{\phi'^2 + \psi'^2}},$$

$$M = \frac{\mathbf{r}_{uv} \cdot (\mathbf{r}_u \times \mathbf{r}_v)}{|\mathbf{r}_u \times \mathbf{r}_v|} = \frac{\begin{vmatrix} x_{uv} & y_{uv} & z_{uv} \\ x_u & y_u & z_u \\ x_v & y_v & z_v \end{vmatrix}}{\sqrt{EG - F^2}} = 0,$$

and

$$N = \frac{\mathbf{r}_{vv} \cdot (\mathbf{r}_u \times \mathbf{r}_v)}{|\mathbf{r}_u \times \mathbf{r}_v|} = \frac{\begin{vmatrix} x_{vv} & y_{vv} & z_{vv} \\ x_u & y_u & z_u \\ x_v & y_v & z_v \end{vmatrix}}{\sqrt{EG - F^2}} = \frac{\psi'\phi}{\sqrt{\phi'^2 + \psi'^2}}.$$

So the second fundamental form is

$$-d\mathbf{n} \cdot d\mathbf{r} = \frac{\psi''\phi' - \phi''\psi'}{\sqrt{\phi'^2 + \psi'^2}}\, du^2 + \frac{\psi'\phi}{\sqrt{\phi'^2 + \psi'^2}}\, dv^2.$$

We see that $M = 0$, which means the coordinate lines are conjugate.[6] Thus the coordinate lines of a surface of revolution are the lines of curvature.

Exercise 5.37. Find the principal curvatures of the surface of revolution.

Exercise 5.38. Find the first and second fundamental forms for (a) the plane, and (b) the sphere.

Exercise 5.39. Find the first and second fundamental forms for each of the following paraboloids: (a) $z = x^2 + y^2$, (b) $z = x^2 - y^2$, (c) $z = x^2$. Find H and K for each of these.

5.11   Natural Equations of a Curve

Suppose we are given functions $k_1(s)$ and $k_2(s)$ of a natural parameter s, continuous on a segment $[s_0, s_1]$, and that $k_1(s)$ is a positive function. We are interested in whether there exists a space curve that has principal curvature $k_1(s)$ and torsion $k_2(s)$. Geometrical considerations show that if such a curve exists then it is uniquely defined up to rigid motions. This means that if we take two such curves, place them in space, and then shift and rotate one of them so that their initial points and moving trihedra at these points coincide, then all points of the curves coincide.

Now we will show that such a curve exists. Thus the form of the curve is defined by $k_1(s)$ and $k_2(s)$ uniquely; this prompts us to call the two functions $k_1 = k_1(s)$ and $k_2 = k_2(s)$ the *natural equations* of the curve.

Let us demonstrate that from $k_1(s)$ and $k_2(s)$ we can find the needed curve. For this, we consider a vector system of differential equations that

---

[6]To any symmetric quadratic form there corresponds something like an inner product of vectors, and hence something akin to orthogonality. Conjugate directions at a point are defined as the directions on the surface $du : dv$ and $\delta u : \delta v$ for which $L\, du\, \delta u + M(du\, \delta v + dv\, \delta u) + N\, dv\, \delta v = 0$. Conjugate coordinate lines are those which are conjugate at each point.

mimics the formula for a tangent vector and the Frenet–Serret equations:

$$\frac{d\mathbf{r}}{ds} = \mathbf{x},$$

$$\frac{d\mathbf{x}}{ds} = k_1(s)\mathbf{y},$$

$$\frac{d\mathbf{y}}{ds} = -k_1(s)\mathbf{x} - k_2(s)\mathbf{z},$$

$$\frac{d\mathbf{z}}{ds} = k_2(s)\mathbf{y}. \tag{5.53}$$

Written in Cartesian components, this is a system of 12 linear ordinary differential equations. ODE theory states that if we define the initial values for all the unknowns (the Cauchy problem), we will have a unique solution to this system.

We can choose arbitrary initial conditions for $\mathbf{r}(s_0)$. The initial values $\mathbf{x}(s_0)$, $\mathbf{y}(s_0)$, $\mathbf{z}(s_0)$ must constitute an arbitrary right-handed orthonormal trihedron $(\mathbf{x}_0, \mathbf{y}_0, \mathbf{z}_0)$.

Thus on $[s_0, s_1]$ there exists a unique solution $(\mathbf{r}(s), \mathbf{x}(s), \mathbf{y}(s), \mathbf{z}(s))$ of the equations (5.53), satisfying some fixed initial conditions. It can be shown that because of skew symmetry of the matrix of the three last equations of (5.53),

$$\begin{vmatrix} 0 & k_1(s) & 0 \\ -k_1(s) & 0 & -k_2(s) \\ 0 & k_2(s) & 0 \end{vmatrix} = 0,$$

the vectors $(\mathbf{x}(s), \mathbf{y}(s), \mathbf{z}(s))$ constitute an orthonormal frame of the same orientation as the initial one on the whole segment $[s_0, s_1]$.

We can treat $\mathbf{r} = \mathbf{r}(s)$ as the equation of the needed curve in natural parametrization. Then the first of the equations (5.53) states that $\mathbf{x}(s)$ is its unit tangent. Comparing the equation $d\mathbf{x}/ds = k_1(s)\mathbf{y}$ with the first of the Frenet–Serret equations for a curve $\mathbf{r} = \mathbf{r}(s)$, we see that $\mathbf{y}(s)$ is its principal normal and $k_1(s)$ is its principal curvature. Similarly we find that $\mathbf{z}(s)$ is the binormal of the curve and $k_2(s)$ is its torsion. This completes the necessary reasoning.

For a plane curve the natural equations reduce to a single equation for the curvature. In the plane we associate the curvature with an algebraic sign. The condition of positivity of $k(s)$ is not necessary.

Natural equation of a curve in the plane

A plane curve has zero torsion. Let us consider how to reconstruct a curve if its curvature $k(s)$ is a given function of the length parameter s. We fix the initial point of the curve, corresponding to $s = s_0$, by the equation

$$\mathbf{r}(s_0) = \mathbf{r}_0. \tag{5.54}$$

We must also fix the direction of the curve at this point. Here it is useful to introduce the angle $\phi = \phi(s)$, measured between the unit tangent to the curve and the vector $\mathbf{i}$ where $\{\mathbf{i}, \mathbf{j}\}$ is an orthonormal basis in the plane. In this way the unit tangent is given by

$$\boldsymbol{\tau} = \mathbf{i}\cos\phi + \mathbf{j}\sin\phi.$$

So to define the initial direction of the curve we introduce

$$\phi(s_0) = \phi_0. \tag{5.55}$$

The equation relating $\boldsymbol{\tau}$ and $\boldsymbol{\nu}$, which is $d\boldsymbol{\tau}/ds = k(s)\boldsymbol{\nu}$, becomes

$$(-\mathbf{i}\sin\phi + \mathbf{j}\cos\phi)\frac{d\phi}{ds} = k(s)\boldsymbol{\nu}.$$

The vector $-\mathbf{i}\sin\phi + \mathbf{j}\cos\phi$ has unit magnitude and is orthogonal to $\boldsymbol{\tau}$. Hence it is parallel to $\boldsymbol{\nu}$, and we conclude that

$$|k(s)| = \left|\frac{d\phi}{ds}\right|.$$

Taking

$$k(s) = \frac{d\phi}{ds},$$

we define a sign of $k(s)$ in the standard manner where $k(s)$ is positive for points at which the curve is concave upwards.

We can integrate the last equation and obtain

$$\phi(s) - \phi_0 = \int_{s_0}^{s} k(t)\, dt.$$

Knowing $\phi(s)$ we can integrate the equation for the unit tangent, which is $d\mathbf{r}/ds = \boldsymbol{\tau}$, rewritten as

$$\frac{d\mathbf{r}}{ds} = \mathbf{i}\cos\phi(s) + \mathbf{j}\sin\phi(s).$$

Integration with respect to s gives us

$$\mathbf{r}(s) - \mathbf{r}_0 = \int_{s_0}^{s} [\mathbf{i}\cos\phi(t) + \mathbf{j}\sin\phi(t)]\, dt. \tag{5.56}$$

By (5.56), the function $k(s)$ and the initial values (5.54) and (5.55) uniquely defined the needed curve $\mathbf{r} = \mathbf{r}(s)$. Changing the initial values of the curve, we get a family of curves with the same $k(s)$ such that all curves have the same shape but different positions with respect to the coordinate axes.

Exercise 5.40. Given the curvature $k(s) = (as)^{-1}$ of a plane curve, find the curve.

5.12 A Word About Rigor

The main goal of this book is the presentation of those tools and formulas of tensor analysis that are needed for applications. Our approach is typical of engineering books; we seldom offer clear statements of the assumptions that guarantee validity of a formula, supposing instead that in applications all the functions, curves, surfaces, etc., should be sufficiently smooth for our purposes. This approach is probably best for any practitioner who must simply get his or her hands on a formula. However, in physics we see surfaces that do not necessarily bound real-world bodies; purely mathematical surfaces can occur, such as those describing the energy of a two-parameter mechanical system. Such a surface can be quite complex, and a physicist may be largely interested in its singular points since these represent the states at which the system changes its behavior crucially. A reader interested in such applications would do well to study a more sophisticated treatment where each major result is stated as the theorem under the weakest possible hypotheses. Many such treatments employ much more advanced (e.g., topological) tools.

However, the reader should be aware that even on the elementary level of our treatment there are questions that require additional explanation. How, for example, should we define the length of a curve or the area of a surface?

Early in our education we learn how to measure the length of a segment or of a more complex set on a straight line. We also learn how to define the circumference of a circle. For this we inscribe an equilateral triangle and calculate its perimeter. Doubling the number of sides of the inscribed triangle, we get an inscribed polygon whose perimeter approximates the circumference. Doubling the number of sides to infinity and calculating the limit of their perimeters, we define this limit as the needed length. Of course, a limit passage of this type is used when we derive the formula

$$\int_a^b |\mathbf{r}'(t)|\, dt \qquad (5.57)$$

for the length of a curve. Here we divide the segment $[a, b]$ into small pieces by the points $t_0 = a, t_1, \ldots, t_n = b$, and draw the vector

$$\Delta \mathbf{r}(t_i + 1) = \mathbf{r}(t_{i+1}) - \mathbf{r}(t_i).$$

In this way we "inscribe" a polygon into the curve. For a continuously differentiable vector function $\mathbf{r}(t)$, the length of a side of a polygon can be found by the mean value theorem:

$$|\mathbf{r}(t_{i+1}) - \mathbf{r}(t_i)| = |\mathbf{r}'(\xi_{i+1})|(t_{i+1} - t_i),$$

where ξ_{i+1} is a point in $[t_{i+1} - t_i]$. Thus the perimeter of the inscribed polygon that approximates the length of the curve is

$$\sum_{i=1}^n |\mathbf{r}'(\xi_{i+1})|(t_{i+1} - t_i).$$

We obtain (5.57) from a limit passage in which the number of partition points tends to infinity while the maximum length of a partition segment tends to zero.

Although this might seem fine, underlying the process is the notion of approximating a small piece of the curve by a many-sided polygon. Our "intuition" tells us that the smaller its sides, the closer is the polygon to the curve, hence its perimeter should approximate the curve length more and more closely. But in elementary geometry one may "demonstrate" that the sum of the legs of any right triangle is equal to its hypotenuse. The construction is as follows. Suppose we are given a right triangle whose legs have lengths a and b. Let us form a curve that looks like the toothed edge of a saw, by placing along the hypotenuse of the given triangle a sequence of small triangles each similar to the given triangle (Fig. 5.4). It is clear that the sum of all the legs of the sawteeth does not depend on the number of teeth and equals $a + b$. But as the number of teeth tends to infinity, the saw edge "coincides" with the hypotenuse of the original triangle. Hence the limiting length is equal to the length of the hypotenuse c, and we have $c = a + b$. So we come to understand that we cannot approximate a curve by a polygon in an arbitrary fashion. Similar remarks apply to the approximation of surfaces to calculate area.

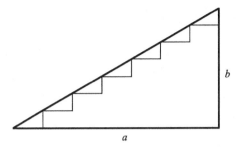

Fig. 5.4   Fallacious estimation of the length of a line.

These remarks should serve as a warning that to apply formulas correctly it is often necessary to understand the restrictions under which they were derived.

5.13   Conclusion

Differential geometry is a well developed subject with many results and formulas. We have presented the main technical formulas that are used in applications. The interested reader could go on to study volumes devoted to the theory of curves, surfaces, manifolds, etc. Such an extensive undertaking falls outside the scope of this book.

5.14   Problems

5.1 Find the parametrization and singular points of the plane curve

$$|x|^{\frac{2}{3}} + |y|^{\frac{2}{3}} = |a|^{\frac{2}{3}},$$

where a is a parameter.

5.2 Find the singular points of the plane curve given by the equations

$$x = a(t - \sin t), \qquad y = a(1 - \cos t),$$

where a is a parameter.

5.3 Find the equation of the tangent to the curve

$$x^2 + y^2 + z^2 = 1, \qquad x^2 + y^2 = x$$

at the point $(0, 0, 1)$.

5.4 Find the length of the astroid

$$x = a\cos^3 t, \qquad y = a\sin^3 t.$$

5.5 Find the length of the part of the cycloid

$$x = a(t - \sin t), \qquad y = a(1 - \cos t)$$

defined by $t \in [0, 2\pi]$.

5.6 Find the length of the cardioid given in polar coordinates by

$$\rho = 2a(1 - \cos\phi).$$

See Fig. 5.5.

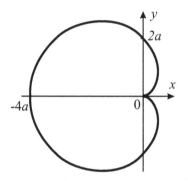

Fig. 5.5   Cardioid.

5.7 Find the length of the portion of the curve

$$x = a\cosh t, \qquad y = a\sinh t, \qquad z = at$$

between the points $t = 0$ and $t = T$.

5.8 Find the curvature of the curve

$$x = t - \sin t, \qquad y = 1 - \cos t, \qquad z = 4\sin\frac{t}{2}.$$

5.9 Find the curvature and torsion of the curve

$$x = a\cosh t, \qquad y = a\sinh t, \qquad z = at$$

at an arbitrary point.

5.10 Find the torsion of the curve

$$x = a \cosh t \cos t, \qquad y = a \cosh t \sin t, \qquad z = at$$

at an arbitrary point.

5.11 Let a triad of orthonormal vectors $e_1(s)$, $e_2(s)$, $e_3(s)$ (i.e., vectors satisfying $e_i \cdot e_j = \delta_{ij}$) be given along a curve. Show that

$$\frac{d}{ds} e_i(s) = d \times e_i(s)$$

where

$$d = -\frac{1}{2}(e_i' \times e_i).$$

Note that these formulas are analogous to the Frenet–Serret equations with the Darboux vector δ.

5.12 Let $Q(s)$ be an orthogonal tensor given along a curve. Verify the formulas

$$Q' = d \times Q, \qquad d = -\frac{1}{2}(Q' \times Q^T)_\times.$$

5.13 Find the second quadratic form of the surface given by the equations

$$x = u \cos v, \qquad y = u \sin v, \qquad z = v.$$

5.14 A surface is defined by the equation $z = f(x, y)$. Show that the coefficients of its second principal form are

$$b_{11} = \frac{f_{xx}}{\sqrt{1 + f_x^2 + f_y^2}}, \qquad b_{12} = b_{21} = \frac{f_{xy}}{\sqrt{1 + f_x^2 + f_y^2}}, \qquad b_{22} = \frac{f_{yy}}{\sqrt{1 + f_x^2 + f_y^2}}.$$

5.15 A surface is defined by the equation $z = f(x, y)$, where f satisfies Laplace's equation $\nabla^2 f = 0$. Demonstrate that its Gaussian curvature satisfies $K \leq 0$.

5.16 Show that the mean curvature of the surface $z = f(x, y)$ is given by

$$H = \text{div} \left(\frac{\text{grad } f}{\sqrt{1 + |\text{grad } f|^2}} \right),$$

where

$$\text{grad} = i_1 \frac{\partial}{\partial x} + i_2 \frac{\partial}{\partial y}.$$

5.17 Show that the mean curvature of a surface is given by $H = -\widetilde{\nabla} \cdot \mathbf{n}/2$.

5.18 Demonstrate that $\widetilde{\nabla} \cdot \mathbf{A} = 2H\mathbf{n}$ where $\mathbf{A} = \mathbf{E} - \mathbf{nn}$.

5.19 Let $\mathbf{X}$ be a second-order tensor. Show that $\mathbf{n} \cdot (\widetilde{\nabla} \times \mathbf{X}) = -\widetilde{\nabla} \cdot (\mathbf{n} \times \mathbf{X})$.

5.20 Find the Gaussian curvature of a surface given by the relation

$$z = f(x) + g(y).$$

5.21 The first principal form of a surface is $A^2 \, du^2 + B^2 \, dv^2$. Determine its Gaussian curvature.

5.22 Let $\mathbf{X}$ be a second-order tensor. Prove the following analog of the Gauss–Ostrogradsky theorem on a surface S having boundary contour Γ:

$$\int_S \left(\widetilde{\nabla} \cdot \mathbf{X} + 2H\mathbf{n} \cdot \mathbf{X} \right) dS = \oint_\Gamma \boldsymbol{\nu} \cdot \mathbf{X} \, ds, \qquad (5.58)$$

where $\boldsymbol{\nu}$ is the outward unit normal to Γ lying in the tangent plane, i.e., $\boldsymbol{\nu} \cdot \mathbf{n} = 0$. Note that when S is a closed surface,

$$\int_S \widetilde{\nabla} \cdot \mathbf{X} \, dS = - \int_S 2H\mathbf{n} \cdot \mathbf{X} \, dS.$$

5.23 Prove that (5.58) holds for a tensor field $\mathbf{X}$ of any order.

5.24 Use the solution of the previous problem to prove that

$$\int_S \left(\widetilde{\nabla}\mathbf{X} + 2H\mathbf{n}\mathbf{X} \right) dS = \oint_\Gamma \boldsymbol{\nu}\mathbf{X} \, ds,$$

$$\int_S \left(\widetilde{\nabla} \times \mathbf{X} + 2H\mathbf{n} \times \mathbf{X} \right) dS = \oint_\Gamma \boldsymbol{\nu} \times \mathbf{X} \, ds,$$

and

$$\int_S \widetilde{\nabla} \times (\mathbf{n}\mathbf{X}) \, dS = \oint_\Gamma \boldsymbol{\tau}\mathbf{X} \, ds,$$

where $\boldsymbol{\tau} = \boldsymbol{\nu} \times \mathbf{n}$ is the unit tangent vector to Γ.

PART 2
Applications in Mechanics

Chapter 6

Linear Elasticity

In this chapter we apply tensor analysis to linear elasticity. Linear elasticity is a powerful tool of engineering design; using general computer programs, engineers can calculate the strains and stresses within a complex elastic body under load. It is useful to learn the principles behind these calculations. Linear elasticity is based on the ideas of classical mechanics, but its technical tools are those of tensor analysis. It treats the small deformations of elastic bodies described by a linear constitutive equation that relates stresses and strains, extending the elementary form of Hooke's law for a spring.

Linear elasticity is the first step (an elementary but not easy step) toward nonlinear mechanics. The latter considers other effects in solids, such as heat propagation, deformation due to piezoelectric or magnetic effects, etc. It therefore incorporates thermodynamics and other areas of physics.

The plan of this chapter is to introduce the principal tools and laws of linear elasticity, to formulate and consider some properties of the boundary value problems of elasticity, and to study some variational principles in applied elasticity.

We start with the idea of the stress tensor, which was introduced by Augustin Louis Cauchy (1789–1857).

6.1 Stress Tensor

The notion of stress is decisive in continuum mechanics. It is an extension of the notion of pressure, and is introduced as the ratio of the value of a force distributed over an elementary surface element to the element area. Consider, for example, a long thin cylindrical bar having cross-sectional area S and stretched by a force f which is uniformly distributed over the cylinder faces (Fig. 6.1). This force is equidistributed over any normal cross-section of the bar. The stress σ is defined as

$$\sigma = f/S.$$

In elementary physics, the pressure is the only force characteristic at a given point in a liquid or gas. But pressure is not sufficient to describe the action of forces inside a three-dimensional solid body. At a particular point in such a body, we find that the direction of the contact force may not be normal to a given area element. Furthermore, as we change the orientation of the area element, the density of the force acting across the element may change in magnitude and direction.

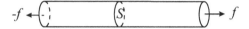

Fig. 6.1   The stress $\sigma = f/S$ in a bar having cross-sectional area S and stretched by a force f.

Forces

Let us discuss the conditions for equilibrium of a deformable body. From classical mechanics, we know that the equilibrium conditions for a rigid body consist of two vector equations. First, the resultant force (i.e., the sum of all forces) acting on the body must be zero. Second, the resultant moment (the sum of the moments of all the forces with respect to some point) must be zero. We write

$$\sum_k \mathbf{f}_k = \mathbf{0}, \qquad \sum_k (\mathbf{r}_k - \mathbf{r}_0) \times \mathbf{f}_k = \mathbf{0}, \qquad (6.1)$$

where $\mathbf{f}_k$ denotes a force applied to a point located by position vector $\mathbf{r}_k$. The position vector $\mathbf{r}_0$ locates an arbitrary but fixed point with respect to which the moments are taken.

To define the equilibrium of a deformable body, we apply the equilibrium equations for a *rigid* body to any portion of the deformable body. We accept this as an axiom, known as the

Solidification principle. *In equilibrium, any part of a deformable body obeys the equilibrium equations as if it were a rigid body under the action of (1) all the external forces, and (2) the force reaction imposed by the remainder of the body on the part under consideration.*

This principle is not a direct consequence of classical mechanics. Rather, it is a kind of axiom which allows us to apply the results of classical mechanics — obtained for non-deformable objects — to deformable objects. We will use the following terminology. A solid body occupies a certain volume in three-dimensional space. The mapping which takes its material points into the spatial points is called a *configuration* of the body; roughly speaking, this describes the geometry of the body in space. If the body is not under load, the configuration is termed the *initial* or *reference configuration*. For a deformed body in equilibrium under a given load, the configuration is termed the *actual configuration*.

Let us apply the solidification principle in the actual configuration of a deformable body in equilibrium. We take an arbitrary portion $\mathcal{P}$ as shown in Fig. 6.2. Two types of forces act on $\mathcal{P}$. First, there are *body forces*. These act on the interior of $\mathcal{P}$ and do not depend on the conditions over the boundary surface of $\mathcal{P}$. An example is the gravitational force. Second, there are *contact* forces. These act on the boundary of $\mathcal{P}$ and represent the reaction forces imposed on $\mathcal{P}$ by the remainder of the body. They arise as follows. Imagine that we isolate $\mathcal{P}$ from the body and replace the effects of the rest of the body by some forces. These reactions are applied only to that part of the boundary of $\mathcal{P}$ that comes into contact with the rest of the body (hence the term "contact forces"). In addition, surface forces may act over the external boundary of the entire body. We will refer to these, together with the body forces, as the *external forces*. By the solidification principle, $\mathcal{P}$ must be in equilibrium under the action of all forces as though it were a rigid body of classic mechanics. (Continuum mechanics permits another approach, in which the reactions of the remaining part arise in the volume as well. However, this approach is not used in modern engineering practice.)

So the total force acting on $\mathcal{P}$ is

$$\mathbf{f}(\mathcal{P}) = \mathbf{f}_B(\mathcal{P}) + \mathbf{f}_C(\mathcal{P}),$$

where the subscripts B and C denote the body and contact forces, respectively.

To characterize the spatial distribution of the forces, we introduce force densities defined by the equalities

$$\mathbf{f}_B(\mathcal{P}) = \int_{V_{\mathcal{P}}} \rho \mathbf{f} \, dV, \qquad \mathbf{f}_C(\mathcal{P}) = \int_{\Sigma_{\mathcal{P}}} \mathbf{t} \, d\Sigma,$$

where $V_{\mathcal{P}}$ is the space volume of $\mathcal{P}$, $\Sigma_{\mathcal{P}} = \partial V_{\mathcal{P}}$ is the boundary of $\mathcal{P}$, and ρ is the (specific) density of the material composing the body. The density $\mathbf{t}$

carries dimensions of force per unit area, whereas **f** carries those of force per unit mass. We emphasize that these are defined in the actual configuration. The quantity **t** is called the *stress vector*.

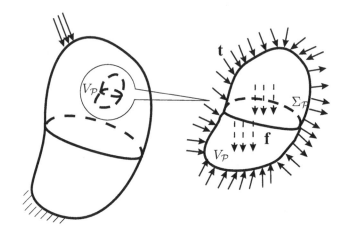

Fig. 6.2   Forces acting on a portion $V_{\mathcal{P}}$ of the body.

Equilibrium equations of a continuum medium

By the solidification principle, the equilibrium equations (6.1) for a rigid body become the following two conditions for a deformable body.

1. The resultant force acting on any portion $\mathcal{P}$ is zero:

$$\int_{V_{\mathcal{P}}} \rho \mathbf{f} \, dV + \int_{\Sigma_{\mathcal{P}}} \mathbf{t} \, d\Sigma = \mathbf{0}. \tag{6.2}$$

2. The resultant moment of all forces acting on $\mathcal{P}$ is zero:

$$\int_{V_{\mathcal{P}}} \{(\mathbf{r} - \mathbf{r}_0) \times \rho \mathbf{f}\} \, dV + \int_{\Sigma_{\mathcal{P}}} \{(\mathbf{r} - \mathbf{r}_0) \times \mathbf{t}\} \, d\Sigma = \mathbf{0}, \tag{6.3}$$

where **r** locates a material point and $\mathbf{r}_0$ locates a fixed reference point. The reader is encouraged to show, using (6.2), that (6.3) does not depend on the choice of $\mathbf{r}_0$.

   Note that $\mathcal{P}$ is not completely arbitrary, but is such that the integration operation over $\mathcal{P}$ makes sense.

Stress tensor

In general, the stress vector depends on the position $\mathbf{r}$ of a particle and on the normal $\mathbf{n}$ to the area element in the body. We will always take the normal outward from the portion of the body under consideration.

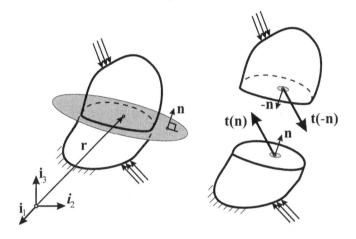

Fig. 6.3 Interaction of two parts of the body.

In continuum mechanics, Newton's third law on action-reaction pairs is called *Cauchy's lemma* and is expressed by the following relation:

$$\mathbf{t}(\mathbf{r}, \mathbf{n}) = -\mathbf{t}(\mathbf{r}, -\mathbf{n}). \tag{6.4}$$

Formula (6.4) describes the interaction of the contacting parts of the body shown in Fig. 6.3. Cauchy's lemma allows us to introduce the stress tensor, which describes the dependence of $\mathbf{t}$ on $\mathbf{n}$, the normal to the area element at a point. We will see this in Cauchy's theorem below.

First, however, we will use (6.2) to obtain the equilibrium equation in differential form. We fix an arbitrary point P in V, the volume of the body, and use it as a vertex of an arbitrary small parallelepiped Π having faces parallel to the Cartesian coordinate planes as shown in Fig. 6.4. Hence the normals to the faces lie along the orthonormal basis vectors $\mathbf{i}_1, \mathbf{i}_2, \mathbf{i}_3$.

Let us expand the stress vector $\mathbf{t}(\mathbf{i}_k)$ on the face having normal $\mathbf{i}_k$:

$$\mathbf{t}(\mathbf{i}_k) = t_{ks}\mathbf{i}_s. \tag{6.5}$$

That is, the t_{ks} are the components of $\mathbf{t}(\mathbf{i}_k)$ in the basis $\mathbf{i}_s$.

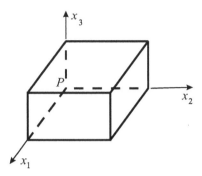

Fig. 6.4   Parallelepiped Π.

Theorem 6.1. *The differential equation*

$$\rho\mathbf{f} + \frac{\partial t_{ks}}{\partial x_k}\mathbf{i}_s = \mathbf{0}, \tag{6.6}$$

known as the **equilibrium equation,** *holds in* V.

Proof.   Equation (6.2) for the parallelepiped Π is

$$\int_{V_\Pi} \rho\mathbf{f}\, dV + \int_{\Sigma_\Pi} \mathbf{t}\, d\Sigma = \mathbf{0}.$$

The surface integral is taken over the faces of Π, each of which is perpendicular to one of the $\mathbf{i}_k$. On the face whose normal is $\mathbf{n} = \mathbf{i}_k$, we have $\mathbf{t}(\mathbf{i}_k) = n_k t_{1s}\mathbf{i}_s$, where $n_k = 1$ and the other two components of $\mathbf{n}$ vanish. On the opposite face the normal is $-\mathbf{i}_k$, so $n_k = -1$ and the remaining components vanish. By Cauchy's lemma (6.4) we have $\mathbf{t}(-\mathbf{i}_k) = -\mathbf{t}(\mathbf{i}_k)$, hence $\mathbf{t}(-\mathbf{i}_k) = n_k t_{1s}\mathbf{i}_s$. Therefore the above equation takes the form

$$\int_{V_\Pi} \rho\mathbf{f}\, dV + \int_{\Sigma_\Pi} n_k t_{ks}\mathbf{i}_s\, d\Sigma = \mathbf{0}.$$

Applying (4.41) to the surface integral, we have

$$\int_{\Sigma_\Pi} n_k t_{ks}\mathbf{i}_s\, d\Sigma = \int_{V_\Pi} \frac{\partial t_{ks}}{\partial x_k}\mathbf{i}_s\, dV \tag{6.7}$$

and it follows that

$$\int_{V_\Pi} \left(\rho\mathbf{f} + \frac{\partial t_{ks}}{\partial x_k}\mathbf{i}_s\right) dV = \mathbf{0}. \tag{6.8}$$

Suppose the integrand in (6.8) is a continuous function. As the vertex P of Π is fixed and Π is arbitrarily small, the differential equation (6.6) must hold at P. Since P is arbitrary, (6.6) holds in V. $\square$

Exercise 6.1. Prove (6.7) by direct integration of the right-hand side.

Now we formulate

Theorem 6.2. *(Cauchy). At any point of a body, the dependence of* **t** *on* **n**, *the normal to an elementary area at a point, is linear:*

$$\mathbf{t} = \mathbf{n} \cdot \boldsymbol{\sigma}.$$

Here $\boldsymbol{\sigma}$ *is a second-order tensor depending on the point; it is the* **Cauchy stress tensor***.*

Proof. We construct a tetrahedron T with vertex O at an arbitrary but fixed point of V as shown in Fig. 6.5. Applying (6.2) to T, we get

$$\int_{V_T} \rho \mathbf{f} \, dV + \int_{\Sigma_T} n_k t_{ks} \mathbf{i}_s \, d\Sigma + \int_{M_1 M_2 M_3} \mathbf{t}(\mathbf{n}) \, d\Sigma = \mathbf{0}, \qquad (6.9)$$

where V_T is the tetrahedron volume, Σ_T is the part of tetrahedron boundary consisting of the faces that are parallel to the coordinate planes, and $M_1 M_2 M_3$ is the inclined face. On the face that is orthogonal to $\mathbf{i}_k$, we will use the representation $\mathbf{t}(\mathbf{i}_k) = n_k t_{1s} \mathbf{i}_s$ from the proof of the previous theorem.

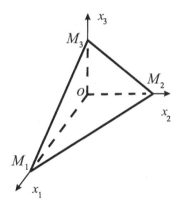

Fig. 6.5   Tetrahedron T.

Using equation (6.6), we transform (6.9) to the following equality:

$$\int_{V_T} \frac{\partial t_{ks}}{\partial x_k} \mathbf{i}_s \, dV = \int_{\Sigma_T} n_k t_{ks} \mathbf{i}_s \, d\Sigma + \int_{M_1 M_2 M_3} \mathbf{t}(\mathbf{n}) \, d\Sigma.$$

A consequence of (4.41) for $V = V_T$ is

$$\int_{V_T} \frac{\partial t_{ks}}{\partial x_k} \mathbf{i}_s \, dV = \int_{\Sigma_T} n_k t_{ks} \mathbf{i}_s \, d\Sigma + \int_{M_1 M_2 M_3} n_k t_{ks} \mathbf{i}_s \, d\Sigma.$$

Comparing the last two equalities, we see that

$$\int_{M_1 M_2 M_3} \left(\mathbf{t}(\mathbf{n}) - n_k t_{ks} \mathbf{i}_s \right) d\Sigma = \mathbf{0}.$$

As the tetrahedron is small and arbitrary at point O, we get the following identity:

$$\mathbf{t}(\mathbf{n}) - n_k t_{ks} \mathbf{i}_s = \mathbf{0}.$$

Because $M_1 M_2 M_3$ can have an arbitrary orientation $\mathbf{n}$, this equality holds for any $\mathbf{n}$ and, moreover, at any point of V.

We have shown that the dependence of $\mathbf{t}$ on $\mathbf{n}$ is linear. As we know (§ 3.2), a linear dependence between two vectors is given by a second-order tensor that we denote by $\boldsymbol{\sigma}$:

$$\mathbf{t}(\mathbf{n}) = \mathbf{n} \cdot \boldsymbol{\sigma}. \tag{6.10}$$

The proof is complete.   □

By the proof of Cauchy's theorem, we see that the components of the matrix (t_{sk}) are the components of the stress tensor in a Cartesian frame:

$$\boldsymbol{\sigma} = t_{sk} \mathbf{i}_k \mathbf{i}_s.$$

To maintain the correspondence between the notations for $\boldsymbol{\sigma}$ and its components, we change t_{sk} to σ_{sk} and write

$$\boldsymbol{\sigma} = \sigma_{sk} \mathbf{i}_k \mathbf{i}_s.$$

Each subscript of σ_{sk} has a certain geometric meaning. The first subscript designates the area element with normal $\mathbf{i}_k$, while the second designates the direction of the projection of the stress vector onto $\mathbf{i}_s$. For example, σ_{31} is the projection of $\mathbf{t}(\mathbf{i}_3)$ onto the axis x_1, and the stress vector $\mathbf{t}(\mathbf{i}_3)$ acts on the elementary surface element having normal parallel to $\mathbf{i}_3$.

Now we return to the equilibrium equation. It is easy to see that the equation

$$\nabla \cdot \boldsymbol{\sigma} + \rho \mathbf{f} = \mathbf{0} \tag{6.11}$$

written in Cartesian coordinates is (6.6). This means we have found the component-free form of the equilibrium equation for the body. Note that the equation does not depend on the properties of the material that makes up the body. We recall that (6.11) is the differential form of the condition that the resultant force applied to an arbitrary part of the body is zero.

So far we have exploited only the force equation (6.2). Now we will formulate the consequences of the moment equation (6.3).

Theorem 6.3. *Let equation* (6.3) *hold for any part of the body. It follows that* $\boldsymbol{\sigma}$ *is a symmetric tensor:* $\boldsymbol{\sigma} = \boldsymbol{\sigma}^T$.

Proof. Using (4.41), we change the surface integral in (6.3) to an integral over $V_{\mathcal{P}}$:

$$
\begin{aligned}
\int_{\Sigma_{\mathcal{P}}} (\mathbf{r} - \mathbf{r}_0) \times \mathbf{t}\, d\Sigma &= \int_{\Sigma_{\mathcal{P}}} (\mathbf{r} - \mathbf{r}_0) \times (\mathbf{n} \cdot \boldsymbol{\sigma})\, d\Sigma \\
&= -\int_{\Sigma_{\mathcal{P}}} \mathbf{n} \cdot \boldsymbol{\sigma} \times (\mathbf{r} - \mathbf{r}_0)\, d\Sigma \\
&= -\int_{V_{\mathcal{P}}} \nabla \cdot [\boldsymbol{\sigma} \times (\mathbf{r} - \mathbf{r}_0)]\, dV.
\end{aligned}
\tag{6.12}
$$

Let us transform the integrand of the last integral. Because $\partial \mathbf{r}/\partial x_k = \mathbf{i}_k$ and $\partial \mathbf{r}_0/\partial x_k = \mathbf{0}$, we have

$$
\begin{aligned}
\nabla \cdot [\boldsymbol{\sigma} \times (\mathbf{r} - \mathbf{r}_0)] &= \nabla \cdot \boldsymbol{\sigma} \times (\mathbf{r} - \mathbf{r}_0) + \mathbf{i}_k \cdot \boldsymbol{\sigma} \times \frac{\partial}{\partial x_k}(\mathbf{r} - \mathbf{r}_0) \\
&= -(\mathbf{r} - \mathbf{r}_0) \times \nabla \cdot \boldsymbol{\sigma} + \mathbf{i}_k \cdot \boldsymbol{\sigma} \times \mathbf{i}_k \\
&= -(\mathbf{r} - \mathbf{r}_0) \times \nabla \cdot \boldsymbol{\sigma} - \sigma_{ks}\mathbf{i}_k \times \mathbf{i}_s.
\end{aligned}
$$

Thus the condition that the resultant moment of all forces acting on $V_{\mathcal{P}}$ is zero brings us to the relation

$$
\int_{V_{\mathcal{P}}} (\mathbf{r} - \mathbf{r}_0) \times (\rho \mathbf{f} + \nabla \cdot \boldsymbol{\sigma})\, dV + \int_{V_{\mathcal{P}}} \sigma_{ks}\mathbf{i}_k \times \mathbf{i}_s\, dV = \mathbf{0}.
\tag{6.13}
$$

The first integral in (6.13) is zero by (6.11). So the second integral in (6.13) is zero for arbitrary $V_{\mathcal{P}}$, and it follows that

$$
\sigma_{ks}\mathbf{i}_k \times \mathbf{i}_s = \mathbf{0} \quad \text{in } V.
$$

This holds if and only if $\boldsymbol{\sigma}$ is symmetric at each point, i.e.,

$$
\boldsymbol{\sigma} = \boldsymbol{\sigma}^T.
$$

Indeed, let us consider the part of the sum $\sigma_{ks}\mathbf{i}_k \times \mathbf{i}_s$ when k, s are 1 or 2. We have

$$
\begin{aligned}
\sigma_{ks}\mathbf{i}_k \times \mathbf{i}_s &= \sigma_{11}\mathbf{i}_1 \times \mathbf{i}_1 + \sigma_{22}\mathbf{i}_2 \times \mathbf{i}_2 + \sigma_{12}\mathbf{i}_1 \times \mathbf{i}_2 + \sigma_{21}\mathbf{i}_2 \times \mathbf{i}_1 \\
&= (\sigma_{12} - \sigma_{21})\mathbf{i}_3 \\
&= \mathbf{0},
\end{aligned}
$$

which implies that $\sigma_{12} = \sigma_{21}$. Similarly we may demonstrate that $\sigma_{23} = \sigma_{32}$ and $\sigma_{13} = \sigma_{31}$. This completes the proof. $\qquad\square$

It is worth noting that in continuum mechanics other types of stresses, such as couple stresses, can be introduced [Cosserat and Cosserat (1909); Eringen (1999)]. For such models, the Cauchy stress tensor is not symmetric in general.

Principal stresses and principal area elements

In a general basis $\mathbf{e}^k$ ($k = 1, 2, 3$), the Cauchy stress tensor $\boldsymbol{\sigma}$ takes the form

$$\boldsymbol{\sigma} = \sigma_{sk}\mathbf{e}^s\mathbf{e}^k,$$

where the matrix σ_{sk} has only six independent components.

Because $\boldsymbol{\sigma}$ is symmetric, there exists the spectral expansion (3.21):

$$\boldsymbol{\sigma} = \sigma_1\mathbf{i}_1\mathbf{i}_1 + \sigma_2\mathbf{i}_2\mathbf{i}_2 + \sigma_3\mathbf{i}_3\mathbf{i}_3. \tag{6.14}$$

Here the eigenvalues σ_k of the matrix (σ_{sk}) are the *principal stresses*, and the normalized eigenvectors $\mathbf{i}_k$ of $\boldsymbol{\sigma}$ are the *principal axes* of $\boldsymbol{\sigma}$. On the principal area element having normal $\mathbf{i}_k$, the tangential stresses are absent. When the σ_k are distinct, the frame $\mathbf{i}_1, \mathbf{i}_2, \mathbf{i}_3$ is orthonormal. For the case of repeated σ_k, the frame of $\mathbf{i}_k$ is not unique; even in this case, however, we can select an orthonormal set $\mathbf{i}_1, \mathbf{i}_2, \mathbf{i}_3$.

6.2 Strain Tensor

Under load, a body changes shape. We will consider how to describe deformation using the strain tensor. We restrict our consideration to very small deformations.

Let us illustrate the notion of strain using a stretched bar as an example. An undeformed bar has length l_0; under load, the length becomes l. The strain is

$$\varepsilon = \frac{l - l_0}{l_0} \equiv \frac{\Delta l}{l_0}.$$

Generalization of this to three dimensions is not straightforward: we should consider changes in shape of the body in all directions.

Let a body initially occupy a volume V in space. Under some external load, it occupies the volume v. The position vectors of a particle in the initial and deformed states are denoted by $\mathbf{r}_0$ and $\mathbf{r}$, respectively. The *displacement vector*

$$\mathbf{u} = \mathbf{r} - \mathbf{r}_0$$

describes the displacement of a particle due to deformation (Fig. 6.6). In this book we restrict ourselves to the case in which $\|\mathbf{u}\| \ll 1$ and all the first derivatives of $\mathbf{u}$ are small in comparison with 1. So we will omit all terms of the second order of smallness in any expression containing first-order terms. Moreover, in the case of small deformations we will not distinguish the initial and actual states of the body; that is, all quantities will be considered to be given in the initial volume V.

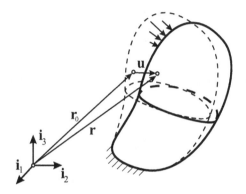

Fig. 6.6   Deformation of a three-dimensional body.

Let us consider the change of an infinitesimal vector-segment $d\mathbf{r}_0$ due to deformation. After deformation the segment is given by the vector $d\mathbf{r}$. We have

$$d\mathbf{r} = d\mathbf{r}_0 \cdot \mathbf{F}, \qquad (6.15)$$

where $\mathbf{F} = \mathbf{E} + \nabla\mathbf{u}$ is the gradient of the deformation.[1]

Exercise 6.2. Derive (6.15) from the relation $\mathbf{r} = \mathbf{r}_0 + \mathbf{u}$.

Next, we consider the change in length of the segment during deformation. Before deformation, the squared length is

$$dS^2 = d\mathbf{r}_0 \cdot d\mathbf{r}_0.$$

---

[1]Some books use $\mathbf{F} = \mathbf{E} + \nabla\mathbf{u}^T$, in which case (6.15) takes the form $d\mathbf{r} = \mathbf{F} \cdot d\mathbf{r}_0$.

After deformation it is

$$ds^2 = d\mathbf{r} \cdot d\mathbf{r} = d\mathbf{r}_0 \cdot \mathbf{F} \cdot \mathbf{F}^T \cdot d\mathbf{r}_0.$$

We have

$$\begin{aligned} ds^2 - dS^2 &= d\mathbf{r}_0 \cdot (\mathbf{F} \cdot \mathbf{F}^T - \mathbf{E}) \cdot d\mathbf{r}_0 \\ &= d\mathbf{r}_0 \cdot \left[(\nabla \mathbf{u} + (\nabla \mathbf{u})^T + (\nabla \mathbf{u}) \cdot \nabla \mathbf{u}^T) \right] \cdot d\mathbf{r}_0. \end{aligned}$$

For small deformations, we may omit all the squared quantities and write

$$ds^2 - dS^2 = d\mathbf{r}_0 \cdot \left[(\nabla \mathbf{u} + (\nabla \mathbf{u})^T) \right] \cdot d\mathbf{r}_0 = 2 d\mathbf{r}_0 \cdot \boldsymbol{\varepsilon} \cdot d\mathbf{r}_0,$$

where

$$\boldsymbol{\varepsilon} = \frac{1}{2} \left(\nabla \mathbf{u} + (\nabla \mathbf{u})^T \right) \tag{6.16}$$

is the *linear strain tensor*. It is clear that $\boldsymbol{\varepsilon}$ is a symmetric tensor.

In a Cartesian frame, $\boldsymbol{\varepsilon}$ is given by

$$\boldsymbol{\varepsilon} = \varepsilon_{mn} \mathbf{i}_m \mathbf{i}_n$$

where

$$\varepsilon_{11} = \frac{\partial u_1}{\partial x_1}, \qquad \varepsilon_{12} = \frac{1}{2} \left(\frac{\partial u_1}{\partial x_2} + \frac{\partial u_2}{\partial x_1} \right),$$

$$\varepsilon_{22} = \frac{\partial u_2}{\partial x_2}, \qquad \varepsilon_{13} = \frac{1}{2} \left(\frac{\partial u_1}{\partial x_3} + \frac{\partial u_3}{\partial x_1} \right),$$

$$\varepsilon_{33} = \frac{\partial u_3}{\partial x_3}, \qquad \varepsilon_{23} = \frac{1}{2} \left(\frac{\partial u_2}{\partial x_3} + \frac{\partial u_3}{\partial x_2} \right).$$

The diagonal components $\varepsilon_{11}, \varepsilon_{22}, \varepsilon_{33}$ describe the changes in the lengths of elementary segments along the $\mathbf{i}_1, \mathbf{i}_2, \mathbf{i}_3$ directions, respectively. The other components ε_{mn} $(m \neq n)$ represent skewing of the body; they characterize the deformational changes in the angles between elementary segments lying initially along the axes.

In arbitrary curvilinear coordinates q^1, q^2, q^3 with basis $\mathbf{r}_k$ $(k = 1, 2, 3)$ and dual basis $\mathbf{r}^k$ $(k = 1, 2, 3)$, the tensor $\boldsymbol{\varepsilon}$ is given by

$$\boldsymbol{\varepsilon} = \varepsilon_{st} \mathbf{r}^s \mathbf{r}^t, \qquad \varepsilon_{st} = \frac{1}{2} \left(\frac{\partial u_s}{\partial q^t} + \frac{\partial u_t}{\partial q^s} \right) - \Gamma_{st}^r u_r.$$

In arbitrary orthogonal curvilinear coordinates, $\boldsymbol{\varepsilon}$ was represented in (4.39). See Appendix A for $\boldsymbol{\varepsilon}$ in the cylindrical and spherical systems.

Equation (6.16) defines ε as a tensorial function of $\mathbf{u}$. The inverse problem, of finding $\mathbf{u}$ when ε is given, has a solution if and only if the compatibility condition

$$\nabla \times (\nabla \times \varepsilon)^T = \mathbf{0} \tag{6.17}$$

holds. In this case, $\mathbf{u}$ can be found using *Cesàro's formula*

$$\mathbf{u} = \mathbf{u}_0 + \boldsymbol{\omega}_0 \times (\mathbf{r} - \mathbf{r}_0) + \int_{M_0}^{M} \{\varepsilon(s) + [\mathbf{r}(s) - \mathbf{r}] \times \nabla \times \varepsilon(s)\} \cdot d\mathbf{r}(s), \tag{6.18}$$

where $\mathbf{u}_0$ and $\boldsymbol{\omega}_0$ are arbitrary but fixed vectors, and the integration path $M_0 M$ joins the points M_0 and M whose position vectors are $\mathbf{r}_0$ and $\mathbf{r}$, respectively. Here $\mathbf{r}(s)$ locates an arbitrary point on $M_0 M$.

The derivations of these formulas can be found in any book on elasticity, e.g., [Green and Zerna (1954); Lurie (2005)].

6.3   Equation of Motion

Using the equilibrium equations (6.11) and d'Alembert's principle of mechanics, we can immediately obtain the equation of motion for a body. The technique is to formally add the inertia forces to the body forces:

$$\mathbf{f} \to \mathbf{f} - \rho \frac{\partial^2 \mathbf{u}}{\partial t^2},$$

where t is the time variable and ρ is the material density. The equation of motion is

$$\nabla \cdot \boldsymbol{\sigma} + \rho \mathbf{f} = \rho \frac{\partial^2 \mathbf{u}}{\partial t^2}. \tag{6.19}$$

The form of (6.19) is simplest in Cartesian coordinates. Putting

$$\boldsymbol{\sigma} = \sigma_{mn} \mathbf{i}_m \mathbf{i}_n, \qquad \mathbf{u} = u_m \mathbf{i}_m,$$

we get

$$\frac{\partial \sigma_{ij}}{\partial x_i} + \rho f_j = \rho \frac{\partial^2 u_j}{\partial t^2} \qquad (j = 1, 2, 3)$$

or explicitly

$$\frac{\partial \sigma_{11}}{\partial x_1} + \frac{\partial \sigma_{21}}{\partial x_2} + \frac{\partial \sigma_{31}}{\partial x_3} + \rho f_1 = \rho \frac{\partial^2 u_1}{\partial t^2},$$

$$\frac{\partial \sigma_{12}}{\partial x_1} + \frac{\partial \sigma_{22}}{\partial x_2} + \frac{\partial \sigma_{32}}{\partial x_3} + \rho f_2 = \rho \frac{\partial^2 u_2}{\partial t^2},$$

$$\frac{\partial \sigma_{13}}{\partial x_1} + \frac{\partial \sigma_{23}}{\partial x_2} + \frac{\partial \sigma_{33}}{\partial x_3} + \rho f_3 = \rho \frac{\partial^2 u_3}{\partial t^2}.$$

With regard for (4.33), in curvilinear coordinates equation (6.19) takes the form

$$\frac{1}{\sqrt{g}} \frac{\partial}{\partial q^i} \left(\sqrt{g} \sigma^{ij} \mathbf{r}_j \right) + \rho \mathbf{f} = \rho \frac{\partial^2 \mathbf{u}}{\partial t^2}. \tag{6.20}$$

In components it is

$$\frac{\partial}{\partial q^i} \left(\sqrt{g} \sigma^{ij} \right) + \Gamma^j_{mn} \sigma^{mn} + \rho \sqrt{g} f^j = \rho \sqrt{g} \frac{\partial^2 u^j}{\partial t^2} \qquad (j = 1, 2, 3). \tag{6.21}$$

The equations of motion can be simplified in orthogonal coordinates. See Appendix A for expressions in cylindrical and spherical coordinates.

6.4   Hooke's Law

Equations (6.16) and (6.19) apply to any small deformation of a continuous medium. However, they do not uniquely define the deformations and stresses in a body. To study a body under load, we should relate the stresses to the strains using the material properties of the body. These relations are called *constitutive equations*. In this book we consider linearly elastic materials for which the constitutive equation represents a linear dependence between $\boldsymbol{\sigma}$ and $\boldsymbol{\varepsilon}$. The simplest version is Hooke's law

$$\sigma = E\varepsilon, \tag{6.22}$$

which describes the elastic properties of a thin rod under tension or compression. Here E is the elastic modulus of the material from which the rod is made; it is known as *Young's modulus*.

Robert Hooke (1635–1703) was the first to establish the linear dependence $f \sim \Delta l$ between the applied force and the elongation of a bar similar to that in Fig. 6.1. Thomas Young (1773–1829) introduced the elastic modulus E as a quantity that does not depend on the cross-sectional area of the bar; rather, E characterizes the material itself. The linear dependence (6.22) holds only in some range $|\varepsilon| < \varepsilon_0$, where ε_0 depends on the material, temperature, and other factors. However, the importance of Hooke's law in engineering cannot be overestimated.

In the general case, a linear dependence between the second-order tensors $\boldsymbol{\sigma}$ and $\boldsymbol{\varepsilon}$ is presented in equation (3.31), known as the *generalized Hooke's law*:

$$\boldsymbol{\sigma} = \mathbf{C} \cdot\cdot \, \boldsymbol{\varepsilon}. \tag{6.23}$$

In a Cartesian basis it is

$$\sigma_{ij} = c_{ijmn}\varepsilon_{mn}.$$

The fourth-order *tensor of elastic moduli*

$$\mathbf{C} = c_{ijmn}\mathbf{i}_i\mathbf{i}_j\mathbf{i}_m\mathbf{i}_n$$

has 81 components; only 36 of these are independent, however, as the symmetries of $\boldsymbol{\sigma}$ and $\boldsymbol{\varepsilon}$ lead to the conditions

$$c_{ijmn} = c_{jimn} = c_{ijnm}.$$

In linear elasticity it is shown [Lurie (2005)] that we can introduce the *strain energy*

$$\frac{1}{2}\int_V W \, dV$$

stored in the elastic body by virtue of its deformation. The integrand $W = W(\boldsymbol{\varepsilon})$, the *strain energy function*, is a quadratic form in $\boldsymbol{\varepsilon}$:

$$W = \frac{1}{2}\boldsymbol{\varepsilon} \cdot\cdot \, \mathbf{C} \cdot\cdot \, \boldsymbol{\varepsilon} = \frac{1}{2}\varepsilon_{ij}c_{ijmn}\varepsilon_{mn}.$$

The fact that W is uniquely defined for any deformation requires $\mathbf{C}$ to possess an additional symmetry property which, in terms of components, is

$$c_{ijmn} = c_{mnij}$$

for any indices i, j, m, n. Indeed, we can represent $\mathbf{C}$ as the sum of two tensors

$$\mathbf{C} = \mathbf{C}' + \mathbf{C}'',$$

where the components of $\mathbf{C}'$ satisfy

$$c'_{ijmn} = c'_{mnij}$$

and those of $\mathbf{C}''$ satisfy

$$c''_{ijmn} = -c''_{mnij}.$$

(This is similar to the representation of a second-order tensor as the sum of symmetric and antisymmetric tensors.) For any symmetric tensors ε_k, we have

$$\varepsilon_1 \cdot\cdot \mathbf{C}'' \cdot\cdot \varepsilon_2 = -\varepsilon_2 \cdot\cdot \mathbf{C}'' \cdot\cdot \varepsilon_1,$$

hence for any strain tensor ε,

$$\varepsilon \cdot\cdot \mathbf{C}'' \cdot\cdot \varepsilon = -\varepsilon \cdot\cdot \mathbf{C}'' \cdot\cdot \varepsilon$$

and

$$\varepsilon \cdot\cdot \mathbf{C}'' \cdot\cdot \varepsilon = 0.$$

It follows that

$$\varepsilon \cdot\cdot \mathbf{C} \cdot\cdot \varepsilon = \varepsilon \cdot\cdot \mathbf{C}' \cdot\cdot \varepsilon.$$

Thus $\mathbf{C}''$ does not affect the values of W, and the constitutive relations should not include it. Setting $\mathbf{C}''$ to zero, we get $\mathbf{C} = \mathbf{C}'$. Thus the components of $\mathbf{C}$ have the following symmetry properties:

$$c_{ijmn} = c_{jimn} = c_{ijnm}, \qquad c_{ijmn} = c_{mnij},$$

for any indices i, j, n, m. Elementary calculation shows that these are 60 equalities. So in Hooke's law, of the 81 components of $\mathbf{C}$ there remain only $81 - 60 = 21$ independent elastic constants.

It follows from Exercise 3.44 that W is a potential for $\boldsymbol{\sigma}$, i.e., that

$$\boldsymbol{\sigma} = W_{,\varepsilon}$$

where $W_{,\varepsilon}$ is the derivative of W with respect to ε.

A dual relation

$$\varepsilon = W_{,\sigma}$$

holds when we express W in terms of $\boldsymbol{\sigma}$, i.e., $W = W(\boldsymbol{\sigma})$. We urge the reader to show this as an exercise.

In engineering analysis, the vector and matrix notations are typically used to describe an elastic body. Because of the symmetry properties of the tensors, all the relations are written in terms of formal six-dimensional "vectors" for the components of $\boldsymbol{\sigma}$ and ε, and 6×6 matrices for $\mathbf{C}$. *Voigt's rule* shows one how to transform the tensor notation to the matrix-vector notation. The rule for changing $c_{ijmn} \rightarrow C_{pq}$ is as follows. The pairs of indices 11, 22, 33 change to 1, 2, 3, the pairs 23 and 32 to 4, the pairs 13 and 31 to 5, and the pairs 12 and 21 to 6. For example, $c_{1122} \rightarrow C_{12}$ and

$c_{1232} \rightarrow C_{64}$. The symmetry property $C_{pq} = C_{qp}$ holds. The components of the stress and strain tensors are transformed by the formulas

$$
\begin{bmatrix} \sigma_{11} \\ \sigma_{22} \\ \sigma_{33} \\ \sigma_{23} \\ \sigma_{31} \\ \sigma_{12} \end{bmatrix} = \begin{bmatrix} \sigma_1 \\ \sigma_2 \\ \sigma_3 \\ \sigma_4 \\ \sigma_5 \\ \sigma_6 \end{bmatrix}, \qquad \begin{bmatrix} \varepsilon_{11} \\ \varepsilon_{22} \\ \varepsilon_{33} \\ 2\varepsilon_{23} \\ 2\varepsilon_{31} \\ 2\varepsilon_{12} \end{bmatrix} = \begin{bmatrix} \varepsilon_1 \\ \varepsilon_2 \\ \varepsilon_3 \\ \varepsilon_4 \\ \varepsilon_5 \\ \varepsilon_6 \end{bmatrix}. \tag{6.24}
$$

In this notation, Hooke's law takes the form

$$
\begin{bmatrix} \sigma_1 \\ \sigma_2 \\ \sigma_3 \\ \sigma_4 \\ \sigma_5 \\ \sigma_6 \end{bmatrix} = \begin{bmatrix} C_{11} & C_{12} & C_{13} & C_{14} & C_{15} & C_{16} \\ C_{12} & C_{22} & C_{23} & C_{24} & C_{25} & C_{26} \\ C_{13} & C_{23} & C_{33} & C_{34} & C_{35} & C_{36} \\ C_{14} & C_{24} & C_{34} & C_{44} & C_{45} & C_{46} \\ C_{15} & C_{25} & C_{35} & C_{45} & C_{55} & C_{56} \\ C_{16} & C_{26} & C_{36} & C_{46} & C_{56} & C_{66} \end{bmatrix} \begin{bmatrix} \varepsilon_1 \\ \varepsilon_2 \\ \varepsilon_3 \\ \varepsilon_4 \\ \varepsilon_5 \\ \varepsilon_6 \end{bmatrix}. \tag{6.25}
$$

The general case in which all 21 elastic constants in (6.25) are independent is not common in applications. In engineering practice, the materials usually possess symmetries that reduce the number of independent constants. Most common are isotropic materials; such a material exhibits complete symmetry of its properties in space so that at each point the material properties do not vary with direction. Mathematically, symmetry is expressed in the language of group theory. Examples of isotropic materials are steel, aluminium, many other metals, polymeric materials, etc.

For an isotropic material, the relation between $\boldsymbol{\sigma}$ and $\boldsymbol{\varepsilon}$ is given by a linear isotropic function. This function, considered in Chapter 3, now takes the form

$$
\boldsymbol{\sigma} = \lambda \mathbf{E} \operatorname{tr} \boldsymbol{\varepsilon} + 2\mu\boldsymbol{\varepsilon}, \tag{6.26}
$$

where λ and μ are *Lamé's moduli*. We may also refer to μ as the shear modulus. For an isotropic material,

$$
\mathbf{C} = \lambda \mathbf{E}\mathbf{E} + \mu(\mathbf{e}_k \mathbf{E}\mathbf{e}^k + \mathbf{I}).
$$

In matrix notation, to $\mathbf{C}$ there corresponds the diagonal matrix

$$\begin{bmatrix} \lambda + 2\mu & 0 & 0 & 0 & 0 & 0 \\ 0 & \lambda + 2\mu & 0 & 0 & 0 & 0 \\ 0 & 0 & \lambda + 2\mu & 0 & 0 & 0 \\ 0 & 0 & 0 & 2\mu & 0 & 0 \\ 0 & 0 & 0 & 0 & 2\mu & 0 \\ 0 & 0 & 0 & 0 & 0 & 2\mu \end{bmatrix}.$$

Other elastic constants were used historically in engineering work, and in terms of these, Hooke's law takes other forms. Common engineering constants include *Young's modulus E*, *Poisson's ratio* ν, the *bulk modulus k*, and the *shear modulus G*. For isotropic materials, these are used in pairs: E and ν, and k and G. The modulus E originated in bar stretching problems; it relates the tension applied to the bar with the resulting strain. The modulus k is used to describe uniform volume deformation of a material, say of a ball under uniform pressure. G describes the shear characteristics of a material. Finally, ν describes the lateral shortening of a stretched band. We will derive some relations between the constants.

Let us split $\boldsymbol{\sigma}$ into a ball tensor and deviator:

$$\boldsymbol{\sigma} = \frac{1}{3}\sigma\mathbf{E} + \operatorname{dev}\boldsymbol{\sigma}.$$

Then (6.26) takes the form

$$\sigma = \frac{1}{3}(3\lambda + 2\mu)\operatorname{tr}\varepsilon, \qquad \operatorname{dev}\boldsymbol{\sigma} = 2\mu\operatorname{dev}\varepsilon.$$

The bulk modulus k relates the mean stress σ with the bulk strain $\operatorname{tr}\varepsilon$ through

$$\sigma = k\operatorname{tr}\varepsilon,$$

and so

$$k = \lambda + 2\mu/3.$$

Shear stresses are related to shear strains by

$$\sigma_{ij} = 2G\varepsilon_{ij} \qquad (i \neq j).$$

The above deviator equation in components is

$$\sigma_{ij} = 2\mu\varepsilon_{ij} \qquad (i \neq j),$$

which yields $G = \mu$.

Now we turn to Young's modulus E. Let us consider a uniaxial homogeneous deformation of a bar; this occurs when the material is uniformly stretched or compressed. Take the unit vector $\mathbf{i}_1$ along the bar axis. Then $\boldsymbol{\sigma} = \sigma_{11}\mathbf{i}_1\mathbf{i}_1$. Relation (6.26) reduces to the three nontrivial component equations

$$\sigma_{11} = \lambda\operatorname{tr}\boldsymbol{\varepsilon} + 2\mu\varepsilon_{11}, \qquad 0 = \lambda\operatorname{tr}\boldsymbol{\varepsilon} + 2\mu\varepsilon_{22}, \qquad 0 = \lambda\operatorname{tr}\boldsymbol{\varepsilon} + 2\mu\varepsilon_{33}.$$

Eliminating ε_{22} and ε_{33} from these, we get

$$\sigma_{11} = E\varepsilon_{11}, \quad \text{where} \quad E = \frac{\mu(3\lambda + 2\mu)}{\lambda + \mu}.$$

When a bar is uniformly stretched by a force and its strain is ε, experiment shows that its transverse dimensions decrease. The transverse strain for this deformation is proportional to the longitudinal strain $-\nu\varepsilon$, the coefficient of proportionality being Poisson ratio. Its relation to the other constants is defined by the following

Exercise 6.3. For the above uniaxial strain state of the bar, find $\varepsilon_{22}/\varepsilon_{11} = -\nu$. Show that $\nu = \lambda/(2\lambda + 2\mu)$. Note that in the uniaxially stretched bar, ν defines the dependence of the lateral strain ε_{22} on the axial strain ε_{11}.

The relations between various pairs of moduli used in the literature are summarized in Appendix A.

Thermodynamic considerations show that the strain energy should be positive:

$$W(\boldsymbol{\varepsilon}) > 0 \quad \text{whenever} \quad \boldsymbol{\varepsilon} \neq \mathbf{0}. \tag{6.27}$$

This puts additional restrictions on the elastic moduli for an isotropic material. We have

$$W(\boldsymbol{\varepsilon}) = \frac{1}{2}\boldsymbol{\sigma}\cdot\cdot\,\boldsymbol{\varepsilon} = \frac{1}{2}\lambda(\operatorname{tr}\boldsymbol{\varepsilon})^2 + \mu\boldsymbol{\varepsilon}\cdot\cdot\,\boldsymbol{\varepsilon} \geq 0.$$

Let us represent $\boldsymbol{\varepsilon}$ as the sum of its ball and deviator terms:

$$\boldsymbol{\varepsilon} = \frac{1}{3}\mathbf{E}\operatorname{tr}\boldsymbol{\varepsilon} + \operatorname{dev}\boldsymbol{\varepsilon}.$$

Because

$$\boldsymbol{\varepsilon}\cdot\cdot\,\boldsymbol{\varepsilon} = \frac{1}{3}(\operatorname{tr}\boldsymbol{\varepsilon})^2 + \operatorname{dev}\boldsymbol{\varepsilon}\cdot\cdot\,\operatorname{dev}\boldsymbol{\varepsilon},$$

we obtain

$$W(\boldsymbol{\varepsilon}) = \frac{1}{6}(3\lambda + 2\mu)(\operatorname{tr}\boldsymbol{\varepsilon})^2 + \mu\operatorname{dev}\boldsymbol{\varepsilon}\cdot\cdot\,\operatorname{dev}\boldsymbol{\varepsilon}.$$

But $\operatorname{tr}\varepsilon$ and $\operatorname{dev}\varepsilon$ are independent quantities, and from (6.27) it follows that

$$3\lambda + 2\mu > 0, \qquad \mu > 0. \tag{6.28}$$

Hence the bulk modulus $k = \lambda + 2/3\mu$ and shear modulus $G = \mu$ are positive. A consequence of (6.28) is explored in

Exercise 6.4. Demonstrate that $E > 0$ and $-1 < \nu \le 1/2$.

For most engineering materials, we have $\nu > 0$. The values of ν for rubbers are close to $1/2$; materials like re-entrant foams can have negative values of ν. In technical reference books, the reader can find data for various materials. For regular steels, the values of E are about 2×10^{11} Pa, and ν lies in the range $0.25 - 0.33$.

Exercise 6.5. For an isotropic body, express ε and W in terms of σ.

6.5 Equilibrium Equations in Displacements

We derived the equation of motion (6.19) and the equilibrium equation (6.11). Equation (6.11) is written in terms of stresses. In component form, (6.11) contains three equations in six independent unknowns σ_{ij}. Using the definition of strain tensor and Hooke's law, we can transform (6.19) and (6.11) to a system of three simultaneous equations with respect to the components of the displacement vector $\mathbf{u}$. So the systems (6.19) and (6.11) reduce to systems involving three equations in the three unknown components of $\mathbf{u}$. This corresponds to the common viewpoint that a well-posed problem should contain the same number of equations as unknowns.

For simplicity, we derive the equilibrium equations for an isotropic homogeneous material defined by (6.26). Suppose λ and μ are constants. First we derive $\nabla \cdot \sigma$ in terms of $\mathbf{u}$:

$$\nabla \cdot \sigma = \nabla \cdot (\lambda \mathbf{E}\operatorname{tr}\varepsilon + 2\mu\varepsilon) = \lambda\nabla\operatorname{tr}\varepsilon + \mu\nabla \cdot \nabla\mathbf{u} + \mu\nabla \cdot (\nabla\mathbf{u})^T.$$

Because

$$\operatorname{tr}\varepsilon = \nabla \cdot \mathbf{u}$$

and

$$\nabla \cdot (\nabla\mathbf{u})^T = \nabla\nabla \cdot \mathbf{u},$$

we get

$$\nabla \cdot \boldsymbol{\sigma} = (\lambda + \mu)\nabla\nabla \cdot \mathbf{u} + \mu\nabla \cdot \nabla\mathbf{u}.$$

So the equilibrium equation takes the form

$$(\lambda + \mu)\nabla\nabla \cdot \mathbf{u} + \mu\nabla \cdot \nabla\mathbf{u} + \rho\mathbf{f} = \mathbf{0}. \tag{6.29}$$

Exercise 6.6. Show that $\nabla \cdot (\nabla\mathbf{u})^T = \nabla\nabla \cdot \mathbf{u}$.

In Cartesian coordinates, equation (6.29) is

$$(\lambda + \mu)\frac{\partial u_i}{\partial x_k \partial x_i} + \mu\frac{\partial^2 u_k}{\partial x_i \partial x_i} + \rho f_k = 0 \qquad (k = 1, 2, 3). \tag{6.30}$$

That is,

$$(\lambda + \mu)\frac{\partial}{\partial x_1}\left(\frac{\partial u_1}{\partial x_1} + \frac{\partial u_2}{\partial x_2} + \frac{\partial u_3}{\partial x_3}\right) + \mu\Delta u_1 + \rho f_1 = 0,$$

$$(\lambda + \mu)\frac{\partial}{\partial x_2}\left(\frac{\partial u_1}{\partial x_1} + \frac{\partial u_2}{\partial x_2} + \frac{\partial u_3}{\partial x_3}\right) + \mu\Delta u_2 + \rho f_2 = 0,$$

$$(\lambda + \mu)\frac{\partial}{\partial x_1}\left(\frac{\partial u_1}{\partial x_1} + \frac{\partial u_2}{\partial x_2} + \frac{\partial u_3}{\partial x_3}\right) + \mu\Delta u_3 + \rho f_3 = 0, \tag{6.31}$$

where

$$\Delta = \frac{\partial^2}{\partial x_1^2} + \frac{\partial^2}{\partial x_2^2} + \frac{\partial^2}{\partial x_3^2}.$$

In curvilinear coordinates, the component representation of (6.29) is

$$(\lambda + \mu)\frac{\partial}{\partial q^k}\varepsilon + \mu g^{mn}\nabla_m\nabla_n u_k + \rho f_k = 0 \qquad (k = 1, 2, 3) \tag{6.32}$$

where

$$\varepsilon = \operatorname{tr}\boldsymbol{\varepsilon} = \frac{1}{\sqrt{g}}\frac{\partial(\sqrt{g}g^{mn}u_n)}{\partial q^m}.$$

Exercise 6.7. Derive (6.32).

See Appendix A for the corresponding relations in cylindrical and spherical coordinates.

Exercise 6.8. Write down the equations of motion in displacements.

6.6 Boundary Conditions and Boundary Value Problems

We have derived the equilibrium equations. To ensure uniqueness of solution, we must supplement the equations with certain conditions on the boundary of the domain where the equations hold. We first pose the main equilibrium problems for elastic bodies. Then we touch on the existence-uniqueness question.

Physical intuition tells us that at a boundary point we should appoint either a displacement vector or a stress vector defined by the contact load. Appointment of both at the same point is impossible. This will be confirmed later. In elasticity, there are three main boundary value problems. Two were mentioned above: one with given displacements, one with given contact forces. The third is a mixed problem with some combination of displacements and contact loads given on the boundary.

In the first boundary value problem of elasticity, we supplement the equilibrium equations in displacements (6.29) or other forms of the equations (6.30)–(6.32) with a given displacement field $\mathbf{u}^0$ on the whole boundary Σ:

$$\mathbf{u}\big|_{\Sigma} = \mathbf{u}^0. \tag{6.33}$$

This is known as the *kinematic boundary condition*.

In the second boundary value problem, we supplement (6.29) with a prescribed contact load $\mathbf{t}^0$ on Σ:

$$\mathbf{n} \cdot \boldsymbol{\sigma}\big|_{\Sigma} = \mathbf{t}^0. \tag{6.34}$$

This is known as a *static boundary condition*. Since no point of the body is fixed, the body can move freely in space. With this condition, the static boundary value problem is well-posed only if the load acting on the body is self-balanced; this means that the resultant force and resultant moment of all the external forces acting on the body must vanish.

The reader familiar with partial differential equations may recognize that these types of boundary conditions are analogous to the Dirichlet and Neumann conditions for Poisson's equation.

The third boundary value problem requires us to supplement (6.29) with *mixed boundary conditions*. That is, we must assign kinematic conditions on some portion Σ_1 of the boundary, and static conditions $\mathbf{t}^0$ on the remainder Σ_2:

$$\mathbf{u}\big|_{\Sigma_1} = \mathbf{u}^0, \qquad \mathbf{n} \cdot \boldsymbol{\sigma}\big|_{\Sigma_2} = \mathbf{t}^0. \tag{6.35}$$

There exist other "mixed problems" of elasticity in which, at each boundary point, we assign three conditions that combine the given displacements and loads in some fashion. Moreover, on the boundary we can combine conditions involving contact with other elastic or inelastic bodies or media. As a rule, such problems require consideration of well-posedness, although engineering intuition commonly permits investigators to propose meaningful boundary conditions of that type. An example of a meaningless problem is the problem with simultaneously given normal displacements and normal stresses on the boundary. These cannot be specified independently.

Henceforth we will assume V is a bounded volume whose boundary Σ is sufficiently regular that we can apply the technique of integration by parts.

6.7 Equilibrium Equations in Stresses

A *boundary value problem in displacements* involves finding a displacement field satisfying the equilibrium equations (6.29) and one of the sets of boundary conditions (6.33), (6.34), or (6.35). It is assumed that the stress tensor in (6.34) or (6.35) is expressed in terms of $\mathbf{u}$ by (6.26) or (6.16). When the stresses are prescribed over the whole boundary, we can try finding the unknown $\boldsymbol{\sigma}$ via (6.11) and (6.34). However, the solution for $\boldsymbol{\sigma}$ is not unique. We must bring in other equations that take into account Hooke's law. We can do this as follows. Suppose we have found a solution $\boldsymbol{\sigma}$ of equations (6.11) supplemented by (6.34). By $\boldsymbol{\sigma}$, through Hooke's law (6.26), we define $\boldsymbol{\varepsilon}$. Calculating the trace of equation (6.26), we get

$$\operatorname{tr}\boldsymbol{\varepsilon} = \frac{1}{3\lambda + 2\mu}\operatorname{tr}\boldsymbol{\sigma}.$$

Substituting this into (6.26), we have

$$\boldsymbol{\varepsilon} = \frac{1}{2\mu}\left[\boldsymbol{\sigma} - \frac{\lambda}{3\lambda + 2\mu}\mathbf{E}\operatorname{tr}\boldsymbol{\sigma}\right] = \frac{1}{2\mu}\left[\boldsymbol{\sigma} - \frac{\nu}{1+\nu}\mathbf{E}\operatorname{tr}\boldsymbol{\sigma}\right]. \qquad (6.36)$$

But we know that the strain tensor makes sense, i.e., by $\boldsymbol{\varepsilon}$ we can find the displacement field, if and only if the compatibility equation (6.17) holds:

$$\nabla \times (\nabla \times \boldsymbol{\varepsilon})^T = \mathbf{0}.$$

Substituting (6.36) into the compatibility equation, we get the equation in terms of $\boldsymbol{\sigma}$:

$$\nabla \times (\nabla \times \boldsymbol{\varepsilon})^T = \frac{1}{2\mu}\left[\nabla \times (\nabla \times \boldsymbol{\sigma})^T - \frac{\nu}{1+\nu}\nabla \times (\nabla \times (\mathbf{E}\operatorname{tr}\boldsymbol{\sigma})^T\right] = \mathbf{0}.$$

Omitting cumbersome calculations, we can reduce this to the *Beltrami–Michell equations*

$$\nabla \cdot \nabla \boldsymbol{\sigma} + \frac{\nu}{1+\nu} \nabla \nabla \operatorname{tr} \boldsymbol{\sigma} + \rho \nabla \mathbf{f} + \rho (\nabla \mathbf{f})^T + \mathbf{E} \frac{\nu}{1-\nu} \rho \nabla \cdot \mathbf{f} = \mathbf{0}. \quad (6.37)$$

These equations, being supplementary to (6.11) and (6.34), present us with the complete setup of the equilibrium problem in stresses. Note that this problem has a solution only if the set of external forces, consisting of the body forces and those acting on the boundary, is self-balanced.

When body forces are absent, (6.37) reduces to

$$\nabla \cdot \nabla \boldsymbol{\sigma} + \frac{\nu}{1+\nu} \nabla \nabla \operatorname{tr} \boldsymbol{\sigma} = \mathbf{0}. \quad (6.38)$$

In Cartesian coordinates, (6.38) implies

$$\Delta \sigma_{11} + \frac{1}{1+\nu} \frac{\partial^2 \sigma}{\partial x_1^2} = 0, \qquad \Delta \sigma_{12} + \frac{1}{1+\nu} \frac{\partial^2 \sigma}{\partial x_1 \partial x_2} = 0,$$

$$\Delta \sigma_{22} + \frac{1}{1+\nu} \frac{\partial^2 \sigma}{\partial x_2^2} = 0, \qquad \Delta \sigma_{23} + \frac{1}{1+\nu} \frac{\partial^2 \sigma}{\partial x_2 \partial x_3} = 0,$$

$$\Delta \sigma_{33} + \frac{1}{1+\nu} \frac{\partial^2 \sigma}{\partial x_3^2} = 0, \qquad \Delta \sigma_{13} + \frac{1}{1+\nu} \frac{\partial^2 \sigma}{\partial x_1 \partial x_3} = 0, \quad (6.39)$$

where

$$\sigma = \operatorname{tr} \boldsymbol{\sigma} = \sigma_{11} + \sigma_{22} + \sigma_{33}.$$

To solve the equilibrium problem completely, we first find $\boldsymbol{\sigma}$ by (6.11), (6.34), and the Beltrami–Michell equations. Then, using Hooke's law, we find $\boldsymbol{\varepsilon}$. Finally, by Cesàro's formula (6.18), we find the displacement field $\mathbf{u}$ that is uniquely defined up to a rigid motion of the body.

We should note that, from the standpoint of classical mathematical physics, the structure of the boundary value problem in stresses is a bit strange. The unknowns are the six components of $\boldsymbol{\sigma}$, satisfying nine equations — three of first order, and six of second order with respect to the components of $\boldsymbol{\sigma}$. These are supplemented with three boundary conditions. By its derivation, this boundary value problem is equivalent to the boundary value problem in displacements. The theory of such "strange" boundary value problems is far from complete.

6.8 Uniqueness of Solution for the Boundary Value Problems of Elasticity

Uniqueness of solution to a boundary value problem is an important indication of its well-posedness. A uniqueness theorem for elasticity problems was established by Gustav R. Kirchhoff (1824–1887).

We will prove uniqueness of solution for equations (6.29) supplemented with boundary conditions (6.35), which include conditions (6.33) as a particular case.

To the contrary, we suppose that there exist two solutions $\mathbf{u}_1$ and $\mathbf{u}_2$ that satisfy (6.29) and (6.35). We denote the corresponding stress tensors by $\boldsymbol{\sigma}_1$ and $\boldsymbol{\sigma}_2$, respectively. So the following two sets of equations hold:

$$\nabla \cdot \boldsymbol{\sigma}_1 + \rho \mathbf{f} = \mathbf{0} \text{ in } V, \quad \mathbf{u}_1\big|_{\Sigma_1} = \mathbf{u}^0, \quad \mathbf{n} \cdot \boldsymbol{\sigma}_1\big|_{\Sigma_2} = \mathbf{t}^0,$$

$$\nabla \cdot \boldsymbol{\sigma}_2 + \rho \mathbf{f} = \mathbf{0} \text{ in } V, \quad \mathbf{u}_2\big|_{\Sigma_1} = \mathbf{u}^0, \quad \mathbf{n} \cdot \boldsymbol{\sigma}_2\big|_{\Sigma_2} = \mathbf{t}^0.$$

Consider the difference $\mathbf{u} = \mathbf{u}_1 - \mathbf{u}_2$ and its corresponding stress field $\boldsymbol{\sigma} = \boldsymbol{\sigma}_1 - \boldsymbol{\sigma}_2$. Subtracting the equations for $\mathbf{u}_1$ and $\mathbf{u}_2$, we see that $\mathbf{u}$ is a solution of the following equilibrium problem:

$$\nabla \cdot \boldsymbol{\sigma} = \mathbf{0} \text{ in } V, \quad \mathbf{u}\big|_{\Sigma_1} = \mathbf{0}, \quad \mathbf{n} \cdot \boldsymbol{\sigma}\big|_{\Sigma_2} = \mathbf{0}.$$

Dot-multiplying the equilibrium equation by $\mathbf{u}$ and integrating over V, we get

$$\int_V (\nabla \cdot \boldsymbol{\sigma}) \cdot \mathbf{u} \, dV = 0.$$

Using the Gauss–Ostrogradsky theorem, we transform this to the equation

$$-\int_V \boldsymbol{\sigma} \cdots (\nabla \mathbf{u})^T \, dV + \int_\Sigma \mathbf{n} \cdot \boldsymbol{\sigma} \cdot \mathbf{u} = 0.$$

At each point of Σ, one of the conditions $\mathbf{u} = \mathbf{0}$ or $\mathbf{n} \cdot \boldsymbol{\sigma} = \mathbf{0}$ holds. So the last surface integral is zero. The tensor $\boldsymbol{\sigma}$ is symmetric. So

$$\boldsymbol{\sigma} \cdots (\nabla \mathbf{u})^T = \boldsymbol{\sigma} \cdots (\nabla \mathbf{u}).$$

This allows us to represent the volume integral as follows:

$$-\int_V \boldsymbol{\sigma} \cdots \boldsymbol{\varepsilon} \, dV = -2 \int_V W(\boldsymbol{\varepsilon}) \, dV = 0.$$

Recall that $W(\boldsymbol{\varepsilon})$ is positive definite. Hence the equality of the integral to zero implies that its integrand vanishes everywhere in V: $W(\boldsymbol{\varepsilon}) = 0$. By positiveness of W, it follows that

$$\boldsymbol{\varepsilon} = \mathbf{0}. \tag{6.40}$$

By Cesàro's formula (6.18), when $\boldsymbol{\varepsilon}(\mathbf{u}) = \mathbf{0}$, the displacement vector takes the form

$$\mathbf{u} = \mathbf{u}_0 + \boldsymbol{\omega}_0 \times (\mathbf{r} - \mathbf{r}_0), \tag{6.41}$$

where $\mathbf{u}_0, \boldsymbol{\omega}_0$ and $\mathbf{r}_0$ are constant vectors. But it is a small displacement field for the rigid volume V. Because $\mathbf{u}|_{\Sigma_1} = \mathbf{0}$, we have $\mathbf{u} = \mathbf{0}$ in V. Thus

$$\mathbf{u}_1 = \mathbf{u}_2.$$

We have proved uniqueness of solution for the first and third boundary value problems of elasticity.

For the second problem, the initial steps of the proof remain valid. We find that $\boldsymbol{\varepsilon} = \mathbf{0}$ in V, hence the two solutions to the problem differ by a rigid-body displacement:

$$\mathbf{u}_1 = \mathbf{u}_2 + \mathbf{u}_0 + \boldsymbol{\omega}_0 \times (\mathbf{r} - \mathbf{r}_0).$$

But this time we cannot conclude that $\mathbf{u}_1 = \mathbf{u}_2$. From physics, the situation is clear. The second problem of elasticity describes a body free of geometric restrictions, so its solution must be defined up to a rigid-body displacement. This is what the last formula states.

6.9 Betti's Reciprocity Theorem

The solutions to the equilibrium problems for an elastic body under two different loads obey *Betti's reciprocal work theorem*.

Let us consider a body under the action of contact forces $\mathbf{t}'$ and body forces $\mathbf{f}'$, and the same body under the action of another pair of external forces $\mathbf{t}''$ and $\mathbf{f}''$. The displacements and other quantities for the two corresponding problems will be denoted similarly, using primes and double-primes, respectively. Now we have two solutions of two different second boundary value problems for the elastic body.

Theorem 6.4. *The solutions $\mathbf{u}'$ and $\mathbf{u}''$ of the second boundary value problem of elasticity that correspond to the respective loading pairs $\mathbf{t}', \mathbf{f}'$ and $\mathbf{t}'', \mathbf{f}''$ satisfy the following relation:*

$$\int_V \rho \mathbf{f}' \cdot \mathbf{u}'' \, dV + \int_\Sigma \mathbf{t}' \cdot \mathbf{u}'' \, d\Sigma = \int_V \rho \mathbf{f}'' \cdot \mathbf{u}' \, dV + \int_\Sigma \mathbf{t}'' \cdot \mathbf{u}' \, d\Sigma. \tag{6.42}$$

This equality is Betti's theorem. First derived for beam theory and later extended to many portions of physics, it is used to derive the equations of the boundary finite element methods. Each of the expressions on the left and right represents the work of a load over a displacement. So we can give a mechanical formulation of Betti's theorem.

The work of the first system of forces over the displacements of the body due to the action of the second system of forces is equal to the work of the second system of forces over the displacements due to the action of the first system of forces.

Proof. The vectors $\mathbf{u}'$ and $\mathbf{u}''$ respectively satisfy the following equilibrium equations and boundary conditions:

$$\nabla \cdot \boldsymbol{\sigma}' + \rho \mathbf{f}' = \mathbf{0} \text{ in } V, \quad \mathbf{n} \cdot \boldsymbol{\sigma}'|_{\Sigma} = \mathbf{t}';$$
$$\nabla \cdot \boldsymbol{\sigma}'' + \rho \mathbf{f}'' = \mathbf{0} \text{ in } V, \quad \mathbf{n} \cdot \boldsymbol{\sigma}''|_{\Sigma} = \mathbf{t}''.$$

Let us consider the left-hand side of (6.42). Writing the forces in terms of the stresses, we get

$$\int_V \rho \mathbf{f}' \cdot \mathbf{u}'' \, dV + \int_{\Sigma} \mathbf{t}' \cdot \mathbf{u}'' \, d\Sigma = -\int_V (\nabla \cdot \boldsymbol{\sigma}') \cdot \mathbf{u}'' \, dV + \int_{\Sigma} \mathbf{n} \cdot \boldsymbol{\sigma}' \cdot \mathbf{u}'' \, d\Sigma.$$

Applying the Gauss–Ostrogradsky theorem to the surface integral, we get

$$-\int_V (\nabla \cdot \boldsymbol{\sigma}') \cdot \mathbf{u}'' \, dV + \int_{\Sigma} \mathbf{n} \cdot \boldsymbol{\sigma}' \cdot \mathbf{u}'' \, d\Sigma$$
$$= \int_V [-(\nabla \cdot \boldsymbol{\sigma}') \cdot \mathbf{u}'' + \nabla \cdot (\boldsymbol{\sigma}' \cdot \mathbf{u}'')] \, dV$$
$$= \int_V \boldsymbol{\sigma}' \cdots (\nabla \mathbf{u}'')^T \, dV$$
$$= \int_V \boldsymbol{\sigma}' \cdots \varepsilon'' \, dV.$$

Similarly,

$$\int_V \rho \mathbf{f}'' \cdot \mathbf{u}' \, dV + \int_{\Sigma} \mathbf{t}'' \cdot \mathbf{u}' \, d\Sigma = \int_V \boldsymbol{\sigma}'' \cdots \varepsilon' \, dV.$$

So the proof reduces to verification of the equality

$$\int_V \boldsymbol{\sigma}' \cdots \varepsilon'' \, dV = \int_V \boldsymbol{\sigma}'' \cdots \varepsilon' \, dV.$$

Using Hooke's law and recalling the symmetry properties of $\mathbf{C}$, we have

$$\boldsymbol{\sigma}' \cdots \varepsilon'' = (\mathbf{C} \cdots \varepsilon') \cdots \varepsilon'' = \varepsilon'' \cdots \mathbf{C} \cdots \varepsilon' = \varepsilon' \cdots \mathbf{C} \cdots \varepsilon'' = \boldsymbol{\sigma}'' \cdots \varepsilon'.$$

This completes the proof. $\square$

Exercise 6.9. Let a portion Σ_1 of the boundary be fixed: $\mathbf{u}|_{\Sigma_1} = \mathbf{0}$. Suppose two boundary value problems include this condition and, moreover, that the body is under the action of one of the two systems of forces $\mathbf{t}', \mathbf{f}'$ or $\mathbf{t}'', \mathbf{f}''$, with $\mathbf{t}', \mathbf{t}''$ given on $\Sigma_2 = \Sigma \setminus \Sigma_1$. Prove that in this case Betti's equality takes the form

$$\int_V \rho \mathbf{f}' \cdot \mathbf{u}'' \, dV + \int_{\Sigma_2} \mathbf{t}' \cdot \mathbf{u}'' \, d\Sigma = \int_V \rho \mathbf{f}'' \cdot \mathbf{u}' \, dV + \int_{\Sigma_2} \mathbf{t}'' \cdot \mathbf{u}' \, d\Sigma.$$

Exercise 6.10. Consider two equilibrium problems for a body under two systems of external forces, as in the previous exercise. This time, however, two distinct displacement fields are prescribed over Σ_1:

$$\mathbf{u}'\big|_{\Sigma_1} = \mathbf{a}', \qquad \mathbf{u}''\big|_{\Sigma_1} = \mathbf{a}''. \tag{6.43}$$

As above, denote the solutions of the corresponding equilibrium problems by $\mathbf{u}'$ and $\mathbf{u}''$. Denote the solutions of the equilibrium problems for a body free of external forces, that satisfy the corresponding conditions (6.43), by $\mathbf{u}'_0$ and $\mathbf{u}''_0$, respectively. Finally, introduce

$$\tilde{\mathbf{u}}' = \mathbf{u}' - \mathbf{u}'_0, \qquad \tilde{\mathbf{u}}'' = \mathbf{u}'' - \mathbf{u}''_0.$$

Prove that Betti's equality takes the form

$$\int_V \rho \mathbf{f}' \cdot \tilde{\mathbf{u}}'' \, dV + \int_{\Sigma_2} \mathbf{t}' \cdot \tilde{\mathbf{u}}'' \, d\Sigma = \int_V \rho \mathbf{f}'' \cdot \tilde{\mathbf{u}}' \, dV + \int_{\Sigma_2} \mathbf{t}'' \cdot \tilde{\mathbf{u}}' \, d\Sigma.$$

6.10 Minimum Total Energy Principle

On a curved surface, a ball in equilibrium takes the lowest position; we call the equilibrium *stable*. In elementary physics, it is said that at such an equilibrium point, the potential energy of the ball takes a minimum value. A stationary point of potential energy corresponds to an equilibrium position of the ball as well, but such an equilibrium may be unstable. An equilibrium that corresponds to a non-minimum point normally is unstable. The stability result for the potential energy of a particle was first extended to classical mechanics by Lagrange. The minimum potential energy principle is now one of the most important principles of physics, holding also for distributed systems like elastic bodies. However, it is not always straightforward to formulate, since we must find an expression for the potential energy and prove that it takes a minimum value in equilibrium.

As the expression for the energy of an elastic body under load, we propose the *total energy*

$$E = \int_V W(\varepsilon)\, dV - \int_V \rho \mathbf{f} \cdot \mathbf{u}\, dV - \int_{\Sigma_2} \mathbf{t}^0 \cdot \mathbf{u}\, d\Sigma.$$

The first term is the strain energy associated with deformation. If we slowly decrease the external forces acting on the body to zero, the strain energy is the maximum work that the body can produce. The other terms, with negative signs, represent the work of external forces on the displacement field of the body. These terms are analogous to the expression for the work of the gravitational force acting on a moving particle, and are so similar to the gravitational potential that they may be called potential energy terms. Again, we must prove that E really takes its minimum value when the body is in equilibrium. Clearly, we should explain what is meant by "the minimum value of E" and introduce the tools necessary to establish the minimum principle.

The reader is familiar with the definition of a local minimum point for a function in n variables: there should exist a neighborhood of the point in which all the values of the function are no less than its value at the point. For a global minimum point, this value should be no greater than the values of the function at any point. These definitions can be extended to quantities like E, as they take values in the set of real numbers. We could call E a function, but because it depends on the vector function $\mathbf{u}$, it is called a *functional*. A functional is a correspondence that takes its argument, which can be a whole function as it is here, to at most one real number.

To extend the idea of minimum to functionals, we should treat the notions of independent variable, the domain of such a variable, and the neighborhood of a "point" (which is now a vector function $\mathbf{u}$ in the domain). For simplicity, we will assume the domain of E consists of vector functions possessing all continuous derivatives in V up to the second order. Moreover, we suppose that for any $\mathbf{u}$, the kinematic boundary conditions hold. Such vector functions will be called *admissible*. To define a neighborhood, we must introduce a metric or norm on the set of admissible vector functions. The reader interested in the formalities should consult standard books on the calculus of variations. Our present approach will be essentially that of the pioneers of the subject, which was to obtain results without worrying too much about formal justification. In fact, we can completely avoid the question about neighborhoods, since for the problems under consideration, in the equilibrium state E takes its global minimum value.

Now we will describe the technical tools that permit us to seek the minimum points of a functional. We wish to derive for E some analogue of Fermat's theorem for a differentiable function in one variable: the first derivative must vanish at a minimum point. Fermat's theorem extends to the theory of functions in many variables: all the first partial derivatives of the function must vanish. Let us review how this extension is accomplished.

Let f be an ordinary real-valued function of a vector variable, and suppose f takes its minimum at a point $\mathbf{x} \in \mathbb{R}^n$. If $\mathbf{a}$ is any vector and τ is a real parameter, then $f(\mathbf{x} + \tau \mathbf{a})$ can be regarded as a function of τ that takes its minimum value at $\tau = 0$. Therefore, at $\tau = 0$ its derivative with respect to τ must vanish:

$$\left. \frac{df(\mathbf{x} + \tau \mathbf{a})}{d\tau} \right|_{\tau=0} = 0. \tag{6.44}$$

At a minimum point $\mathbf{x}$, this must hold for any $\mathbf{a}$. As the reader is aware, the expression on the left is the directional derivative of f at $\mathbf{x}$ along the direction of $\mathbf{a}$. Moreover, if we formally set $\mathbf{a} = d\mathbf{x}$, then the left member takes the form of the first differential df at point $\mathbf{x}$:

$$df = \left. \frac{df(\mathbf{x} + \tau \, d\mathbf{x})}{d\tau} \right|_{\tau=0} = \nabla f|_{\mathbf{x}} \cdot d\mathbf{x}.$$

Putting $\mathbf{a} = \mathbf{i}_k$ in (6.44), we obtain Fermat's theorem for a function in n variables: the partial derivative of f with respect to x_k at $\mathbf{x}$ must vanish if $\nabla f|_{\mathbf{x}}$ exists and $\mathbf{x}$ is a minimum point of f. We emphasize that this is merely a necessary condition. If at some point $\mathbf{x}$ equation (6.44) holds, then $\mathbf{x}$ may not be a minimum (or a maximum) point. Any $\mathbf{x}$ satisfying (6.44) is called a *stationary point* of f. It could be a minimum point, a maximum point, or a saddle point.

This procedure extends to functionals in a straightforward manner. Suppose $\mathbf{u}$ is a minimum point of E. Let us fix some $\delta\mathbf{u}$ such that $\mathbf{u} + \tau \, \delta\mathbf{u}$ is an admissible displacement for all small τ. Substitute this into E. Then $E(\mathbf{u} + \tau \, \delta\mathbf{u})$ is a function of the real variable τ that takes its minimum value at $\tau = 0$. If the derivative of this function with respect to τ exists, then it vanishes at $\tau = 0$:

$$\delta E = \left. \frac{d}{d\tau} E(\mathbf{u} + \tau \, \delta\mathbf{u}) \right|_{\tau=0} = 0$$

for any $\delta\mathbf{u}$ such that $\mathbf{u} + \tau \, \delta\mathbf{u}$ is admissible for any small τ.

We call δE the *first variation* of the functional E. Its zeros are called *stationary points* of E; they are analogous to the stationary points of an ordinary function f. The equation we obtained is a necessary condition for $\mathbf{u}$ to be a minimum point of E.

We assumed the displacements $\mathbf{u} + \tau\,\delta\mathbf{u}$ were admissible for all small τ. Clearly, if all derivatives of $\mathbf{u}$ and $\delta\mathbf{u}$ up to order two are continuous in V, the same is true of $\mathbf{u} + \tau\,\delta\mathbf{u}$. We required $\mathbf{u}$ to satisfy the kinematic restrictions on Σ_1. Hence if $\delta\mathbf{u}$ satisfies

$$\delta\mathbf{u}\big|_{\Sigma_1} = \mathbf{0},$$

then $\mathbf{u} + \tau\,\delta\mathbf{u}$ also satisfies the kinematic restrictions for any τ. In this case $\delta\mathbf{u}$ is called a *virtual displacement*.

We note that the definition of δE can be applied to any sufficiently smooth E. (We leave this on an intuitive level, as a formal definition would require a background in functional analysis.) It is called the *Gâteaux derivative* of the functional. In the notation δE, we regard the symbol δ as some action on E given by a formula similar to the one used to find the differential of a function. By $\delta\mathbf{u}$ we denote a virtual displacement; this could be denoted by $\mathbf{v}$ as well, but we would have to keep clarifying that it is a *virtual* displacement. That is, in this case the symbol δ is merely a notation; it is not an operation and is used for historical reasons. When we apply the operation δ to an ordinary function, it coincides with the operation of taking its first differential. The differentials of variables in this case are written with the use of the symbol δ instead of d; for example, we have $\delta(x^2) = 2x\,\delta x$.

Using the definition, let us find δE for our problem:

$$\delta E = \int_V \delta W(\boldsymbol{\varepsilon})\,dV - \int_V \rho\mathbf{f} \cdot \delta\mathbf{u}\,dV - \int_{\Sigma_2} \mathbf{t}^0 \cdot \delta\mathbf{u}\,d\Sigma.$$

Calculating δW in the integrand and recalling that for intermediate steps we can use the formulas for the first differential, we get

$$\delta W(\boldsymbol{\varepsilon}) = \frac{1}{2}\delta(\boldsymbol{\varepsilon} \cdot\cdot \mathbf{C} \cdot\cdot \boldsymbol{\varepsilon}) = \boldsymbol{\varepsilon} \cdot\cdot \mathbf{C} \cdot\cdot \delta\boldsymbol{\varepsilon} = \boldsymbol{\sigma} \cdot\cdot \delta\boldsymbol{\varepsilon},$$

where

$$\delta\boldsymbol{\varepsilon} = \boldsymbol{\varepsilon}(\delta\mathbf{u}) = \frac{1}{2}\left(\nabla\delta\mathbf{u} + (\nabla\delta\mathbf{u})^T\right).$$

It follows that

$$\delta E = \int_V \boldsymbol{\sigma} \cdot\cdot \delta\boldsymbol{\varepsilon}\,dV - \int_V \rho\mathbf{f} \cdot \delta\mathbf{u}\,dV - \int_{\Sigma_2} \mathbf{t}^0 \cdot \delta\mathbf{u}\,d\Sigma.$$

Now we wish to find a relationship between a solution to the equilibrium problem for an elastic body and the minimum problem for E. We start with the following theorem.

Theorem 6.5. *A stationary point* $\mathbf{u}$ *of* E *on the set of admissible displacements satisfies the equilibrium equations of the elastic body in the volume* V *and the boundary condition* $\mathbf{n} \cdot \boldsymbol{\sigma}|_{\Sigma_2} = \mathbf{t}^0$, *and conversely.*

Proof. First we prove that $\mathbf{u}$, a solution of the equilibrium problem for an elastic body under load, is a stationary point of E; that is,

$$\delta E = 0$$

when $\mathbf{u}$ is a solution and $\delta\mathbf{u}$ is an arbitrary virtual displacement. Assume $\mathbf{u}$ satisfies

$$\nabla \cdot \boldsymbol{\sigma} + \rho\mathbf{f} = \mathbf{0} \text{ in } V, \qquad \mathbf{u}\big|_{\Sigma_1} = \mathbf{u}^0, \quad \mathbf{n} \cdot \boldsymbol{\sigma}\big|_{\Sigma_2} = \mathbf{t}^0. \qquad (6.45)$$

Dot-multiply the equilibrium equation by an admissible $\delta\mathbf{u}$ and integrate over V. Then apply the Gauss–Ostrogradsky theorem:

$$0 = \int_V [(\nabla \cdot \boldsymbol{\sigma}) \cdot \delta\mathbf{u} + \rho\mathbf{f} \cdot \delta\mathbf{u}] \, dV$$

$$= \int_V [-\boldsymbol{\sigma} \cdot\cdot (\nabla\delta\mathbf{u})^T + \rho\mathbf{f} \cdot \delta\mathbf{u}] \, dV + \int_\Sigma \mathbf{n} \cdot \boldsymbol{\sigma} \cdot \delta\mathbf{u} \, d\Sigma$$

$$= \int_V [-\boldsymbol{\sigma} \cdot\cdot \delta\boldsymbol{\varepsilon} + \rho\mathbf{f} \cdot \delta\mathbf{u}] \, dV + \int_{\Sigma_2} \mathbf{t}^0 \cdot \delta\mathbf{u} \, d\Sigma$$

$$= -\delta E.$$

We used the fact that $\delta\mathbf{u} = \mathbf{0}$ on Σ_1 and $\mathbf{n} \cdot \boldsymbol{\sigma} = \mathbf{t}^0$ on Σ_2. Hence $\mathbf{u}$ is a stationary point of E.

Now we prove the converse statement. Let $\mathbf{u}$ be a stationary point of E, so that it satisfies the equation $\delta E = 0$ for any virtual $\delta\mathbf{u}$ being sufficiently smooth, and let $\mathbf{u}|_{\Sigma_1} = \mathbf{u}^0$. We will show that $\mathbf{u}$ satisfies the equilibrium equations in V as well as the contact condition $\mathbf{n} \cdot \boldsymbol{\sigma} = \mathbf{t}^0$ on Σ_2. Indeed, doing the above calculations in reverse order, we get

$$
\begin{aligned}
\delta E &= \int_V \boldsymbol{\sigma} \cdot\cdot \, \delta\boldsymbol{\varepsilon} \, dV - \int_V \rho \mathbf{f} \cdot \delta\mathbf{u} \, dV - \int_{\Sigma_2} \mathbf{t}^0 \cdot \delta\mathbf{u} \, d\Sigma \\
&= \int_V \boldsymbol{\sigma} \cdot\cdot \, (\nabla \delta\mathbf{u})^T \, dV - \int_V \rho \mathbf{f} \cdot \delta\mathbf{u} \, dV - \int_{\Sigma_2} \mathbf{t}^0 \cdot \delta\mathbf{u} \, d\Sigma \\
&= -\int_V [(\nabla \cdot \boldsymbol{\sigma}) \cdot \delta\mathbf{u} + \rho\mathbf{f} \cdot \delta\mathbf{u}] \, dV + \int_\Sigma \mathbf{n} \cdot \boldsymbol{\sigma} \cdot \delta\mathbf{u} \, d\Sigma - \int_{\Sigma_2} \mathbf{t}^0 \cdot \delta\mathbf{u} \, d\Sigma \\
&= -\int_V [\nabla \cdot \boldsymbol{\sigma} + \rho\mathbf{f}] \cdot \delta\mathbf{u} \, dV \\
&\quad + \int_{\Sigma_1} \mathbf{n} \cdot \boldsymbol{\sigma} \cdot \delta\mathbf{u} \, d\Sigma - \int_{\Sigma_2} [\mathbf{n} \cdot \boldsymbol{\sigma} - \mathbf{t}^0] \cdot \delta\mathbf{u} \, d\Sigma.
\end{aligned}
$$

Because $\delta\mathbf{u} = 0$ on Σ_1, we finally obtain

$$
\delta E = -\int_V [\nabla \cdot \boldsymbol{\sigma} + \rho\mathbf{f}] \cdot \delta\mathbf{u} \, dV - \int_{\Sigma_2} [\mathbf{n} \cdot \boldsymbol{\sigma} - \mathbf{t}^0] \cdot \delta\mathbf{u} \, d\Sigma = 0. \qquad (6.46)
$$

We assumed $\mathbf{u}$ to be smooth enough that the above integrands are continuous. As admissible $\delta\mathbf{u}$ is arbitrary, from this equality it follows that

$$
\nabla \cdot \boldsymbol{\sigma} + \rho\mathbf{f} = \mathbf{0} \text{ in } V, \qquad \mathbf{n} \cdot \boldsymbol{\sigma}\big|_{\Sigma_2} = \mathbf{t}^0. \qquad (6.47)
$$

The justification of this last step requires standard material from the calculus of variations. We postpone this discussion until the end of the present section. □

Now we would like to prove

Theorem 6.6. *Let* $\mathbf{u}$ *be a solution of the boundary problem* (6.45). *Then it is a point of global minimum of* E.

Proof. The statement is a consequence of the fact that W is a positive quadratic form with respect to $\boldsymbol{\varepsilon}$. Indeed, let $\mathbf{v}$ be an arbitrary admissible displacement so that it is smooth enough and satisfies the kinematic condition $\mathbf{v}|_{\Sigma_1} = \mathbf{u}^0$. Consider the difference

$$
\Delta E = E(\mathbf{v}) - E(\mathbf{u}).
$$

We have

$$\Delta E = \int_V W(\varepsilon(\mathbf{v}))\, dV - \int_V \rho \mathbf{f} \cdot \mathbf{v}\, dV - \int_{\Sigma_2} \mathbf{t}^0 \cdot \mathbf{v}\, d\Sigma$$
$$- \int_V W(\varepsilon(\mathbf{u}))\, dV + \int_V \rho \mathbf{f} \cdot \mathbf{u}\, dV + \int_{\Sigma_2} \mathbf{t}^0 \cdot \mathbf{u}\, d\Sigma$$
$$= \int_V [W(\varepsilon(\mathbf{v})) - W(\varepsilon(\mathbf{u}))]\, dV$$
$$- \int_V \rho \mathbf{f} \cdot (\mathbf{v} - \mathbf{u})\, dV - \int_{\Sigma_2} \mathbf{t}^0 \cdot (\mathbf{v} - \mathbf{u})\, d\Sigma.$$

Let $\mathbf{w} = \mathbf{v} - \mathbf{u}$. Because $\mathbf{u}$ and $\mathbf{v}$ coincide on Σ_1, we have $\mathbf{w}|_{\Sigma_1} = \mathbf{0}$. Next,

$$2\,[W(\varepsilon(\mathbf{v})) - W(\varepsilon(\mathbf{u}))] = \varepsilon(\mathbf{v}) \cdots \mathbf{C} \cdots \varepsilon(\mathbf{v}) - \varepsilon(\mathbf{u}) \cdots \mathbf{C} \cdots \varepsilon(\mathbf{u})$$
$$= \varepsilon(\mathbf{w}) \cdots \mathbf{C} \cdots \varepsilon(\mathbf{w}) + 2\varepsilon(\mathbf{u}) \cdots \mathbf{C} \cdots \varepsilon(\mathbf{w}).$$

Therefore

$$\Delta E = \frac{1}{2} \int_V \varepsilon(\mathbf{w}) \cdots \mathbf{C} \cdots \varepsilon(\mathbf{w})\, dV$$
$$+ \int_V \varepsilon(\mathbf{u}) \cdots \mathbf{C} \cdots \varepsilon(\mathbf{w})\, dV - \int_V \rho \mathbf{f} \cdot \mathbf{w}\, dV - \int_{\Sigma_2} \mathbf{t}^0 \cdot \mathbf{w}\, d\Sigma.$$

The sum of the terms in the second line constitutes δE at $\mathbf{u}$ with $\delta \mathbf{u} = \mathbf{w}$, which is a virtual displacement, so this sum is zero. Thus we have

$$\Delta E = \frac{1}{2} \int_V \varepsilon(\mathbf{w}) \cdots \mathbf{C} \cdots \varepsilon(\mathbf{w})\, dV = \int_V W(\varepsilon(\mathbf{w}))\, dV. \qquad (6.48)$$

As W is a positive definite form, $\Delta E \geq 0$ for any admissible $\mathbf{v}$. Hence $\mathbf{u}$ is a global minimizer of E. □

The two preceding theorems can be summarized as *Lagrange's variational principle*.

Theorem 6.7. *The solution of the equilibrium problem for an elastic body is equivalent to the problem of minimizing the total energy functional E over the set of all kinematically admissible smooth displacement fields.*

To complete the proof of Theorem 6.5, we must show how to obtain (6.47) from (6.46). It will suffice to do this for one of the component equations. Denoting the kth component of the expressions in the brackets in (6.46) by F and f respectively, the kth component of $\delta \mathbf{u}$ by δu, and setting the remaining components of $\delta\mathbf{u}$ to zero, we reduce equation (6.46) to

$$\int_V F \, \delta u \, dV + \int_{\Sigma_2} f \, \delta u \, d\Sigma = 0. \tag{6.49}$$

Equation (6.47) follows from the next theorem.

Theorem 6.8. *Let F be continuous in V and let f be continuous on Σ_2. Suppose (6.49) holds for all sufficiently smooth functions δu that vanish on Σ_1. Then*

$$F = 0 \; in \; V, \qquad f = 0 \; on \; \Sigma_2.$$

Proof. The proof is done in two steps. We begin by restricting the set of all admissible δu to those that vanish on the whole boundary Σ. Then (6.49) becomes

$$\int_V F \, \delta u \, dV = 0.$$

Suppose, contrary to the theorem statement, that $F(\mathbf{x}_*) = a \neq 0$ at some point $\mathbf{x}_*$. Without loss of generality we take $\mathbf{x}_*$ to be an interior point of V and $a > 0$. By continuity of F there is an open ball B_r with center $\mathbf{x}_*$ and radius $r > 0$ that lies inside V such that $F(\mathbf{x}) > a/2 > 0$ on B_r. Now if we take a smooth function δu_0 that is positive on B_r and zero outside B_r, we get

$$\int_V F \, \delta u_0 \, dV > 0,$$

which contradicts the above equality that must hold for all δu that vanish on the boundary of V. It is clear that such a function δu_0 exists; the reader can find examples of "bell-shaped functions" in textbooks on the calculus of variations, cf. [Lebedev and Cloud (2003)]. So $F = 0$ in V.

The second step is to prove that $f = 0$ on Σ_2. For this we return to (6.49). Because $F = 0$, the first integral is zero for any admissible δu. So we find that

$$\int_{\Sigma_2} f \, \delta u \, d\Sigma = 0$$

holds for all admissible δu that take arbitrary values on Σ_2. But now we are in a position similar to the proof for F. We can repeat that proof, but on a two-dimensional domain Σ_2, so $f = 0$ on Σ_2. $\qquad\square$

A reader familiar with the calculus of variations may recognize the above procedure as leading to one version of the "main lemma" of that subject.

It is worth noting that in the theory of elasticity there are several variational principles. Some are of minimax type; others are of stationary type. We can mention principles that carry the names of Castigliano, Reissner, Tonti, Hamilton, etc.

6.11 Ritz's Method

The minimum total energy principle is of great physical importance. It is also the basis for introducing the generalized setup of the equilibrium problems, which in turn defines *weak solutions* in elasticity. These solutions have finite strain energies. The principle is even more important from a practical standpoint. The variational equation $\delta E = 0$ is the basis for various numerical methods, including the Ritz and Galerkin methods, the variational finite difference methods, and the finite element methods. Minimality of the total energy on the solution warrants stability of the algorithms of these methods. For details, the reader should consult specialized literature. Here we present Ritz's method, which spawned the other methods mentioned above.

The minimum total energy principle states that a solution of the equilibrium problem

$$\nabla \cdot \boldsymbol{\sigma} + \rho \mathbf{f} = \mathbf{0} \text{ in } V, \qquad \mathbf{u}\big|_{\Sigma_1} = \mathbf{u}^0, \qquad \mathbf{n} \cdot \boldsymbol{\sigma}\big|_{\Sigma_2} = \mathbf{t}^0,$$

expressed in terms of $\mathbf{u}$, is a point of global minimum of the total energy functional

$$E = \int_V W(\boldsymbol{\varepsilon}) \, dV - \int_V \rho \mathbf{f} \cdot \mathbf{u} \, dV - \int_{\Sigma_2} \mathbf{t}^0 \cdot \mathbf{u} \, d\Sigma$$

on the set of admissible displacements. The converse statement is also valid, hence by the uniqueness theorem this point of minimum is unique. Walter Ritz (1878–1909) thought that by finding a minimum point on some subset of admissible displacements, we can get an approximation to the solution. The decisive idea was how to select this subset in such a way that solving the approximate minimum problem would be relatively easy. Ritz proposed to minimize E on the set of displacements of the form

$$\mathbf{u}_N = \sum_{k=1}^N u_k \boldsymbol{\varphi}_k + \mathbf{u}^*, \qquad \mathbf{u}^*\big|_{\Sigma_1} = \mathbf{u}^0, \qquad \boldsymbol{\varphi}_n\big|_{\Sigma_1} = \mathbf{0} \qquad (6.50)$$

with some fixed $\boldsymbol{\varphi}_k$ and $\mathbf{u}^*$, by determining the numerical coefficients u_k. The first step is to find a sufficiently smooth vector function $\mathbf{u}^*$ that satisfies the condition $\mathbf{u}^*|_{\Sigma_1} = \mathbf{u}^0$. Then we choose some set of basis elements $\boldsymbol{\varphi}_k$ that vanish on Σ_1. Clearly, any $\mathbf{u}_N$ satisfies the kinematic condition $\mathbf{u}_N|_{\Sigma_1} = \mathbf{u}^0$.

In Ritz's time, when calculations where done manually, the selection of the basis was extremely important; an engineer needed to find a few coefficients u_k in a reasonable time. Now, with powerful computers, the number of basis elements can be large. One requirement for the basis elements is that they constitute a linearly independent set. Another is that using this basis, we can actually approximate a solution with satisfactory precision. We shall not pursue this issue, as it requires a serious excursion into mathematics.

Let us formulate *Ritz's minimization problem for the Nth approximation*:

Minimize E over the set of all $\mathbf{u}_N$ satisfying (6.50).

That is, find

$$\tilde{\mathbf{u}}_N = \sum_{k=1}^{N} \tilde{u}_k \boldsymbol{\varphi}_k + \mathbf{u}^*$$

such that

$$E(\tilde{\mathbf{u}}_N) \leq E(\mathbf{u}_N) \text{ for any } \mathbf{u}_N \text{ from the set } (6.50),$$

where

$$E(\mathbf{u}_N) = \int_V W(\boldsymbol{\varepsilon}(\mathbf{u}_N))\,dV - \int_V \rho \mathbf{f} \cdot \mathbf{u}_N\,dV - \int_{\Sigma_2} \mathbf{t}^0 \cdot \mathbf{u}_N\,d\Sigma.$$

Because $\mathbf{u}^*$ is fixed, we should minimize $E(\mathbf{u}_N)$ on the N-dimensional linear space spanned by the elements $\boldsymbol{\varphi}_1, \ldots, \boldsymbol{\varphi}_N$. Hence we must find the values of the real coefficients $\tilde{u}_1, \ldots, \tilde{u}_N$. From the minimum theorem on the whole set of admissible displacements, it follows that $E(\mathbf{u}_N)$ has a minimum point in the spanned space.

Now we need a practical way of finding the coefficients of the minimizer of E. Let us write out the expression for $E(\mathbf{u}_N)$ in detail. We introduce the following notation:

$$\langle \mathbf{u}, \mathbf{v} \rangle = \int_V \boldsymbol{\varepsilon}(\mathbf{u}) \cdots \mathbf{C} \cdots \boldsymbol{\varepsilon}(\mathbf{v})\,dV. \tag{6.51}$$

Note that $\langle \mathbf{u}, \mathbf{v} \rangle = \langle \mathbf{v}, \mathbf{u} \rangle$, and that $\langle \mathbf{u}, \mathbf{u} \rangle$ is twice the strain energy for the displacement field $\mathbf{u}$. This form $\langle \mathbf{u}, \mathbf{v} \rangle$ can be considered as an inner product on any linear space of sufficiently smooth $\mathbf{u}$ such that $\mathbf{u}|_{\Sigma_1} = \mathbf{0}$, for which the expression $\langle \mathbf{u}, \mathbf{v} \rangle$ makes sense.

Next we introduce

$$f_m = \int_V \rho \mathbf{f} \cdot \boldsymbol{\varphi}_m \, dV + \int_{\Sigma_2} \mathbf{t}^0 \cdot \boldsymbol{\varphi}_m \, d\Sigma - \langle \mathbf{u}^*, \boldsymbol{\varphi}_m \rangle$$

and

$$F = \int_V \rho \mathbf{f} \cdot \mathbf{u}^* \, dV + \int_{\Sigma_2} \mathbf{t}^0 \cdot \mathbf{u}^* \, d\Sigma - \frac{1}{2} \langle \mathbf{u}^*, \mathbf{u}^* \rangle.$$

In this notation we have

$$E(\mathbf{u}_N) = \frac{1}{2} \sum_{m=1}^{N} \sum_{n=1}^{N} u_m u_n \langle \boldsymbol{\varphi}_m, \boldsymbol{\varphi}_n \rangle - \sum_{n=1}^{N} f_n u_n - F.$$

So $E(\mathbf{u}_N)$ is a quadratic function in the N variables u_n, and we can apply standard tools from calculus.

At the point of minimum of $E(\mathbf{u}_N)$, the simultaneous equations

$$\frac{\partial E(\mathbf{u}_N)}{\partial u_m} = 0 \qquad (m = 1, \ldots, N)$$

must hold. Writing these as

$$\sum_{n=1}^{N} u_n \langle \boldsymbol{\varphi}_n, \boldsymbol{\varphi}_m \rangle = f_m \qquad (m = 1, \ldots, N), \tag{6.52}$$

we have a system of linear algebraic equations. The matrix of the system is

$$\mathbf{A} = \begin{pmatrix} \langle \boldsymbol{\varphi}_1, \boldsymbol{\varphi}_1 \rangle & \langle \boldsymbol{\varphi}_2, \boldsymbol{\varphi}_1 \rangle & \langle \boldsymbol{\varphi}_3, \boldsymbol{\varphi}_1 \rangle & \cdots & \langle \boldsymbol{\varphi}_N, \boldsymbol{\varphi}_1 \rangle \\ \langle \boldsymbol{\varphi}_1, \boldsymbol{\varphi}_2 \rangle & \langle \boldsymbol{\varphi}_2, \boldsymbol{\varphi}_2 \rangle & \langle \boldsymbol{\varphi}_3, \boldsymbol{\varphi}_2 \rangle & \cdots & \langle \boldsymbol{\varphi}_N, \boldsymbol{\varphi}_2 \rangle \\ \langle \boldsymbol{\varphi}_1, \boldsymbol{\varphi}_3 \rangle & \langle \boldsymbol{\varphi}_2, \boldsymbol{\varphi}_3 \rangle & \langle \boldsymbol{\varphi}_3, \boldsymbol{\varphi}_3 \rangle & \cdots & \langle \boldsymbol{\varphi}_N, \boldsymbol{\varphi}_3 \rangle \\ \vdots & \vdots & \vdots & \ddots & \vdots \\ \langle \boldsymbol{\varphi}_1, \boldsymbol{\varphi}_N \rangle & \langle \boldsymbol{\varphi}_2, \boldsymbol{\varphi}_N \rangle & \langle \boldsymbol{\varphi}_3, \boldsymbol{\varphi}_N \rangle & \cdots & \langle \boldsymbol{\varphi}_N, \boldsymbol{\varphi}_N \rangle \end{pmatrix}.$$

Its determinant $\det(\mathbf{A})$ is Gram's determinant. In linear algebra, it is shown that for a linearly independent system of elements the Gram determinant is nonzero and conversely. But the elements $\boldsymbol{\varphi}_1, \ldots, \boldsymbol{\varphi}_N$ are linearly independent by assumption, hence (6.52) has a unique solution.

Now let us touch on the question of convergence for Ritz's approximations. Our use of the term "approximations" does not, by itself, answer the question whether $\mathbf{u}_N$ is really close to the solution.

We shall reformulate the problem. Let $\mathbf{u}$ be a solution to the equilibrium problem under consideration. Then the Nth approximation of Ritz's method minimizes the functional $E(\mathbf{u}_N) - E(\mathbf{u})$ as well. But by the formula (6.48) in the proof of Theorem 6.6, we have

$$E(\mathbf{u}_N) - E(\mathbf{u}) = \int_V W(\varepsilon(\mathbf{u}_N - \mathbf{u})) \, dV. \qquad (6.53)$$

Because W is a positive definite form, the minimization procedure seems to be convergent if we can approximate *any* admissible displacement $\mathbf{u}$ with $\mathbf{u}_N$. The sense in which we should approximate the displacements is given by the above limit $E(\mathbf{u}_N) - E(\mathbf{u}) \to 0$ as $N \to \infty$. So to get an approximation, it suffices to have the set $\boldsymbol{\varphi}_m$ be *complete* in the sense that for any admissible $\mathbf{v}$ such that $\mathbf{v}|_{\Sigma_1} = \mathbf{0}$ and $\varepsilon > 0$, we can find a $\tilde{\mathbf{u}}_n$ such that

$$\int_V W(\varepsilon(\tilde{\mathbf{u}}_n - \mathbf{v})) \, dV < \varepsilon.$$

In practice we can find complete systems, but proof of completeness is not easy.

Having a complete basis set of $\boldsymbol{\varphi}_m$, it seems we could say the following. Because the Nth Ritz approximation is the best approximation of the solution from all the possible approximations, by completeness of the $\boldsymbol{\varphi}_m$ we immediately obtain convergence of the Ritz approximations $\mathbf{u}_N$ to $\mathbf{u}$ in the sense that

$$\int_V W(\varepsilon(\mathbf{u}_N - \mathbf{u})) \, dV \to 0 \quad \text{as } N \to \infty.$$

This argument is, unfortunately, merely plausible.

We can show that for Ritz's approximations the following holds:

$$\int_V W(\varepsilon(\mathbf{u}_N) - \varepsilon(\mathbf{u}_M)) \, dV \to 0 \quad \text{whenever } N, M \to \infty. \qquad (6.54)$$

So $\{\mathbf{u}_N\}$ seems to be a Cauchy sequence in the integral energy sense. Unfortunately, using only classical calculus we cannot conclude that a limit element exists and determine its properties. These issues are discussed in the theory of Sobolev spaces. Engineers who use Ritz's method should understand that they work with smooth approximations that are ordinary functions, but that the limit functions and convergence questions are functional analytic issues; see, for example, [Lebedev and Cloud (2003)].

Often engineers attempt to compensate for a lack of theoretical justification by means of experiment. However, experiments are quite restrictive and cannot provide real justification in all practical cases. They can only support opinion regarding the applicability of a method.

Korn's inequality

For various qualitative questions about solutions of elasticity problems, *Korn's inequality* plays a significant role. It states that a displacement with finite energy belongs to the space $W^{1,2}(V)$; that is, all of its Cartesian components have square-integrable first derivatives. It can also yield an estimate for the displacement field through the expression for the strain energy.

One form of Korn's inequality is

$$\int_V (\mathbf{u} \cdot \mathbf{u} + \nabla \mathbf{u} \cdots \nabla \mathbf{u}^T)\, dV \leq c \int_V W(\boldsymbol{\varepsilon}(\mathbf{u}))\, dV \qquad (6.55)$$

for any sufficiently smooth $\mathbf{u}$ with a constant c that depends only on V and Σ_1, the part of the boundary on which $\mathbf{u}|_{\Sigma_1} = \mathbf{0}$. Because $W(\boldsymbol{\varepsilon}(\mathbf{u}))$ is a positive definite quadratic form in $\boldsymbol{\varepsilon}$ and we are not interested in the exact value of c, we can change $W(\boldsymbol{\varepsilon}(\mathbf{u}))$ to $\mathrm{tr}(\boldsymbol{\varepsilon} \cdot \boldsymbol{\varepsilon})$. A general proof for $\Sigma_1 \neq \Sigma$ is difficult [Ciarlet (1988)], so we establish the inequality when $\mathbf{u}|_{\Sigma} = \mathbf{0}$. We will prove

$$\int_V \mathrm{tr}(\nabla \mathbf{u} \cdot \nabla \mathbf{u}^T)\, dV \leq 2 \int_V \mathrm{tr}(\boldsymbol{\varepsilon} \cdot \boldsymbol{\varepsilon})\, dV, \qquad (6.56)$$

from which (6.55) follows. In Cartesian coordinates this is

$$\int_V \sum_{i,j=1}^{3} \left(\frac{\partial u_i}{\partial u_j} \right)^2 dV \leq 2 \int_V \left(\sum_{i,j=1}^{3} \varepsilon_{ij}^2 \right) dV.$$

Another part of the inequality for u_i follows from *Friedrich's inequality*

$$\int_V \sum_{i=1}^{3} |u_i|^2\, dV \leq c_1 \int_V \sum_{i,j=1}^{3} \left(\frac{\partial u_i}{\partial u_j} \right)^2 dV,$$

which holds with a constant c_1 independent of u_i for any smooth function u_i that vanishes on the boundary. See [Lebedev and Cloud (2003)] for a proof.

So we prove (6.56). We have

$$\int_V \text{tr}(\varepsilon \cdot \varepsilon)\, dV = \frac{1}{2} \int_V \left[\text{tr}(\nabla \mathbf{u} \cdot (\nabla \mathbf{u})^T) + \text{tr}(\nabla \mathbf{u} \cdot \nabla \mathbf{u}) \right] dV$$

$$= \frac{1}{2} \|\nabla \mathbf{u}\|^2 + \frac{1}{2} \int_V \text{tr}(\nabla \mathbf{u} \cdot \nabla \mathbf{u})\, dV.$$

On the boundary, $\mathbf{u} = \mathbf{0}$. Using the Gauss–Ostrogradsky theorem, let us transform the last integral:

$$\int_V \text{tr}(\nabla \mathbf{u} \cdot \nabla \mathbf{u})\, dV = \int_V \frac{\partial u_j}{\partial x_i} \frac{\partial u_i}{\partial x_j}\, dV = -\int_V u_j \frac{\partial^2 u_i}{\partial x_i \partial x_j}\, dV$$

$$= \int_V \frac{\partial u_i}{\partial x_i} \frac{\partial u_j}{\partial x_j}\, dV = \int_V (\nabla \cdot \mathbf{u})^2\, dV.$$

So

$$2\|\varepsilon\|^2 = \|\nabla \mathbf{u}\|^2 + \int_V (\nabla \cdot \mathbf{u})^2\, dV.$$

Inequality (6.56) follows from the positivity of the second term.

6.12 Rayleigh's Variational Principle

The minimum total energy principle, also known as Lagrange's principle, is formulated for equilibrium problems; it cannot be applied to dynamics problems. But there is one important dynamics problem for which there exists a variational principle based on a similar minimization idea. This is the eigenoscillation problem for an elastic body. In this problem we seek solutions to a dynamic homogeneous problem in displacements in the form

$$\mathbf{u} = \mathbf{u}(\mathbf{r}, t) = \mathbf{w}(\mathbf{r})e^{i\omega t}. \tag{6.57}$$

The equations of the dynamical problem are

$$\nabla \cdot \boldsymbol{\sigma} = \rho \ddot{\mathbf{u}} \text{ in } V, \qquad \mathbf{u}\big|_{\Sigma_1} = \mathbf{0}, \quad \mathbf{n} \cdot \boldsymbol{\sigma}\big|_{\Sigma_2} = \mathbf{0}. \tag{6.58}$$

Substituting (6.57) into (6.58) expressed in displacements, and canceling the factor $e^{i\omega t}$, we get

$$\nabla \cdot \boldsymbol{\sigma} = -\rho \omega^2 \mathbf{w} \text{ in } V, \qquad \mathbf{w}\big|_{\Sigma_1} = \mathbf{0}, \quad \mathbf{n} \cdot \boldsymbol{\sigma}\big|_{\Sigma_2} = \mathbf{0}. \tag{6.59}$$

Here $\boldsymbol{\sigma}$ is given by

$$\boldsymbol{\sigma} = \mathbf{C} \cdot\cdot\, \varepsilon, \qquad \varepsilon = \varepsilon(\mathbf{w}) = \frac{1}{2} \left(\nabla \mathbf{w} + (\nabla \mathbf{w})^T \right).$$

Equations (6.59) constitute an *eigenvalue problem*: we must find positive values ω, known as *eigenfrequencies* of the elastic body, for which (6.59) has a nontrivial solution $\mathbf{w}$, called an *eigenoscillation*.

It can be shown that the problem has only nonnegative eigenfrequencies, for which $\mathbf{w}$ has only real components, and that the set of eigenfrequencies is countable. To demonstrate that the set of eigenfrequencies is countable, and to show that a complete set of linearly independent eigenmodes constitutes a basis in the space of modes having finite energy, we need techniques that fall outside the scope of this book. Note that these proofs require V to be a compact volume with a sufficiently smooth boundary. The term "sufficiently smooth" covers the needs of engineering practice: the boundary cannot have cusps, but V can be a pyramid or a cone.

Now we show that any eigenfrequency ω is real and, moreover, nonnegative. Suppose to the contrary ω is a complex number so that its corresponding mode $\mathbf{w}$ is also complex-valued. Dot-multiplying the first equation of (6.59) by $\overline{\mathbf{w}}$ and integrating over V, we get

$$\int_V (\nabla \cdot \boldsymbol{\sigma}) \cdot \overline{\mathbf{w}} \, dV = -\omega^2 \int_V \rho \mathbf{w} \cdot \overline{\mathbf{w}} \, dV.$$

Applying the Gauss–Ostrogradsky theorem to the left side, we get

$$\int_V \boldsymbol{\sigma} \cdot\cdot \, \overline{\boldsymbol{\varepsilon}} \, dV = \omega^2 \int_V \rho \mathbf{w} \cdot \overline{\mathbf{w}} \, dV. \tag{6.60}$$

Because $\boldsymbol{\sigma} \cdot\cdot \, \overline{\boldsymbol{\varepsilon}} \geq 0$ and $\rho \mathbf{w} \cdot \overline{\mathbf{w}} \geq 0$, from (6.60) it follows that ω^2 is a real nonnegative number, hence so is ω. As the eigenfrequency equation is linear with real-valued coefficients, the real and imaginary parts of an eigensolution are eigensolutions as well. This means that we can consider only real eigenmodes.

Engineers are mostly interested in some range of eigenfrequencies, say the lowest few, or those belonging to some finite range.

Let us note that for the first and third boundary value problems, the minimum eigenfrequency is positive. Indeed, for $\omega = 0$ the problem (6.59) is described by the equilibrium equations, and by uniqueness we have only the zero solution $\mathbf{w} = \mathbf{0}$. For the second boundary value problem, where $\Sigma_2 = \Sigma$, to the value $\omega = 0$ there corresponds a nontrivial solution that represents a displacement of the body as a rigid whole:

$$\mathbf{w} = \mathbf{w}_0 + \boldsymbol{\omega}_0 \times (\mathbf{r} - \mathbf{r}_0)$$

with arbitrary but constant vectors $\mathbf{w}_0, \boldsymbol{\omega}_0$, and $\mathbf{r}_0$. We restrict ourselves to the case of positive ω.

By linearity of the problem, if $\mathbf{w}$ is an eigensolution, then so is $a\mathbf{w}$ for any scalar a. We typically choose a so that $a\mathbf{w}$ has unit L^2 norm. Such an eigensolution is called an *oscillation mode*.

We prove the following theorem.

Theorem 6.9. *For oscillation eigenmodes* $\mathbf{w}_1$ *and* $\mathbf{w}_2$ *corresponding to distinct eigenfrequencies* ω_1 *and* ω_2 *respectively, the relation*

$$\int_V \rho \mathbf{w}_1 \cdot \mathbf{w}_2 \, dV = 0 \qquad (6.61)$$

holds. Moreover,

$$\langle \mathbf{w}_1, \mathbf{w}_2 \rangle = 0. \qquad (6.62)$$

Equality (6.61) *is the* **orthogonality relation,** *and* (6.62) *is the* **generalized orthogonality relation** *between* $\mathbf{w}_1$ *and* $\mathbf{w}_2$.

Proof. The proof mimics that of Theorem 3.2. Let $\mathbf{w}_1$ and $\mathbf{w}_2$ satisfy

$$\nabla \cdot \boldsymbol{\sigma}_1 = -\rho \omega_1^2 \mathbf{w}_1 \text{ in } V, \qquad \mathbf{w}_1\big|_{\Sigma_1} = \mathbf{0}, \qquad \mathbf{n} \cdot \boldsymbol{\sigma}_1\big|_{\Sigma_2} = \mathbf{0},$$

and

$$\nabla \cdot \boldsymbol{\sigma}_2 = -\rho \omega_2^2 \mathbf{w}_2 \text{ in } V, \qquad \mathbf{w}_2\big|_{\Sigma_1} = \mathbf{0}, \qquad \mathbf{n} \cdot \boldsymbol{\sigma}_2\big|_{\Sigma_2} = \mathbf{0},$$

where $\boldsymbol{\sigma}_k = \mathbf{C} \cdots \boldsymbol{\varepsilon}_k$ and $\boldsymbol{\varepsilon}_k = \boldsymbol{\varepsilon}(\mathbf{w}_k)$. Dot-multiply the first equation in V by $\mathbf{w}_2$ and integrate this over V. Applying the Gauss–Ostrogradsky theorem, we get

$$-\int_V \boldsymbol{\sigma}_1 \cdots \boldsymbol{\varepsilon}_2 \, dV + \omega_1^2 \int_V \rho \mathbf{w}_1 \cdot \mathbf{w}_2 \, dV = 0. \qquad (6.63)$$

Similarly,

$$-\int_V \boldsymbol{\sigma}_2 \cdots \boldsymbol{\varepsilon}_1 \, dV + \omega_2^2 \int_V \rho \mathbf{w}_2 \cdot \mathbf{w}_1 \, dV = 0.$$

Subtracting these two integral equalities and using $\boldsymbol{\sigma}_1 \cdots \boldsymbol{\varepsilon}_2 = \boldsymbol{\sigma}_2 \cdots \boldsymbol{\varepsilon}_1$, we get

$$(\omega_1^2 - \omega_2^2) \int_V \rho \mathbf{w}_1 \cdot \mathbf{w}_2 \, dV = 0.$$

Relation (6.61) follows from the fact that $\omega_1 \neq \omega_2$. Substituting this into the above equality, we obtain (6.62). $\qquad \square$

Mathematically, Theorem 6.9 states that the oscillation eigenmodes corresponding to distinct eigenfrequencies are orthogonal in the space $L_2(V)$ with weight ρ. From (6.63) it follows that they are also orthogonal with respect to the energy inner product $\langle \cdot, \cdot \rangle$. If to some ω there correspond a few linearly independent eigenmodes, we can always construct an orthonormal set of corresponding eigenmodes using the Gram–Schmidt procedure. See [Yosida (1980)].

This orthogonality is the basis for practical application of the following variational Rayleigh principle.

Theorem 6.10. *Eigenmodes are stationary points of the energy functional*

$$E_0(\mathbf{w}) = \frac{1}{2} \int_V W(\varepsilon(\mathbf{w})) \, dV$$

on the set of displacements satisfying the boundary conditions $\mathbf{w}|_{\Sigma_1} = \mathbf{0}$ *and subject to the constraint*

$$\frac{1}{2} \int_V \rho \mathbf{w} \cdot \mathbf{w} \, dV = 1. \tag{6.64}$$

Conversely, all the stationary points of $E_0(\mathbf{w})$ *on the above set of displacements are eigenmodes of the body that correspond to its eigenfrequencies.*

Proof. Let us write out the stationarity condition for $E_0(\mathbf{w})$. Using the same reasoning as in the proof of Theorem 6.3, we get

$$\delta E_0 = -\int_V (\nabla \cdot \boldsymbol{\sigma}) \cdot \delta \mathbf{w} \, dV - \int_{\Sigma_2} \mathbf{n} \cdot \boldsymbol{\sigma} \cdot \delta \mathbf{w} \, d\Sigma = 0. \tag{6.65}$$

This should be considered on the set of displacements with the restrictions described in the theorem statement. So $\delta \mathbf{w}$ is not independent as it was in Theorem 6.3, but must satisfy the condition

$$\delta \left(\frac{1}{2} \int_V \rho \mathbf{w} \cdot \mathbf{w} \, dV - 1 \right) = \int_V \rho \mathbf{w} \cdot \delta \mathbf{w} \, dV = 0. \tag{6.66}$$

In calculus, the minimization of a function with constraints is solved using Lagrange multipliers. We will adapt this method to our problem for a functional. To $E_0(\mathbf{w})$ let us add

$$-\lambda \left(\frac{1}{2} \int_V \rho \mathbf{w} \cdot \mathbf{w} \, dV - 1 \right)$$

where λ is an indefinite multiplier (we shall call it a Lagrange multiplier).

Let us introduce a functional $\tilde{E}_0(\mathbf{w}, \lambda)$ depending on the variables $\mathbf{w}$ and λ, as

$$\tilde{E}_0(\mathbf{w}, \lambda) = E_0(\mathbf{w}) - \lambda \left(\frac{1}{2} \int_V \rho \mathbf{w} \cdot \mathbf{w} \, dV - 1 \right).$$

Now $\mathbf{w}$ is not subjected to the constraint. We will show that $\mathbf{w}$ from a stationary point $(\mathbf{w}, \lambda)$ of $\tilde{E}_0$ is a stationary point of $E_0(\mathbf{w})$ under the constraint (6.64). Indeed, $\delta \tilde{E}_0 = 0$ is

$$\delta \tilde{E}_0 = \delta E_0 - \lambda \int_V \rho \mathbf{w} \cdot \delta \mathbf{w} \, dV - (\delta \lambda) \left(\frac{1}{2} \int_V \rho \mathbf{w} \cdot \mathbf{w} \, dV - 1 \right) = 0,$$

where δE_0 is given by (6.65). Because $\mathbf{w}$ and λ are independent, $\delta \tilde{E}_0 = 0$ implies two simultaneous equations:

$$\delta E_0 - \lambda \int_V \rho \mathbf{w} \cdot \delta \mathbf{w} \, dV = 0, \qquad \frac{1}{2} \int_V \rho \mathbf{w} \cdot \mathbf{w} \, dV - 1 = 0. \qquad (6.67)$$

This shows the required property possessed by stationary points of $\tilde{E}_0$.

So we consider a stationary point $\tilde{E}_0(\mathbf{w}, \lambda)$ without constraints on $\mathbf{w}$. From the first equation in (6.67) we get

$$-\int_V [(\nabla \cdot \boldsymbol{\sigma}) \cdot \delta \mathbf{w} + \lambda \rho \mathbf{w} \cdot \delta \mathbf{w}] \, dV - \int_{\Sigma_2} \mathbf{n} \cdot \boldsymbol{\sigma} \cdot \delta \mathbf{w} \, d\Sigma = 0. \qquad (6.68)$$

As in the derivation of the minimum total energy principle, it follows from this integral equality for arbitrary $\delta \mathbf{w}$ that

$$\nabla \cdot \boldsymbol{\sigma} = -\rho \lambda \mathbf{w}, \qquad \mathbf{n} \cdot \boldsymbol{\sigma}\big|_{\Sigma_2} = \mathbf{0}.$$

If we change λ to ω^2, this equation coincides with (6.59). This explains the meaning of the Lagrange multiplier: it is equal to the squared eigenvalue ω whose non-negativeness was proved above. Thus the stationarity condition for $E_0(\mathbf{w})$ is valid on eigensolutions of the problem (6.59).

Now we prove the converse. Let $\mathbf{w}$ be a solution of the equation

$$\nabla \cdot \boldsymbol{\sigma} = -\rho \omega^2 \mathbf{w}$$

in V for some ω. We dot-multiply it by $\mathbf{v}$, integrate over V, and apply the Gauss–Ostrogradsky theorem to get

$$0 = \int_V [(\nabla \cdot \boldsymbol{\sigma}) \cdot \mathbf{v} + \omega^2 \rho \mathbf{w} \cdot \mathbf{v}] \, dV$$

$$= \int_V [-\boldsymbol{\sigma} \cdots \boldsymbol{\varepsilon}(\mathbf{v}) + \omega^2 \rho \mathbf{w} \cdot \mathbf{v}] \, dV + \int_{\Sigma_2} \mathbf{n} \cdot \boldsymbol{\sigma} \cdot \mathbf{v} \, d\Sigma, \qquad (6.69)$$

where

$$\varepsilon(\mathbf{v}) = \frac{1}{2}(\nabla\mathbf{v} + (\nabla\mathbf{v})^T).$$

Let us select $\mathbf{v}$ such that

$$\int_V \rho\mathbf{w} \cdot \mathbf{v}\, dV = 0.$$

Denoting $\mathbf{v} = \delta\mathbf{w}$, we transform the second line of (6.69) to $-\delta E_0$ from (6.65). Thus δE_0, which is calculated at the eigenmode, is equal to zero. This completes the proof. $\square$

We will reformulate Rayleigh's principle for ease of application. In the new formulation, one need not stipulate separate integral restrictions on the set of $\mathbf{w}$.

Theorem 6.11. *On the set of admissible vector-functions satisfying the condition* $\mathbf{w}|_{\Sigma_1} = \mathbf{0}$, *the oscillation eigenmodes are stationary points of the functional*

$$R(\mathbf{w}) = \frac{E_0}{K}, \quad where \quad K = \frac{1}{2}\int_V \rho\mathbf{w} \cdot \mathbf{w}\, dV.$$

$R(\mathbf{w})$ *is called* **Rayleigh's quotient.** *Conversely, a normalized stationary point of* $R(\mathbf{w})$ *is an eigenmode that corresponds to some eigenfrequency; the value of* $R(\mathbf{w})$ *on an eigenmode is the squared eigenfrequency.*

Proof.   The proof of the first part follows from the previous proof. Indeed, R does not change when we multiply $\mathbf{w}$ by a constant factor. Selecting such a factor that

$$\frac{1}{2}\int_V \rho\mathbf{w} \cdot \mathbf{w}\, dV = 1,$$

we see that $R = E_0$.

Now we show that for the eigenmode $\mathbf{w}$ corresponding to the eigenfrequency ω, we have

$$R(\mathbf{w}) = \omega^2. \tag{6.70}$$

We dot-multiply (6.59) by $\mathbf{w}$, integrate over V, and apply the Gauss–Ostrogradsky theorem. We get

$$\int_V \boldsymbol{\sigma} \cdots \boldsymbol{\varepsilon}\, dV = \omega^2 \int_V \rho\mathbf{w} \cdot \mathbf{w}\, dV,$$

from which (6.70) follows. $\square$

Equation (6.70) is widely used in mechanics for finding approximate values of eigenfrequencies. After using eigenmode orthogonality to approximate an eigenmode by $\tilde{\mathbf{w}}$, we can approximate the eigenfrequency through the equality

$$\tilde{\omega} = \sqrt{R(\tilde{\mathbf{w}})}.$$

A finite-dimensional analogue of Rayleigh's quotient gave birth to a class of numerical methods for large sparse systems of linear algebraic equations. The interested reader should consult specialized literature.

6.13 Plane Waves

It is harder to obtain an exact solution for a dynamics problem than for an equilibrium problem. However, there is an important dynamics problem that can be solved analytically. This is the problem of plane waves in an unbounded homogeneous anisotropic space that is free of body forces.

Such solutions describe propagation of sound far from the source, say for distant earthquakes or explosions, so they are of significant interest in applications. On the other hand, their theory is relatively simple, and the investigation reduces to an algebraic eigenvalue problem.

A solution of the form

$$\mathbf{u} = U(\mathbf{k} \cdot \mathbf{r} - \omega t)\mathbf{a}$$

is called a *plane wave*. Here $\mathbf{k}$ is a constant vector called the *wave vector*, ω is the frequency, $\mathbf{a}$ is a constant vector, and $U = U(x)$ is an unknown function in one variable $x = \mathbf{k} \cdot \mathbf{r} - \omega t$. The equation

$$\mathbf{k} \cdot \mathbf{r} - \omega t = \text{const}$$

represents a plane propagating in the space $\mathbb{R}^3$. At any instant, the normal to this plane is parallel to $\mathbf{k}$. The propagation velocity of the plane is $c = \omega/|\mathbf{k}|$; this is the *wave velocity*. A solution of this form is constant on each propagating plane. The vector $\mathbf{k}$ defines the direction of wave propagation, whereas $\mathbf{a}$ is the direction of the displacement. For $\mathbf{u}$ we have

$$\nabla\mathbf{u} = U'(\mathbf{k}\cdot\mathbf{r}-\omega t)\mathbf{k}\mathbf{a}, \quad \varepsilon = \frac{1}{2}U'(\mathbf{k}\cdot\mathbf{r}-\omega t)(\mathbf{k}\mathbf{a}+\mathbf{a}\mathbf{k}), \quad \ddot{\mathbf{u}} = \omega^2 U''(\mathbf{k}\cdot\mathbf{r}-\omega t)\mathbf{a},$$

where the prime denotes differentiation with respect to x, where $x = \mathbf{k} \cdot \mathbf{r} - \omega t$, and the overdot denotes differentiation with respect to t.

Using the symmetry properties of $\mathbf{C}$, we write out Hooke's law for this displacement field:

$$\boldsymbol{\sigma} = \mathbf{C} \cdots (\mathbf{ka})U'.$$

Substituting $\ddot{\mathbf{u}}$ and $\boldsymbol{\sigma}$ into the equations of motion, we get

$$\mathbf{k} \cdot \mathbf{C} \cdots (\mathbf{ka})U'' = \rho\omega^2 U'' \mathbf{a}.$$

This equation has the uninteresting trivial solution $U'' = 0$. The necessary nontrivial solutions are defined by the following algebraic problem:

Find a nontrivial solution of the equation

$$\mathbf{A}(\mathbf{k}) \cdot \mathbf{a} = \rho\omega^2 \mathbf{a}, \quad where \ \mathbf{A}(\mathbf{k}) = \mathbf{k} \cdot \mathbf{C} \cdot \mathbf{k}. \tag{6.71}$$

*We call $\mathbf{A}$ the **acoustic tensor**.*

We recall that $\mathbf{C}$ has some symmetry properties, and that $\boldsymbol{\varepsilon} \cdots \mathbf{C} \cdots \boldsymbol{\varepsilon}$ is positive for any $\boldsymbol{\varepsilon} \neq 0$. It is easy to verify that $\mathbf{A}(\mathbf{k})$ is a positive definite symmetric tensor. Considering $\rho\omega^2$ as an eigenvalue of $\mathbf{A}(\mathbf{k})$, we have an eigenvalue problem (6.71) that possesses three positive eigenvalues. In other words, from (6.71) it follows that for any $\mathbf{k}$ there exist three values of ω. A plane wave corresponds to each of these.

Let us consider the plane wave problem for an isotropic medium in more detail. Now

$$\mathbf{C} = \lambda \mathbf{EE} + \mu(\mathbf{e}_k \mathbf{E} \mathbf{e}^k + \mathbf{I}),$$

where $\mathbf{E} = \mathbf{e}_k \mathbf{e}^k$ is the unit tensor and $\mathbf{I} = \mathbf{e}_k \mathbf{e}_m \mathbf{e}^k \mathbf{e}^m$. The acoustic tensor becomes

$$\mathbf{A} = \mathbf{k} \cdot \mathbf{C} \cdot \mathbf{k} = (\lambda + \mu)\mathbf{kk} + \mu(\mathbf{k} \cdot \mathbf{k})\mathbf{E}. \tag{6.72}$$

Let us take a Cartesian frame with basis vectors $\mathbf{i}_1$, $\mathbf{i}_2$, $\tilde{\mathbf{k}} = \mathbf{k}/|\mathbf{k}|$, so $|\mathbf{i}_1| = |\mathbf{i}_2| = |\tilde{\mathbf{k}}| = 1$ and $\mathbf{i}_1 \cdot \mathbf{i}_2 = \mathbf{i}_1 \cdot \tilde{\mathbf{k}} = \mathbf{i}_2 \cdot \tilde{\mathbf{k}} = 0$. In this basis, the matrix $\mathbf{A}(\mathbf{k})$ is diagonal:

$$\mathbf{A}(\mathbf{k}) = \begin{pmatrix} \mu|\mathbf{k}|^2 & 0 & 0 \\ 0 & \mu|\mathbf{k}|^2 & 0 \\ 0 & 0 & (2\mu + \lambda)|\mathbf{k}|^2 \end{pmatrix}.$$

The eigenvalues and eigenvectors from (6.71) take the form

$$\omega_1 = \sqrt{\frac{\mu}{\rho}}|\mathbf{k}| \qquad\qquad \text{for } \mathbf{a} = \mathbf{i}_1,$$

$$\omega_2 = \sqrt{\frac{\mu}{\rho}}|\mathbf{k}| \qquad\qquad \text{for } \mathbf{a} = \mathbf{i}_2,$$

$$\omega_3 = \sqrt{\frac{\lambda + 2\mu}{\rho}}|\mathbf{k}| \qquad\qquad \text{for } \mathbf{a} = \tilde{\mathbf{k}}.$$

The first two solutions describe *transverse waves*; their directions of displacement are those **a** that are perpendicular to **k**, the direction of wave propagation. This type of wave is also called a *shear* or *S wave*. The third equation describes a *longitudinal wave*; its displacement is along the direction of propagation (Fig. 6.7). This type of wave is also called a *dilatational*, *pressure*, or *P wave*. Such solutions are used in acoustics and seismology.

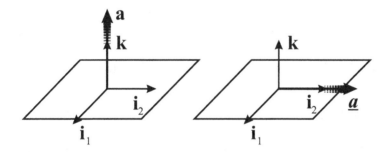

Fig. 6.7   Plane waves in an isotropic medium. Longitudinal wave is on the left. Transverse wave is on the right.

Let us note that for an arbitrary anisotropic medium, the spectral problem describes waves that may be neither transverse nor longitudinal.

Exercise 6.11. Verify (6.72).

Exercise 6.12. Prove that $\mathbf{A}(\mathbf{k})$ is a symmetric tensor.

Exercise 6.13. Show that $\mathbf{A}(\mathbf{k})$ is positive definite.

6.14 Plane Problems of Elasticity

Two important classes of deformation for which elastic problems can be significantly simplified are the *plane strain* and *plane stress* deformations. The problems for these deformations are called the *plane problems of elasticity*.

First we consider the plane deformation problem. A deformation is called plane if all the displacements of the body points are parallel to a plane and depend only on the coordinates in this plane. With reasonable precision, this is the case for a central portion of a long cylindrical or prismatic body stretched by axial forces.

The displacement vector for plane deformation takes the form

$$\mathbf{u} = u_1(x_1, x_2)\mathbf{i}_1 + u_2(x_1, x_2)\mathbf{i}_2.$$

The body and contact forces should be of similar form:

$$\mathbf{f} = f_1(x_1, x_2)\mathbf{i}_1 + f_2(x_1, x_2)\mathbf{i}_2, \qquad \mathbf{t}^0 = t_1(x_1, x_2)\mathbf{i}_1 + t_2(x_1, x_2)\mathbf{i}_2.$$

In a Cartesian frame, the strain tensor components with indices 3, corresponding to $\mathbf{u}$, are zero: $\varepsilon_{13} = \varepsilon_{23} = \varepsilon_{33} = 0$, so

$$\boldsymbol{\varepsilon} = \varepsilon_{11}\mathbf{i}_1\mathbf{i}_1 + \varepsilon_{12}(\mathbf{i}_1\mathbf{i}_2 + \mathbf{i}_2\mathbf{i}_1) + \varepsilon_{22}\mathbf{i}_2\mathbf{i}_2.$$

Using Hooke's law, we find the components of the stress tensor:

$$\sigma_{11} = \lambda \operatorname{tr}\boldsymbol{\varepsilon} + 2\mu\varepsilon_{11}, \qquad \sigma_{22} = \lambda \operatorname{tr}\boldsymbol{\varepsilon} + 2\mu\varepsilon_{22}, \qquad \sigma_{33} = \lambda \operatorname{tr}\boldsymbol{\varepsilon},$$

$$\sigma_{12} = \mu\varepsilon_{12}, \qquad \sigma_{23} = 0, \qquad \sigma_{13} = 0,$$

where $\operatorname{tr}\boldsymbol{\varepsilon} = \varepsilon_{11} + \varepsilon_{22}$. Because $\varepsilon_{33} = 0$, we can express σ_{33} in terms of σ_{11} and σ_{22}:

$$\sigma_{33} = \frac{\lambda}{2(\lambda + \mu)}(\sigma_{11} + \sigma_{22}) = \nu(\sigma_{11} + \sigma_{22}).$$

All the components of $\boldsymbol{\sigma}$ do not depend on x_3; this allows us to simplify the equilibrium equations:

$$\frac{\partial \sigma_{11}}{\partial x_1} + \frac{\partial \sigma_{21}}{\partial x_2} + \rho f_1 = 0, \qquad \frac{\partial \sigma_{12}}{\partial x_1} + \frac{\partial \sigma_{22}}{\partial x_2} + \rho f_2 = 0.$$

Let us consider a particular case when body forces are absent. Introducing the *Airy stress function* Φ via the relations

$$\sigma_{11} = \frac{\partial^2 \Phi}{\partial x_2^2}, \qquad \sigma_{22} = \frac{\partial^2 \Phi}{\partial x_1^2}, \qquad \sigma_{12} = -\frac{\partial^2 \Phi}{\partial x_1 \partial x_2}, \qquad (6.73)$$

we identically satisfy the equilibrium equations. From the set of Beltrami–Michell equations, in the plane problem there remain only three nontrivial equations:

$$\Delta\sigma_{11} + \frac{1}{1+\nu}\frac{\partial^2\sigma}{\partial x_1^2} = 0,$$

$$\Delta\sigma_{22} + \frac{1}{1+\nu}\frac{\partial^2\sigma}{\partial x_2^2} = 0,$$

$$\Delta\sigma_{12} + \frac{1}{1+\nu}\frac{\partial^2\sigma}{\partial x_1\partial x_2} = 0, \tag{6.74}$$

where

$$\sigma = \operatorname{tr}\boldsymbol{\sigma} = (1+\nu)(\sigma_{11} + \sigma_{22}).$$

Substitute (6.73) into (6.74). Since

$$\sigma = (1+\nu)\Delta\Phi,$$

we reduce the first of the equations (6.74) to the biharmonic equation for Φ:

$$0 = \Delta\sigma_{11} + \frac{1}{1+\nu}\frac{\partial^2\sigma}{\partial x_1^2} = \Delta\frac{\partial^2\Phi}{\partial x_2^2} + \frac{\partial^2}{\partial x_1^2}\Delta\Phi = \Delta^2\Phi.$$

We can reduce the second equation from (6.74) to the same biharmonic equation, whereas the third equation from (6.74) holds identically. Thus Φ satisfies

$$\Delta^2\Phi = 0. \tag{6.75}$$

In the plane problem, this plays the role of the compatibility equation; it is the equation we have to solve.

Similar simplifications can be done for the plane stress problem. This type of deformation occurs in a thin plate with planar faces free of load. We appoint the x_3 direction along the normal to the plate faces. Now the stress components σ_{13}, σ_{23}, and σ_{33} are small, so we set them to zero: $\sigma_{13} = \sigma_{23} = \sigma_{33} = 0$. The equality $\sigma_{33} = 0$ distinguishes the plane strain deformation from the above plane deformation where $\sigma_{33} \neq 0$. For the plane stress deformation, the analysis of the equilibrium equations can be performed as above.

Exercise 6.14. Find the components of $\boldsymbol{\varepsilon}$ for the plane stress deformation.

Exercise 6.15. Find the components of the strain tensor for a plane stress state.

The results of Exercises 6.14 and 6.15 show that the equations of the plane strain and plain stress problems differ only in the values of the corresponding elastic moduli λ.

Special tools have been developed for use in plane elasticity theory. Some are based on the theory of functions of a complex variable [Muskhelishvili (1966)]. Using the complex potential method, we can find exact solutions to rather complex problems. However, computer techniques and the finite element method have significantly reduced interest in these.

6.15 Problems

6.1 Write out the boundary conditions for the problem of plane elasticity on the square ABCD shown in Fig. 6.8.

6.2 Write out the boundary conditions for the plane elasticity boundary value problem on the triangle ABC shown in Fig. 6.9.

6.3 Write out the boundary conditions for the plane elasticity problem on the triangle shown in Fig. 6.10. Assume the forces are normal to the sides and equal to p_1, p_2, p_3, respectively.

6.4 Write out the boundary conditions for the plane elasticity problem on the portion of a ring depicted in Fig. 6.11.

6.5 Write out the boundary conditions for the elasticity problem on the visible portions of the cubes shown in Fig. 6.12.

6.6 At a point of an elastic body the principal stresses are $\sigma_1 = 50$ MPa, $\sigma_2 = -50$ MPa, and $\sigma_3 = 75$ MPa. Referring to Fig. 6.13, find the stress vector on ABC that is equi-inclined to the principal axes.

6.7 Referring to Fig. 6.14, write out the boundary conditions for the elasticity problem on the visible portions of the cylindrical bodies.

6.8 Let γ, λ, k be given parameters. For the following displacement vectors, find the expressions for the strain tensors.

(a) $\mathbf{u} = \gamma x_2 \mathbf{i}_1$;

(b) $\mathbf{u} = \lambda x_1 \mathbf{i}_1$;

(c) $\mathbf{u} = \lambda \mathbf{r}$;

(d) $\mathbf{u} = u(r)\mathbf{e}_r + kz\mathbf{e}_z$ (use cylindrical coordinates);

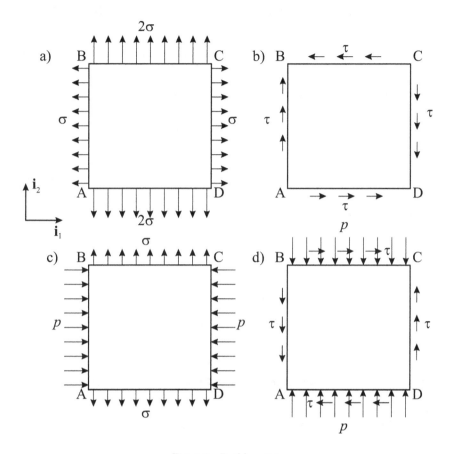

Fig. 6.8   Problem 6.1.

(e) $\mathbf{u} = u(r)\mathbf{e}_\phi + kz\mathbf{e}_z$ (use cylindrical coordinates);

(f) $\mathbf{u} = (u(r) + kz)\mathbf{e}_z$ (use cylindrical coordinates);

(g) $\mathbf{u} = u(r)\mathbf{e}_r$ (use spherical coordinates).

6.9 Let φ be an arbitrary biharmonic function. Verify that the expression

$$2\mu\mathbf{u} = 2(1 - \nu)\nabla^2\varphi - \nabla\nabla\cdot\varphi$$

is a solution of Lamé's equations (6.29) at $\mathbf{f} = \mathbf{0}$. It is the *Boussinesq–Galerkin* solution of the equilibrium problem.

6.10 Let ψ be an arbitrary harmonic vector function and ψ_0 an arbitrary

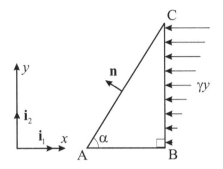

Fig. 6.9   Problem 6.2.

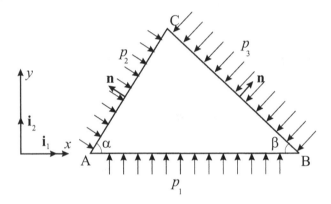

Fig. 6.10   Problem 6.3.

harmonic function. Verify that the expression

$$2\mu\mathbf{u} = 4(1 - \nu)\boldsymbol{\psi} - \nabla(\mathbf{r} \cdot \boldsymbol{\psi} + \psi_0)$$

is a solution of the equilibrium equations (6.29) at $\mathbf{f} = \mathbf{0}$. It is called the *Papkovich–Neuber* solution.

6.11 Let $\boldsymbol{\psi}$ be an arbitrary harmonic vector function, and suppose the function η satisfies the equation

$$\nabla^2\eta = 2\nabla \cdot \boldsymbol{\psi}.$$

Verify that the expression

$$2\mu\mathbf{u} = 4(1 - \nu)\boldsymbol{\psi} - \nabla\eta$$

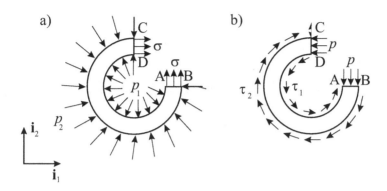

Fig. 6.11   Problem 6.4.

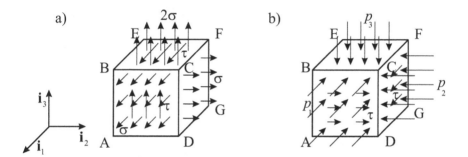

Fig. 6.12   Problem 6.5.

is a solution to (6.29) at $\mathbf{f} = \mathbf{0}$.

6.12 Let ψ be an arbitrary harmonic vector function, and suppose the function ξ satisfies the equation

$$\frac{\partial \xi}{\partial z} = \frac{1}{4\nu - 3} \nabla \cdot \psi.$$

Verify that the expression

$$2\mu \mathbf{u} = \nabla \psi + z \nabla \xi$$

represents a solution to equations (6.29) at $\mathbf{f} = \mathbf{0}$.

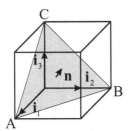

Fig. 6.13   Problem 6.6.

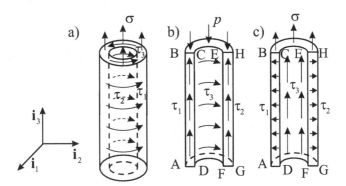

Fig. 6.14   Problem 6.7.

Chapter 7

Linear Elastic Shells

Thin-walled structures are common in engineering practice. Examples include ship hulls, tank beds, rocket and aircraft airframes, and various tubular structures such as arteries, membranes, and vases. A *shell* resembles a surface having thickness. Mathematical models for such objects were developed along with the theory of three-dimensional elasticity. However, shell theory is still under development; it is an approximation to describe the behavior of three-dimensional objects using two-dimensional models given on a surface. Clearly, any shell theory can be accurate only in some restricted number of situations. A few popular shell models exist. We will present some linear models that are used in applications.

Historically, the first shell model was based on the Kirchhoff–Love assumptions. Models are also attributed to Timoshenko, Reissner, Cosserat, Naghdi, and others. In shell theory, we encounter the names of Euler, Lagrange, Poisson, Kelvin, etc. A particular case of shell theory is *plate theory*, where the object under consideration is akin to the top of a desk. In this chapter, some features of shell theory will be demonstrated only for plate models: our goal is an introduction to the theory and a demonstration of how tensor analysis is applied in its construction.

A mathematical model of a shell takes the form of a boundary value problem for simultaneous partial differential equations given on a base surface, usually the midsurface of the shell. Since the quantities describing shell behavior are functions of the surface coordinates, shell theory is closely related to the theory of surfaces from differential geometry (Chapter 5). The decisive point in any version of shell theory is the method of eliminating the third spatial coordinate — along the normal to the shell surface — from the three-dimensional model.

Before the invention of the computer, the numerical solution of three-dimensional elasticity problems was nearly impossible. Much interest was directed toward shell theories, where problems could be solved manually to required accuracy. Modern computer solution has not changed the situation for thin-walled problems. This is because numerical calculations of strains in shell-like bodies harbor difficulties related to huge mesh sizes in discrete models of a shell as a three-dimensional body and the ill-posedness of the corresponding three-dimensional problems. It is interesting to note that the application of powerful computers has increased interest in more accurate shell models which allow engineers to apply two-dimensional finite elements that incorporate a more realistic strain distribution along the shell thickness coordinate.

There are two approaches to the development of two-dimensional shell models. The first is based on direct approximation of a given three-dimensional boundary value problem for a shell-like body in terms of a boundary value problem given on the base surface of the body. This is done in a few ways. One is the *hypothesis method*, in which we assume the form of the dependence of the shell displacement and stress fields on the thickness coordinate. We will present an example of this type of model, based on the classical Kirchhoff–Love hypotheses. Another method involves asymptotic expansion of the solution of a three-dimensional problem with respect to a natural small parameter ε, the ratio of the shell thickness to a characteristic shell size such as the diameter of a circular plate. There are other principles for expanding a solution in a series with respect to ε. These models commonly use as a starting point the equilibrium equations for the shell as a three-dimensional body. Afterwards, the expansions reduce the equations to equations given on the shell surface. The models typically result in a system of equations on the base surface whose order exceeds the order of the initial system for a three-dimensional body.

Another approach to deriving shell models can be called the *direct approach*. In this case a shell is considered as a material surface having additional mechanical properties. So it is a surface that resists deformation, possesses strain energy, can have distributed mass, etc. The dynamical or equilibrium equations for such a surface are formulated directly through fundamental laws of mechanics such as the law of impulse conservation. In the direct approach, the equations of three-dimensional elasticity are not used; hence constitutive laws similar to Hooke's law should be formulated independently.

In this chapter, we will demonstrate how tensor calculus can be used to derive a few linear shell and plate models for the case of small strains. We will also discuss some properties of the boundary value problems of shell theory. We start with the Kirchhoff–Love model.

7.1 Some Useful Formulas of Surface Theory

In this section we collect some tensorial formulas needed to formulate shell relations. Many are taken from Chapter 5, with certain changes in notation.

Let Σ be a sufficiently smooth surface in $\mathbb{R}^3$ (Fig. 7.1). The position vector of a point on Σ is denoted by $\boldsymbol{\rho}$. We introduce[1] coordinates q^1 and q^2 over Σ, which is then described by the equation $\boldsymbol{\rho} = \boldsymbol{\rho}(q^1, q^2)$.

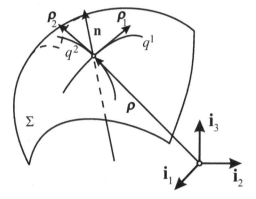

Fig. 7.1 Surface Σ with curvilinear coordinate lines (q^1, q^2).

The basis vectors on Σ are tangent to the coordinate lines at a point; they are denoted by $\boldsymbol{\rho}_1$ and $\boldsymbol{\rho}_2$. The vectors $\boldsymbol{\rho}^1$ and $\boldsymbol{\rho}^2$ constitute the dual basis. We recall that

$$\boldsymbol{\rho}_1 = \frac{\partial \boldsymbol{\rho}}{\partial q^1}, \qquad \boldsymbol{\rho}_2 = \frac{\partial \boldsymbol{\rho}}{\partial q^2}, \qquad \boldsymbol{\rho}_\alpha \cdot \boldsymbol{\rho}^\beta = \delta_\alpha^\beta \quad (\alpha, \beta = 1, 2).$$

From now on, Greek indices take the values $1, 2$, whereas Roman indices take the values $1, 2, 3$. The normal $\mathbf{n}$ to Σ at a point is defined by the relation

$$\mathbf{n} = \frac{\boldsymbol{\rho}_1 \times \boldsymbol{\rho}_2}{|\boldsymbol{\rho}_1 \times \boldsymbol{\rho}_2|}.$$

---

[1]In Chapter 5, following the tradition of differential geometry, we denoted the coordinates by u^1 and u^2. To avoid confusion with the components of the displacement vector $\mathbf{u}$ in this chapter, we rename the coordinates as indicated.

Let us introduce the metric tensor $\mathbf{A}$ on the surface:

$$\mathbf{A} = \boldsymbol{\rho}^\alpha \boldsymbol{\rho}_\alpha = \mathbf{E} - \mathbf{nn},$$

where $\mathbf{E}$ is the metric tensor in $\mathbb{R}^3$. It plays the role of the unit tensor on Σ at a point: if a vector $\mathbf{v}$ lies in the tangent plane at the point so that $\mathbf{v} \cdot \mathbf{n} = 0$, then $\mathbf{A} \cdot \mathbf{v} = \mathbf{v}$.

Exercise 7.1. Let $\mathbf{v} \cdot \mathbf{n} = 0$. Show that $\mathbf{A} \cdot \mathbf{v} = \mathbf{v}$.

Exercise 7.2. Show that $\mathbf{A} = a_{\alpha\beta}\boldsymbol{\rho}^\alpha\boldsymbol{\rho}^\beta = a^{\alpha\beta}\boldsymbol{\rho}_\alpha\boldsymbol{\rho}_\beta$, where $a_{\alpha\beta} = \boldsymbol{\rho}_\alpha \cdot \boldsymbol{\rho}_\beta$ and $a^{\alpha\beta} = \boldsymbol{\rho}^\alpha \cdot \boldsymbol{\rho}^\beta$.

Exercise 7.2 states that the components of $\mathbf{A}$ are the coefficients of the first fundamental form of Σ. This is why $\mathbf{A}$ is also called the *first fundamental tensor* of Σ.

At a point of Σ, the vectors $(\boldsymbol{\rho}_1, \boldsymbol{\rho}_2, \mathbf{n})$ constitute a basis of $\mathbb{R}^3$. The dual basis is $(\boldsymbol{\rho}^1, \boldsymbol{\rho}^2, \mathbf{n})$. This means that an arbitrary vector field $\mathbf{v}$ given on Σ can be represented as

$$\mathbf{v} = \mathbf{v}(q^1, q^2) = v_1(q^1, q^2)\boldsymbol{\rho}^1 + v_2(q^1, q^2)\boldsymbol{\rho}^2 + v_3(q^1, q^2)\mathbf{n}$$
$$= v^1(q^1, q^2)\boldsymbol{\rho}_1 + v^2(q^1, q^2)\boldsymbol{\rho}_2 + v^3(q^1, q^2)\mathbf{n}.$$

Note that $v_3 = v^3$.

The gradient operator on Σ is given by the formula

$$\widetilde{\nabla} = \boldsymbol{\rho}^\alpha \frac{\partial}{\partial q^\alpha}.$$

To apply the differential operations, we need the derivatives of the basis vectors $\boldsymbol{\rho}_1$, $\boldsymbol{\rho}_2$, $\mathbf{n}$, $\boldsymbol{\rho}^1$, $\boldsymbol{\rho}^2$. We recall that the derivatives of the normal $\mathbf{n}$ are (see equation (5.39))

$$\frac{\partial \mathbf{n}}{\partial q^\alpha} = -b_{\alpha\beta}\boldsymbol{\rho}^\beta.$$

We use the components $b_{\alpha\beta}$ to construct the tensor

$$\mathbf{B} = b_{\alpha\beta}\boldsymbol{\rho}^\alpha\boldsymbol{\rho}^\beta.$$

Using the definition of $\widetilde{\nabla}$, we see that

$$\mathbf{B} = -\widetilde{\nabla}\mathbf{n}.$$

The symmetric tensor $\mathbf{B}$ is called the *curvature tensor* of Σ or its *second fundamental tensor*; the components of $\mathbf{B}$ are the coefficients of the second fundamental form of Σ.

Exercise 7.3. Show that $\mathbf{B} = b_{\alpha\beta}\boldsymbol{\rho}^\alpha\boldsymbol{\rho}^\beta = b^{\alpha\beta}\boldsymbol{\rho}_\alpha\boldsymbol{\rho}_\beta$. Also show that the expressions for $b_{\alpha\beta}$ coincide with the coefficients of the second fundamental form (§ 5.5).

The derivatives of $\boldsymbol{\rho}_\alpha$ and $\boldsymbol{\rho}^\alpha$ are given by the formulas

$$\frac{\partial\boldsymbol{\rho}_\alpha}{\partial q^\beta} = \Gamma^\gamma_{\alpha\beta}\boldsymbol{\rho}_\gamma + b_{\alpha\beta}\mathbf{n}, \qquad \frac{\partial\boldsymbol{\rho}^\alpha}{\partial q^\beta} = -\Gamma^\alpha_{\beta\gamma}\boldsymbol{\rho}^\gamma + b^\alpha_\beta\mathbf{n},$$

where $\Gamma^\gamma_{\alpha\beta}$ is the Christoffel symbol introduced in § 4.5.

Let us derive the expression for the gradient $\widetilde{\nabla}\mathbf{v}$ of a vector field $\mathbf{v}$ given on Σ. We split $\mathbf{v}$ into two components; one, $\tilde{\mathbf{v}}$, is tangent to Σ such that $\tilde{\mathbf{v}} \cdot \mathbf{n} = 0$, and the other is normal to Σ:

$$\mathbf{v} = \tilde{\mathbf{v}} + w\mathbf{n}, \quad \tilde{\mathbf{v}} = v_1(q^1, q^2)\boldsymbol{\rho}^1 + v_2(q^1, q^2)\boldsymbol{\rho}^2, \quad w = v_3 = v^3 = \mathbf{v} \cdot \mathbf{n}.$$

Then

$$\widetilde{\nabla}\mathbf{v} = (\widetilde{\nabla}\tilde{\mathbf{v}}) \cdot \mathbf{A} - w\mathbf{B} + (\widetilde{\nabla}w + \mathbf{B} \cdot \tilde{\mathbf{v}})\mathbf{n}. \tag{7.1}$$

We also will need the expression for the tensor divergence:

$$\widetilde{\nabla} \cdot \mathbf{T} = \boldsymbol{\rho}^\gamma \frac{\partial}{\partial q^\gamma} \cdot (T^{\alpha\beta}\boldsymbol{\rho}_\alpha\boldsymbol{\rho}_\beta + T^{3\beta}\mathbf{n}\boldsymbol{\rho}_\beta + T^{\alpha 3}\boldsymbol{\rho}_\alpha\mathbf{n} + T^{33}\mathbf{n}\mathbf{n})$$

$$= \frac{\partial T^{\alpha\beta}}{\partial q^\alpha}\boldsymbol{\rho}_\beta + T^{\alpha\beta}\boldsymbol{\rho}^\gamma \cdot \frac{\partial\boldsymbol{\rho}_\alpha}{\partial q^\gamma}\boldsymbol{\rho}_\beta + T^{\alpha\beta}\frac{\partial\boldsymbol{\rho}_\beta}{\partial q^\alpha} + T^{3\beta}\boldsymbol{\rho}^\gamma \cdot \frac{\partial\mathbf{n}}{\partial q^\gamma}\boldsymbol{\rho}_\beta$$

$$+ T^{33}\boldsymbol{\rho}^\gamma \cdot \frac{\partial\mathbf{n}}{\partial q^\gamma}\mathbf{n} + \frac{\partial T^{\alpha 3}}{\partial q^\alpha}\mathbf{n} + T^{\alpha 3}\boldsymbol{\rho}^\gamma \cdot \frac{\partial\boldsymbol{\rho}_\alpha}{\partial q^\gamma}\mathbf{n} + T^{\alpha 3}\frac{\partial\mathbf{n}}{\partial q^\alpha}$$

$$= \frac{\partial T^{\alpha\beta}}{\partial q^\alpha}\boldsymbol{\rho}_\beta + T^{\alpha\beta}\Gamma^\gamma_{\alpha\gamma}\boldsymbol{\rho}_\beta + T^{\alpha\beta}\Gamma^\gamma_{\beta\alpha}\boldsymbol{\rho}_\gamma + T^{\alpha\beta}b_{\alpha\beta}\mathbf{n} - T^{3\beta}b^\gamma_\gamma\boldsymbol{\rho}_\beta$$

$$- T^{33}b^\gamma_\gamma\mathbf{n} + \frac{\partial T^{\alpha 3}}{\partial q^\alpha}\mathbf{n} + T^{\alpha 3}\Gamma^\gamma_{\alpha\gamma}\mathbf{n} - T^{\alpha 3}b^\beta_\alpha\boldsymbol{\rho}_\beta. \tag{7.2}$$

Exercise 7.4. Establish (7.1).

Exercise 7.5. Use (7.1) to show that $\widetilde{\nabla} \cdot \mathbf{v} = \widetilde{\nabla} \cdot \tilde{\mathbf{v}} - w\operatorname{tr}\mathbf{B}$.

Exercise 7.6. Let $\mathbf{v} = \mathbf{v}(\phi, \theta)$ be a vector given on a sphere with spherical coordinates ϕ and θ. Find $\widetilde{\nabla}\mathbf{v}$.

Exercise 7.7. Let $\mathbf{v} = \mathbf{v}(\phi, z)$ be a vector given on a cylindrical surface with cylindrical coordinates ϕ and z. Find $\widetilde{\nabla}\mathbf{v}$.

Exercise 7.8. Let $\mathbf{T}$ be a second-order tensor given on Σ, such that $\mathbf{T}\cdot\mathbf{n} = \mathbf{n} \cdot \mathbf{T} = 0$. Show that the equation $\widetilde{\nabla} \cdot \mathbf{T} = 0$ implies the algebraic relation $T^{\alpha\beta}b_{\alpha\beta} = 0$.

Exercise 7.9. Let $\mathbf{T}$ be a second-order tensor given on a sphere with spherical coordinates ϕ, θ. Find $\widetilde{\nabla} \cdot \mathbf{T}$.

Exercise 7.10. Let $\mathbf{T}$ be a second-order tensor given on a cylindrical surface with cylindrical coordinates ϕ, z. Find $\widetilde{\nabla} \cdot \mathbf{T}$.

7.2 Kinematics in a Neighborhood of Σ

To describe a spatial domain that surrounds Σ, we will use the coordinates and quantities of the previous section. A shell is a three-dimensional body occupying some neighborhood of the base surface Σ, which is used to describe shell kinematics as well as displacements and strains under load.

The coordinates of a point Q in a neighborhood of Σ are introduced as follows. Let $\mathbf{n}$ be the normal to Σ through a point Q. Let the base point of $\mathbf{n}$ be P, whose position vector is $\boldsymbol{\rho}(q^1, q^2)$ so that its coordinates on Σ are q^1, q^2. We appoint q^1, q^2 as the two first coordinates of Q, and the distance z from P to Q as the third coordinate (Fig. 7.2). Note that z is taken positive when $\overrightarrow{PQ}$ is co-directed with $\mathbf{n}$ and negative when oppositely directed.

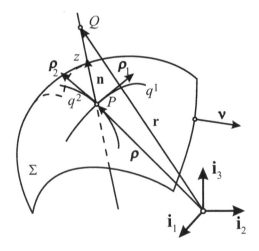

Fig. 7.2   Kinematics in a neighborhood of Σ.

The position vector $\mathbf{r}$ of Q is

$$\mathbf{r} = \mathbf{r}(q^1, q^2, z) = \boldsymbol{\rho}(q^1, q^2) + z\mathbf{n}.$$

The volume occupied by the shell is defined by the inequalities

$$-h/2 \leq z \leq h/2,$$

where h is the *shell thickness*. In general, $h = h(q^1, q^2)$ can vary from point to point. At a point Q near Σ, the basis vectors are

$$\mathbf{r}_\alpha = \frac{\partial \mathbf{r}}{\partial q^\alpha} = \boldsymbol{\rho}_\alpha + z \frac{\partial \mathbf{n}}{\partial q^\alpha} = (\mathbf{A} - z\mathbf{B}) \cdot \boldsymbol{\rho}_\alpha, \qquad \mathbf{r}_3 = \mathbf{n}. \qquad (7.3)$$

The dual basis is given by

$$\mathbf{r}^\alpha = (\mathbf{A} - z\mathbf{B})^{-1} \cdot \boldsymbol{\rho}^\alpha, \qquad \mathbf{r}^3 = \mathbf{n}. \qquad (7.4)$$

Let us explain the notation $(\mathbf{A} - z\mathbf{B})^{-1}$. It is seen that $(\mathbf{A} - z\mathbf{B}) \cdot \mathbf{n} = \mathbf{0}$, so $\mathbf{A} - z\mathbf{B}$ degenerates as a three-dimensional tensor and its inverse does not exist. Here we restrict $\mathbf{A} - z\mathbf{B}$ to an operator acting in the two-dimensional subspace tangent to Σ. This subspace has basis $\boldsymbol{\rho}^1, \boldsymbol{\rho}^2$, and the tensor $\mathbf{A}$ plays the role of the unit operator on it. In what follows, we suppose that $h \|\mathbf{B}\|$ is very small. Now, using the Banach contraction principle [Lebedev and Cloud (2003)], we can prove the existence of a unique inverse to $\mathbf{A} - z\mathbf{B}$ for small $|z| \leq h/2$ on the tangent space. So we have

$$(\mathbf{A} - z\mathbf{B})^{-1} \cdot (\mathbf{A} - z\mathbf{B}) = (\mathbf{A} - z\mathbf{B}) \cdot (\mathbf{A} - z\mathbf{B})^{-1} = \mathbf{A}. \qquad (7.5)$$

We illustrate how to derive $(\mathbf{A} - z\mathbf{B})^{-1}$ via two important examples.

(1) Let Σ be a cylinder of radius R. In cylindrical coordinates

$$\mathbf{n} = \mathbf{e}_r,$$
$$\mathbf{A} = \mathbf{E} - \mathbf{nn} = \mathbf{e}_\phi \mathbf{e}_\phi + \mathbf{e}_z \mathbf{e}_z,$$
$$\mathbf{B} = -\widetilde{\nabla} \mathbf{e}_r = -\mathbf{e}_\phi \mathbf{e}_\phi / R.$$

Then

$$\mathbf{A} - z\mathbf{B} = \left(1 + \frac{z}{R}\right) \mathbf{e}_\phi \mathbf{e}_\phi + \mathbf{e}_z \mathbf{e}_z.$$

It is easy to check that

$$(\mathbf{A} - z\mathbf{B})^{-1} = \left(1 + \frac{z}{R}\right)^{-1} \mathbf{e}_\phi \mathbf{e}_\phi + \mathbf{e}_z \mathbf{e}_z$$

satisfies (7.5) and hence is the needed inverse tensor.

(2) Now we define $(\mathbf{A} - z\mathbf{B})^{-1}$ for a sphere of radius R. In spherical coordinates,

$$\mathbf{n} = \mathbf{e}_r,$$
$$\mathbf{A} = \mathbf{E} - \mathbf{nn} = \mathbf{e}_\phi \mathbf{e}_\phi + \mathbf{e}_\theta \mathbf{e}_\theta,$$
$$\mathbf{B} = -\widetilde{\nabla} \mathbf{e}_r = -(\mathbf{e}_\phi \mathbf{e}_\phi + \mathbf{e}_\theta \mathbf{e}_\theta)/R.$$

Then

$$\mathbf{A} - z\mathbf{B} = \left(1 + \frac{z}{R}\right)(\mathbf{e}_\phi \mathbf{e}_\phi + \mathbf{e}_\theta \mathbf{e}_\theta).$$

Again, a direct check of formula (7.5) for

$$(\mathbf{A} - z\mathbf{B})^{-1} = \left(1 + \frac{z}{R}\right)^{-1}(\mathbf{e}_\phi \mathbf{e}_\phi + \mathbf{e}_\theta \mathbf{e}_\theta)$$

demonstrates that it is the needed inverse tensor.

These examples show that $(\mathbf{A} - z\mathbf{B})^{-1}$ cannot be found for all values of z. In both cases, the inverse tensor exists for all $|z| \leq h/2$ if $h/2R < 1$. This is the domain in which each point is uniquely defined by the coordinate system introduced above. We shall use the definition of $(\mathbf{A} - z\mathbf{B})^{-1}$ in this restricted sense.

Finally, using the dual basis and (7.4), we find the following representation of the spatial nabla operator:

$$\begin{aligned}
\nabla &= \mathbf{r}^\alpha \frac{\partial}{\partial q^\alpha} + \mathbf{n}\frac{\partial}{\partial z} \\
&= (\mathbf{A} - z\mathbf{B})^{-1} \cdot \boldsymbol{\rho}^\alpha \frac{\partial}{\partial q^\alpha} + \mathbf{n}\frac{\partial}{\partial z} \\
&= (\mathbf{A} - z\mathbf{B})^{-1} \cdot \tilde{\nabla} + \mathbf{n}\frac{\partial}{\partial z}.
\end{aligned}$$

7.3   Shell Equilibrium Equations

Let us derive the two-dimensional equations for shell equilibrium. These will be a direct consequence of the three-dimensional equilibrium equations for the shell as a spatial body occupying a domain V in $\mathbb{R}^3$ (see Fig. 7.3). V is bounded by two surfaces (faces) S_- and S_+, each at distance $h/2$ from the midsurface Σ, and by the lateral ruled surface S_ν. S_ν is formed by the motion of the normal $\mathbf{n}$ to Σ along the boundary of Σ. In other words, V is the set of all spatial points given by the position vector

$$\mathbf{r} = \boldsymbol{\rho}(q^1, q^2) + z\mathbf{n},$$

where $\boldsymbol{\rho}$ is the position vector of a point on Σ and $-h/2 \leq z \leq h/2$. For simplicity, we suppose h is constant.

On the faces $S_\pm$, the boundary conditions are

$$\mathbf{n}_+ \cdot \boldsymbol{\sigma}\big|_{z=h/2} = \mathbf{t}_+^0, \qquad \mathbf{n}_- \cdot \boldsymbol{\sigma}\big|_{z=-h/2} = \mathbf{t}_-^0, \tag{7.6}$$

where $\mathbf{n}_\pm$ are the normals to $S_\pm$ (Fig. 7.3) and $\mathbf{t}_\pm^0$ are the surface loads on $S_\pm$, respectively. In the general case, on S_ν we will assign the mixed boundary conditions of the third problem of elasticity.

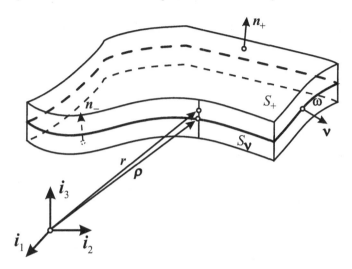

Fig. 7.3   A shell-like body.

Equation (6.20) with $\ddot{\mathbf{u}} = \mathbf{0}$ describes the equilibrium of a three-dimensional elastic body occupying volume V:

$$\frac{1}{\sqrt{g}} \frac{\partial}{\partial q^i} \left(\sqrt{g} \sigma^{ij} \mathbf{r}_j \right) + \rho \mathbf{f} = \mathbf{0}. \tag{7.7}$$

Using q^1, q^2, $q^3 = z$ as the coordinates, in equation (7.7) we can isolate the differentiation with respect to z. Let us denote

$$\boldsymbol{\sigma}^\alpha = \mathbf{r}^\alpha \cdot \boldsymbol{\sigma} = \sigma^{\alpha j} \mathbf{r}_j, \qquad \boldsymbol{\sigma}^3 = \mathbf{n} \cdot \boldsymbol{\sigma} = \sigma^{3j} \mathbf{r}_j.$$

We get

$$\frac{1}{\sqrt{g}} \left(\frac{\partial(\sqrt{g} \boldsymbol{\sigma}^\alpha)}{\partial q^\alpha} + \frac{\partial(\sqrt{g} \boldsymbol{\sigma}^3)}{\partial z} \right) + \rho \mathbf{f} = \mathbf{0}. \tag{7.8}$$

We recall that g is the determinant of the matrix $g_{ij} = \mathbf{r}_i \cdot \mathbf{r}_j$. We can represent g in terms of the parameters of Σ. Indeed,

$$g_{\alpha\beta} = \boldsymbol{\rho}_\alpha \cdot (\mathbf{A} - z\mathbf{B})^2 \cdot \boldsymbol{\rho}_\beta, \qquad g_{13} = 0, \qquad g_{23} = 0, \qquad g_{33} = 1.$$

So g takes the form

$$g = aG^2, \qquad a = a_{11}a_{22} - a_{12}^2, \qquad G = \det(\mathbf{A} - z\mathbf{B}), \tag{7.9}$$

where $a_{\alpha\beta}$ are equal to the values of $g_{\alpha\beta}$ taken on Σ. Here $\det(\mathbf{A} - z\mathbf{B})$ is calculated as the determinant of a two-dimensional tensor:

$$\det \mathbf{X} = X_1^1 X_2^2 - \left(X_1^2\right)^2,$$

where X_α^β are the mixed components of $\mathbf{X} = \mathbf{A} - z\mathbf{B}$. Also observe that in the new notation (7.6) is

$$\sigma^3\big|_{z=\pm h/2} = \mathbf{t}_\pm^0.$$

Let us rewrite (7.8) in the form

$$\frac{1}{\sqrt{a}}\frac{\partial(\sqrt{a}G\boldsymbol{\sigma}^\alpha)}{\partial q^\alpha} + \frac{\partial(G\boldsymbol{\sigma}^3)}{\partial z} + \rho G\mathbf{f} = \mathbf{0}. \tag{7.10}$$

Now we can derive two-dimensional equilibrium equations for the shell. We integrate (7.10) over the thickness, i.e., along the normal coordinate z from $-h/2$ to $h/2$. Taking into account the boundary conditions (7.6), we get

$$\frac{1}{\sqrt{a}}\frac{\partial}{\partial q^\alpha}\left(\sqrt{a}\,[\![\boldsymbol{\sigma}^\alpha]\!]\right) + G_+\mathbf{t}_+^0 - G_-\mathbf{t}_-^0 + [\![\rho\mathbf{f}]\!] = \mathbf{0}, \tag{7.11}$$

where

$$G_\pm = G\big|_{z=\pm h/2}.$$

Here we introduced the notation $[\![\cdots]\!]$. For any quantity f, this denotes the definite integral of Gf over thickness:

$$[\![f]\!] = \int_{-h/2}^{h/2} Gf\,dz.$$

Let us transform (7.11) to component-free form. The quantity

$$\mathbf{T} = \boldsymbol{\rho}_\alpha[\![\boldsymbol{\sigma}^\alpha]\!]$$

can be represented as follows:

$$\begin{aligned}
\mathbf{T} &= \boldsymbol{\rho}_\alpha[\![\mathbf{r}^\alpha \cdot \boldsymbol{\sigma}]\!] \\
&= \boldsymbol{\rho}_\alpha[\![\boldsymbol{\rho}^\alpha \cdot (\mathbf{A} - z\mathbf{B})^{-1} \cdot \boldsymbol{\sigma}]\!] \\
&= \boldsymbol{\rho}_\alpha\boldsymbol{\rho}^\alpha \cdot [\![(\mathbf{A} - z\mathbf{B})^{-1} \cdot \boldsymbol{\sigma}]\!] \\
&= \mathbf{A} \cdot [\![(\mathbf{A} - z\mathbf{B})^{-1} \cdot \boldsymbol{\sigma}]\!] \\
&= [\![(\mathbf{A} - z\mathbf{B})^{-1} \cdot \boldsymbol{\sigma}]\!].
\end{aligned}$$

We call $\mathbf{T}$ the *stress resultant tensor*. By definition, it is clear that $\mathbf{n}\cdot\mathbf{T} = \mathbf{0}$. The components of $\mathbf{T}\cdot\mathbf{A}$ are the *stress resultants* acting in the tangent plane of the shell. $\mathbf{T} \cdot \mathbf{n}$ is the vector of *transverse shear stress resultants*. So (7.11) becomes

$$\widetilde{\nabla} \cdot \mathbf{T} + \mathbf{q} = \mathbf{0}, \tag{7.12}$$

where

$$\mathbf{q} = G_+ \mathbf{t}_+^0 - G_- \mathbf{t}_-^0 + [\![\rho \mathbf{f}]\!].$$

The vector $\mathbf{q}$ is the distributed load on the shell. Its component $\mathbf{q} \cdot \mathbf{A}$ acts in the tangent plane to Σ, and $\mathbf{q} \cdot \mathbf{n}$ is the transverse load. So we have derived the first shell equilibrium equation.

To derive the second equilibrium equation, we cross-multiply the terms of (7.8) by $z\mathbf{n}$ from the left:

$$\frac{1}{\sqrt{a}} z\mathbf{n} \times \frac{\partial(\sqrt{a} G \boldsymbol{\sigma}^\alpha)}{\partial q^\alpha} + z\mathbf{n} \times \frac{\partial(G \boldsymbol{\sigma}^3)}{\partial z} + \rho G z\mathbf{n} \times \mathbf{f} = \mathbf{0}. \tag{7.13}$$

Moving $\mathbf{n}$ through the derivative sign, we get

$$\frac{1}{\sqrt{a}} \frac{\partial}{\partial q^\alpha}(\mathbf{n} \times \sqrt{a} G z \boldsymbol{\sigma}^\alpha) - \frac{\partial \mathbf{n}}{\partial q^\alpha} \times G z \boldsymbol{\sigma}^\alpha + z \frac{\partial}{\partial z}(G\mathbf{n} \times \boldsymbol{\sigma}^3) + \rho G z\mathbf{n} \times \mathbf{f} = \mathbf{0}.$$

Because

$$\mathbf{r}_\alpha = \boldsymbol{\rho}_\alpha + z\frac{\partial \mathbf{n}}{\partial q^\alpha}$$

so that

$$z\frac{\partial \mathbf{n}}{\partial q^\alpha} = \mathbf{r}_\alpha - \boldsymbol{\rho}_\alpha,$$

we obtain

$$\frac{\partial \mathbf{n}}{\partial q^\alpha} \times z\boldsymbol{\sigma}^\alpha = (\mathbf{r}_\alpha - \boldsymbol{\rho}_\alpha) \times \boldsymbol{\sigma}^\alpha = -\mathbf{n} \times \boldsymbol{\sigma}^3 - \boldsymbol{\rho}_\alpha \times \boldsymbol{\sigma}^\alpha.$$

Since

$$\begin{aligned}
\mathbf{r}_i \times \boldsymbol{\sigma}^i &= \mathbf{r}_i \times (\mathbf{r}^i \cdot \boldsymbol{\sigma}) \\
&= \mathbf{r}_i \times (\mathbf{r}^i \cdot \sigma^{kj} \mathbf{r}_k \mathbf{r}_j) \\
&= \mathbf{r}_i \times \mathbf{r}_j \sigma^{ij} \\
&= -\mathbf{r}_j \times \mathbf{r}_i \sigma^{ij} \\
&= -\mathbf{r}_j \times \mathbf{r}_i \sigma^{ji} \\
&= -\mathbf{r}_i \times \mathbf{r}_j \sigma^{ij} \\
&= -\mathbf{r}_i \times \boldsymbol{\sigma}^i,
\end{aligned}$$

it follows that

$$\mathbf{r}_i \times \boldsymbol{\sigma}^i = \mathbf{0}.$$

From

$$\mathbf{r}_i \times \boldsymbol{\sigma}^i = \mathbf{r}_\alpha \times \boldsymbol{\sigma}^\alpha + \mathbf{n} \times \boldsymbol{\sigma}^3 = \mathbf{0}$$

it follows that

$$\mathbf{r}_\alpha \times \boldsymbol{\sigma}^\alpha = -\mathbf{n} \times \boldsymbol{\sigma}^3.$$

Equation (7.13) becomes

$$\frac{1}{\sqrt{a}} \frac{\partial}{\partial q^\alpha} (\mathbf{n} \times \sqrt{a} G z \boldsymbol{\sigma}^\alpha) + \boldsymbol{\rho}_\alpha \times G \boldsymbol{\sigma}^\alpha + \frac{\partial}{\partial z} (G \mathbf{n} \times z \boldsymbol{\sigma}^3) + \rho G z \mathbf{n} \times \mathbf{f} = \mathbf{0}.$$

Integrating with respect to z from $-h/2$ to $h/2$ and using (7.6), we get the second vector equilibrium equation for the shell:

$$\frac{1}{\sqrt{a}} \frac{\partial}{\partial q^\alpha} \left(\sqrt{a} [\![\mathbf{n} \times z \boldsymbol{\sigma}^\alpha]\!] \right) + \boldsymbol{\rho}_\alpha \times [\![\boldsymbol{\sigma}^\alpha]\!]$$

$$+ \frac{hG_+}{2} \mathbf{n} \times \mathbf{t}_+^0 + \frac{hG_-}{2} \mathbf{n} \times \mathbf{t}_-^0 + [\![\rho z \mathbf{n} \times \mathbf{f}]\!] = \mathbf{0}. \qquad (7.14)$$

Now we represent it in component-free form. Let us introduce the *stress couple tensor*

$$\mathbf{M} = \boldsymbol{\rho}_\alpha [\![\mathbf{n} \times z \boldsymbol{\sigma}^\alpha]\!],$$

which we represent as

$$\mathbf{M} = -\boldsymbol{\rho}_\alpha [\![\mathbf{r}^\alpha \cdot z \boldsymbol{\sigma} \times \mathbf{n}]\!]$$

$$= -\boldsymbol{\rho}_\alpha [\![\boldsymbol{\rho}^\alpha \cdot (\mathbf{A} - z\mathbf{B})^{-1} \cdot z \boldsymbol{\sigma} \times \mathbf{n}]\!]$$

$$= -\boldsymbol{\rho}_\alpha \boldsymbol{\rho}^\alpha \cdot [\![(\mathbf{A} - z\mathbf{B})^{-1} \cdot z \boldsymbol{\sigma} \times \mathbf{n}]\!]$$

$$= -\mathbf{A} \cdot [\![(\mathbf{A} - z\mathbf{B})^{-1} \cdot z \boldsymbol{\sigma} \times \mathbf{n}]\!]$$

$$= -[\![(\mathbf{A} - z\mathbf{B})^{-1} \cdot z \boldsymbol{\sigma} \times \mathbf{n}]\!].$$

By definition of $\mathbf{M}$, we see that $\mathbf{n} \cdot \mathbf{M} = \mathbf{M} \cdot \mathbf{n} = \mathbf{0}$. Let us denote

$$\mathbf{m} = \frac{1}{2} hG_+ \mathbf{n} \times \mathbf{t}_+^0 + \frac{1}{2} hG_- \mathbf{n} \times \mathbf{t}_-^0 + [\![\rho z \mathbf{n} \times \mathbf{f}]\!],$$

which represents the external bending couples applied to the shell surface. By the definition of $\mathbf{m}$, we see that $\mathbf{m}$ belongs to the tangent plane of Σ; that is, $\mathbf{m} \cdot \mathbf{n} = 0$.

Finally, we introduce a vectorial invariant of $\mathbf{T}$, which is

$$\mathbf{T}_\times = \boldsymbol{\rho}_\alpha \times [\![\boldsymbol{\sigma}^\alpha]\!];$$

this notation is called the *Gibbsian cross*. Now we can rewrite (7.14) in component-free form as follows:

$$\widetilde{\nabla} \cdot \mathbf{M} + \mathbf{T}_\times + \mathbf{m} = \mathbf{0}. \tag{7.15}$$

Note that (7.12) and (7.15) are *exact* consequences of the equilibrium equations for the three-dimensional continuum of the shell. They express the equality to zero of the sum of all forces and moments acting on a shell element.

So we have derived two two-dimensional equilibrium equations (7.12) and (7.15) for a shell. The unknowns $\mathbf{T}$ and $\mathbf{M}$ are defined on Σ. In what follows, Σ will be called the shell surface, or the *midsurface* or *base surface* of the shell.

7.4 Shell Deformation and Strains; Kirchhoff's Hypotheses

Kirchhoff formulated his famous hypotheses on the laws of plate deformation that extend Bernoulli's assumptions for bending of a beam. Since *Kirchhoff's hypotheses* were extended by Love to shell theory, they are often called the *Kirchhoff–Love hypotheses*.

Kirchhoff's kinematic hypothesis. Any shell cross section that is perpendicular to the midsurface before deformation remains perpendicular to the deformed midsurface. The transverse shear strains are negligibly small. Mathematically, this means that $\mathbf{n} \cdot \boldsymbol{\varepsilon} \cdot \boldsymbol{\tau} = 0$ for all $\boldsymbol{\tau}$ such that $\boldsymbol{\tau} \cdot \mathbf{n} = 0$.

Kirchhoff's static hypothesis. The normal stress is negligible in comparison with the other stresses, so $\sigma_{33} = \mathbf{n} \cdot \boldsymbol{\sigma} \cdot \mathbf{n} = 0$.

Experimentally, the behavior of the normal often reflects the real situation quite well. In this theory, we ignore shear strains and transverse compression in the shell. In the literature we can find different versions of Kirchhoff's hypotheses. We can also find discussions regarding their interpretations, applications, etc. See [Ciarlet (1997); Ciarlet (2000); Chróścielewski, Makowski, and Pietraszkiewicz (2004); Donnell (1976); Goldenveizer (1976); Libai and Simmonds (1998); Novozhilov, Chernykh, and Mikhailovskiy (1991); Timoshenko (1985); Vorovich (1999); Zhilin (2006); Zubov (1982)]. These topics fall outside the scope of the present book.

Kirchhoff's static hypothesis allows us to eliminate $\varepsilon_{33} \equiv \mathbf{n} \cdot \boldsymbol{\varepsilon} \cdot \mathbf{n}$ from Hooke's law. This is done in a manner similar to the elimination of ε_{33} in plane elasticity (cf., Exercises 6.14 and 6.15). By Kirchhoff's hypothesis,

$$\sigma_{33} \equiv \mathbf{n} \cdot \boldsymbol{\sigma} \cdot \mathbf{n} = 0.$$

Let the shell material be isotropic. Because

$$\sigma_{33} = \lambda \operatorname{tr} \boldsymbol{\varepsilon} + 2\mu\varepsilon_{33} = 0$$

it follows that

$$\varepsilon_{33} = -\frac{\lambda}{\lambda + 2\mu}(\varepsilon_{11} + \varepsilon_{22}).$$

So for the stress and strain tensors of the shell we get the representations

$$\boldsymbol{\sigma} = \tilde{\boldsymbol{\sigma}} + \sigma_{\alpha 3}(\mathbf{r}^{\alpha}\mathbf{n} + \mathbf{n}\mathbf{r}^{\alpha}),$$

$$\boldsymbol{\varepsilon} = \tilde{\boldsymbol{\varepsilon}} + \varepsilon_{\alpha 3}(\mathbf{r}^{\alpha}\mathbf{n} + \mathbf{n}\mathbf{r}^{\alpha}) - \frac{\lambda}{\lambda + 2\mu}(\varepsilon_{11} + \varepsilon_{22})\mathbf{n}\mathbf{n},$$

where $\tilde{\boldsymbol{\sigma}}$ and $\tilde{\boldsymbol{\varepsilon}}$ satisfy the relations

$$\mathbf{n} \cdot \tilde{\boldsymbol{\sigma}} = 0, \qquad \tilde{\boldsymbol{\sigma}} \cdot \mathbf{n} = 0, \qquad \mathbf{n} \cdot \tilde{\boldsymbol{\varepsilon}} = 0, \qquad \tilde{\boldsymbol{\varepsilon}} \cdot \mathbf{n} = 0.$$

Hooke's law takes the form

$$\boldsymbol{\sigma} = 2\mu \left[\boldsymbol{\varepsilon} + \frac{\lambda}{\lambda + 2\mu}\mathbf{E}\operatorname{tr}\tilde{\boldsymbol{\varepsilon}} \right] = \frac{E}{1+\nu}\left[\boldsymbol{\varepsilon} + \frac{\nu}{1-\nu}\mathbf{E}\operatorname{tr}\tilde{\boldsymbol{\varepsilon}} \right]. \qquad (7.16)$$

In component representation, this is

$$\sigma_1^1 = \frac{E}{1-\nu^2}(\varepsilon_1^1 + \nu\varepsilon_2^2), \qquad\qquad \sigma_2^2 = \frac{E}{1-\nu^2}(\varepsilon_2^2 + \nu\varepsilon_1^1),$$

$$\sigma_1^2 = \frac{E}{1+\nu}\varepsilon_1^2, \qquad\qquad\qquad \sigma_1^3 = \frac{E}{1+\nu}\varepsilon_1^3,$$

$$\sigma_2^3 = \frac{E}{1+\nu}\varepsilon_2^3, \qquad\qquad\qquad\qquad\qquad (7.17)$$

where

$$\sigma_\alpha^\beta = \mathbf{r}_\alpha \cdot \boldsymbol{\sigma} \cdot \mathbf{r}^\beta, \qquad \sigma_\alpha^3 = \mathbf{r}_\alpha \cdot \boldsymbol{\sigma} \cdot \mathbf{n}.$$

The components of $\boldsymbol{\varepsilon}$ are denoted similarly.

Exercise 7.11. Derive (7.16) and (7.17).

Now we turn to the kinematic hypothesis. By this hypothesis, any cross-section perpendicular to the midsurface before deformation remains perpendicular after deformation and, moreover, remains planar. Let us fix a point q^1, q^2 on Σ. Any cross-section through the point is planar after deformation if and only if the dependence of the displacement of the points on the normal at q^1, q^2 is linear in z. So we conclude that the general form of the displacement vector is

$$\mathbf{u}(q^1, q^2, z) = \tilde{\mathbf{v}}(q^1, q^2) + w(q^1, q^2)\mathbf{n} - z\boldsymbol{\vartheta}(q^1, q^2), \qquad (7.18)$$

where

$$\boldsymbol{\vartheta}(q^1, q^2) = \tilde{\boldsymbol{\vartheta}} + \vartheta_n \mathbf{n}, \qquad \mathbf{n} \cdot \tilde{\boldsymbol{\vartheta}} = 0.$$

Let us derive the expression for the strain tensor corresponding to (7.18). We have

$$\begin{aligned}
\nabla \mathbf{u} &= \left(\tilde{\nabla} + \mathbf{n} \frac{\partial}{\partial z} \right) (\tilde{\mathbf{v}} + w\mathbf{n} - z\boldsymbol{\vartheta}) \\
&= \tilde{\nabla}\tilde{\mathbf{v}} + (\tilde{\nabla}w)\mathbf{n} - w\mathbf{B} - z\tilde{\nabla}\tilde{\boldsymbol{\vartheta}} + z\vartheta_n \mathbf{B} - \mathbf{n}\boldsymbol{\vartheta} \qquad (7.19)
\end{aligned}$$

and

$$(\nabla \mathbf{u})^T = (\tilde{\nabla}\tilde{\mathbf{v}})^T + \mathbf{n}\tilde{\nabla}w - w\mathbf{B} - z(\tilde{\nabla}\tilde{\boldsymbol{\vartheta}})^T + z\vartheta_n \mathbf{B} - \boldsymbol{\vartheta}\mathbf{n}.$$

So the strain tensor is

$$\begin{aligned}
2\boldsymbol{\varepsilon} &= \nabla \mathbf{u} + \nabla \mathbf{u}^T \\
&= \tilde{\nabla}\tilde{\mathbf{v}} + (\tilde{\nabla}\tilde{\mathbf{v}})^T - 2w\mathbf{B} - z\left(\tilde{\nabla}\tilde{\boldsymbol{\vartheta}} + (\tilde{\nabla}\tilde{\boldsymbol{\vartheta}})^T \right) + 2z\vartheta_n \mathbf{B} \\
&\quad + (\tilde{\nabla}w)\mathbf{n} - \mathbf{n}\boldsymbol{\vartheta} + \mathbf{n}\tilde{\nabla}w - \boldsymbol{\vartheta}\mathbf{n}. \qquad (7.20)
\end{aligned}$$

We recall that by (7.1),

$$\tilde{\nabla}\tilde{\mathbf{v}} = (\tilde{\nabla}\tilde{\mathbf{v}}) \cdot \mathbf{A} + (\mathbf{B} \cdot \tilde{\mathbf{v}})\mathbf{n}, \qquad \tilde{\nabla}\tilde{\boldsymbol{\vartheta}} = (\tilde{\nabla}\tilde{\boldsymbol{\vartheta}}) \cdot \mathbf{A} + (\mathbf{B} \cdot \tilde{\boldsymbol{\vartheta}})\mathbf{n}.$$

Substituting these into (7.20), we get

$$\begin{aligned}
2\boldsymbol{\varepsilon} &= (\tilde{\nabla}\tilde{\mathbf{v}}) \cdot \mathbf{A} + \mathbf{A} \cdot (\tilde{\nabla}\tilde{\mathbf{v}})^T - 2w\mathbf{B} - z\left((\tilde{\nabla}\tilde{\boldsymbol{\vartheta}}) \cdot \mathbf{A} + \mathbf{A} \cdot (\tilde{\nabla}\tilde{\boldsymbol{\vartheta}})^T \right) \\
&\quad + (\mathbf{B} \cdot \tilde{\mathbf{v}})\mathbf{n} + \mathbf{n}(\mathbf{B} \cdot \tilde{\mathbf{v}}) - z(\mathbf{B} \cdot \tilde{\boldsymbol{\vartheta}})\mathbf{n} - z\mathbf{n}(\mathbf{B} \cdot \tilde{\boldsymbol{\vartheta}}) + 2z\vartheta_n \mathbf{B} \\
&\quad + (\tilde{\nabla}w)\mathbf{n} - \mathbf{n}\boldsymbol{\vartheta} + \mathbf{n}\tilde{\nabla}w - \boldsymbol{\vartheta}\mathbf{n}. \qquad (7.21)
\end{aligned}$$

Let us derive some other consequences of Kirchhoff's kinematic hypothesis. By the hypothesis we have

$$\mathbf{n} \cdot \boldsymbol{\varepsilon} \cdot \boldsymbol{\tau} = 0, \qquad (7.22)$$

where $\boldsymbol{\tau}$ is an arbitrary vector orthogonal to $\mathbf{n}$; in components this is $\varepsilon_{31} = \varepsilon_{32} = 0$. Using (7.21) we find that

$$2\mathbf{n} \cdot \boldsymbol{\varepsilon} = \mathbf{B} \cdot \tilde{\mathbf{v}} - \tilde{\boldsymbol{\vartheta}} + \tilde{\nabla}w - z\mathbf{B} \cdot \tilde{\boldsymbol{\vartheta}} - 2\vartheta_n \mathbf{n}.$$

All terms on the right-hand side except $2\vartheta_n \mathbf{n}$ are orthogonal to $\mathbf{n}$. It follows from (7.22), which holds for any $\boldsymbol{\tau}$ orthogonal to $\mathbf{n}$, that

$$\mathbf{B} \cdot \tilde{\mathbf{v}} - \tilde{\boldsymbol{\vartheta}} + \widetilde{\nabla} w - z\mathbf{B} \cdot \tilde{\boldsymbol{\vartheta}} = \mathbf{0}.$$

From this we find $\tilde{\boldsymbol{\vartheta}}$:

$$\tilde{\boldsymbol{\vartheta}} = (\mathbf{A} + z\mathbf{B})^{-1} \cdot (\widetilde{\nabla} w + \mathbf{B} \cdot \tilde{\mathbf{v}}).$$

In linear elasticity, the displacements and their derivatives should be very small. For a sufficiently thin and smooth shell, we also can suppose that $h\mathbf{B}$ is small. We formalize this as

Assumption S. Suppose $h \, \|\mathbf{B}\| \ll 1$.

So we assume that the terms $z\mathbf{B}$ are negligibly small in comparison with unity. Note that by (7.9), from Assumption S it follows that $G = \det(\mathbf{A} - z\mathbf{B})$ is very close to 1. Under Assumption S we find that

$$\tilde{\boldsymbol{\vartheta}} = \widetilde{\nabla} w + \mathbf{B} \cdot \tilde{\mathbf{v}}. \tag{7.23}$$

With this, we have

$$2\mathbf{n} \cdot \boldsymbol{\varepsilon} \cdot \boldsymbol{\tau} = -z\boldsymbol{\tau} \cdot \mathbf{B} \cdot \tilde{\boldsymbol{\vartheta}}.$$

On the left side of this equality we see $\mathbf{n} \cdot \boldsymbol{\varepsilon} \cdot \boldsymbol{\tau}$ which, according to Kirchhoff's kinematic hypothesis, is zero. On the right, by Assumption S, we see terms of a higher order of smallness than $\tilde{\boldsymbol{\vartheta}}$. So the above choice of $\tilde{\boldsymbol{\vartheta}}$ makes Kirchhoff's kinematic hypothesis sufficiently accurate. Note that the quantity ϑ_n is still undefined.

Substituting (7.23) into (7.21), we find that

$$2\boldsymbol{\varepsilon} = (\widetilde{\nabla}\tilde{\mathbf{v}}) \cdot \mathbf{A} + \mathbf{A} \cdot (\widetilde{\nabla}\tilde{\mathbf{v}})^T - 2w\mathbf{B} - z\left((\widetilde{\nabla}\tilde{\boldsymbol{\vartheta}}) \cdot \mathbf{A} + \mathbf{A} \cdot (\widetilde{\nabla}\tilde{\boldsymbol{\vartheta}})^T\right)$$
$$- 2\vartheta_n \mathbf{nn} + \underline{2z\vartheta_n\mathbf{B} - z(\mathbf{B} \cdot \tilde{\boldsymbol{\vartheta}})\mathbf{n} - z\mathbf{n}(\mathbf{B} \cdot \tilde{\boldsymbol{\vartheta}})}. \tag{7.24}$$

By Assumption S, we set the underlined terms in (7.24) to zero. Finally, we get the expression for the strain tensor in the form

$$2\boldsymbol{\varepsilon} = (\widetilde{\nabla}\tilde{\mathbf{v}}) \cdot \mathbf{A} + \mathbf{A} \cdot (\widetilde{\nabla}\tilde{\mathbf{v}})^T - 2w\mathbf{B}$$
$$- z\left((\widetilde{\nabla}\tilde{\boldsymbol{\vartheta}}) \cdot \mathbf{A} + \mathbf{A} \cdot (\widetilde{\nabla}\tilde{\boldsymbol{\vartheta}})^T\right) - 2\vartheta_n\mathbf{nn}$$
$$= 2\tilde{\boldsymbol{\varepsilon}} - 2\vartheta_n\mathbf{nn}, \tag{7.25}$$

where

$$2\tilde{\boldsymbol{\varepsilon}} = (\widetilde{\nabla}\tilde{\mathbf{v}}) \cdot \mathbf{A} + \mathbf{A} \cdot (\widetilde{\nabla}\tilde{\mathbf{v}})^T - 2w\mathbf{B} - z\left((\widetilde{\nabla}\tilde{\boldsymbol{\vartheta}}) \cdot \mathbf{A} + \mathbf{A} \cdot (\widetilde{\nabla}\tilde{\boldsymbol{\vartheta}})^T\right).$$

We rewrite the last equation as

$$\tilde{\varepsilon} = \epsilon + z\text{æ}, \qquad (7.26)$$

where

$$2\epsilon = (\widetilde{\nabla}\tilde{\mathbf{v}}) \cdot \mathbf{A} + \mathbf{A} \cdot (\widetilde{\nabla}\tilde{\mathbf{v}})^T - 2w\mathbf{B},$$

$$\text{æ} = -\frac{1}{2}\left((\widetilde{\nabla}\tilde{\boldsymbol{\vartheta}}) \cdot \mathbf{A} + \mathbf{A} \cdot (\widetilde{\nabla}\tilde{\boldsymbol{\vartheta}})^T\right).$$

The tensor ϵ is the *tangential strain measure* or the *midsurface strain tensor*. It describes the shell deformation in the tangent plane. The tensor æ describes the shell deformation due to bending, and is called the *bending strain measure* or the *tensor of changes of curvature*. The simplest interpretation for these quantities will be formulated for a plate, a particular case of the shell, later. Various simplifications of the strain measures appear in the literature. They are based on the assumptions of smallness of certain terms in ϵ and æ. We will not consider these particular cases here. The above deformation measures are used, for example, by W.T. Koiter [Koiter (1970)].

Let us note that in the framework of Kirchhoff's hypotheses, the quantity ϑ_n does not appear in the expressions for ϵ and æ. It is only involved in ε_{33}. Thus, in the Kirchhoff–Love theory, ϑ_n does not affect the shell deformation and stress characteristics: indeed, if we set $\vartheta_n = 0$ we get the above expressions for ϵ and æ. The quantity ϑ_n describes the rotation of the shell cross section, which is tangent to the midsurface, about the normal. From a physical standpoint, at least for a homogeneous shell, these rotations are very small in comparison with the other rotations. This is related to the fact that the shell twisting stiffness is much bigger than its bending stiffness. Thus, in the Kirchhoff–Love theory, a consideration of ϑ_n and the rotations of the cross sections about the normal related to ϑ_n is impossible. In much of the literature, it is assumed that ϑ_n is zero; alternatively some analysis of its smallness, similar to the above, is proposed. See [Novozhilov, Chernykh, and Mikhailovskiy (1991); Vorovich (1999); Zhilin (2006)]. The reader can find a rigorous asymptotic proof of the smallness of ϑ_n, under certain assumptions on the loads, the type of boundary conditions, and shell homogeneity, in [Goldenveizer (1976)].

Let us turn to Hooke's law. Substitute ε into equation (7.17). The expression for $\tilde{\sigma}$ becomes

$$\tilde{\sigma} = \frac{E}{1+\nu}\left[\tilde{\varepsilon} + \frac{\nu}{1-\nu}\mathbf{A}\operatorname{tr}\tilde{\varepsilon}\right]. \qquad (7.27)$$

In component form it is

$$\sigma_\alpha^\beta = \frac{E}{1+\nu}\left[\varepsilon_\alpha^\beta + \frac{\nu}{1-\nu}\delta_\alpha^\beta \operatorname{tr}\tilde{\varepsilon}\right] \tag{7.28}$$

or, in more detail,

$$\sigma_1^1 = \frac{E}{1-\nu^2}(\epsilon_1^1 + \nu\epsilon_2^2) + \frac{Ez}{1-\nu^2}(\mathit{æ}_1^1 + \nu\mathit{æ}_2^2),$$

$$\sigma_2^2 = \frac{E}{1-\nu^2}(\epsilon_2^2 + \nu\epsilon_1^1) + \frac{Ez}{1-\nu^2}(\mathit{æ}_2^2 + \nu\mathit{æ}_1^1),$$

$$\sigma_1^2 = \frac{E}{1+\nu}\epsilon_1^2 + \frac{Ez}{1+\nu}\mathit{æ}_1^2.$$

Using these, we can find the resultants $\mathbf{T}\cdot\mathbf{A}$ and $\mathbf{M}$ via the formulas

$$\mathbf{T} = [\![(\mathbf{A}-z\mathbf{B})^{-1}\cdot\boldsymbol{\sigma}]\!], \qquad \mathbf{M} = -[\![(\mathbf{A}-z\mathbf{B})^{-1}\cdot z\boldsymbol{\sigma}\times\mathbf{n}]\!].$$

In the context of Kirchhoff's theory, knowing the deformation, we cannot recover the transverse shear stress resultants $\mathbf{T}\cdot\mathbf{n}$; indeed by Kirchhoff's hypothesis, the shear strains $\varepsilon_1^{\cdot 3}$ and $\varepsilon_2^{\cdot 3}$ in the shell are zero. But the vector $\mathbf{T}\cdot\mathbf{n}$, which represents the shear stress components, is not zero. Until now it has been left indefinite, but $\mathbf{T}\cdot\mathbf{n}$ can be found via (7.12) and (7.15).

Assumption S allows us to overcome the difficulties of integrating $(\mathbf{A}-z\mathbf{B})^{-1}$ over the thickness. This simplifies the formulas for the stress resultants and stress couples to

$$\mathbf{T}\cdot\mathbf{A} = [\![\mathbf{A}\cdot\tilde{\boldsymbol{\sigma}}]\!], \qquad \mathbf{M} = -[\![\mathbf{A}\cdot z\tilde{\boldsymbol{\sigma}}\times\mathbf{n}]\!]. \tag{7.29}$$

The simplification (7.29) implies that the two-dimensional equilibrium equations (7.12) and (7.15) with these versions of $\mathbf{T}$ and $\mathbf{M}$ are not exact consequences of the three-dimensional equilibrium equations as earlier. Rather, they are only first approximations to the equilibrium equations that differ in terms of the second order of smallness. Integrating with respect to z over the thickness, we get

$$\mathbf{T}\cdot\mathbf{A} = \frac{Eh}{1+\nu}\left[\boldsymbol{\epsilon} + \frac{\nu}{1-\nu}\mathbf{A}\operatorname{tr}\boldsymbol{\epsilon}\right] \tag{7.30}$$

and

$$\mathbf{M} = -\frac{Eh^3}{12(1+\nu)}\left[\mathit{æ} + \frac{\nu}{1-\nu}\mathbf{A}\operatorname{tr}\mathit{æ}\right]\times\mathbf{n}. \tag{7.31}$$

In the basis $\boldsymbol{\rho}^\alpha$, we have

$$\mathbf{T}\cdot\mathbf{A} = T_{\alpha\beta}\boldsymbol{\rho}^\alpha\boldsymbol{\rho}^\beta$$

and

$$\epsilon = \epsilon_{\alpha\beta} \rho^\alpha \rho^\beta.$$

Then the component representation of the stress resultant tensor is

$$T_{\alpha\beta} = \frac{Eh}{1+\nu} \left[\epsilon_{\alpha\beta} + \frac{\nu}{1-\nu} a_{\alpha\beta} \epsilon_\gamma^\gamma \right]. \qquad (7.32)$$

Comparing this with (7.28), we see that

$$T_{\alpha\beta} = [\![\sigma_{\alpha\beta}]\!] \equiv \int_{-h/2}^{h/2} \sigma_{\alpha\beta} \, dz. \qquad (7.33)$$

Here we have used the fact that, under Assumption S, we can take $G = 1$; hence we redefined

$$[\![f]\!] \equiv \int_{-h/2}^{h/2} f(z) \, dz.$$

Similarly, we find the components of the stress couple tensor:

$$M_{\alpha\beta} = [\![z\sigma_{\alpha\beta}]\!] \equiv \int_{-h/2}^{h/2} z\sigma_{\alpha\beta} \, dz. \qquad (7.34)$$

These are related to the components of æ through the formulas

$$M_{\alpha\beta} = \frac{Eh}{1+\nu} \left[æ_{\alpha\beta} + \frac{\nu}{1-\nu} a_{\alpha\beta} æ_\gamma^\gamma \right]. \qquad (7.35)$$

Comparing (7.33) and (7.34) with (7.29), we see that

$$\mathbf{T} \cdot \mathbf{A} = T_{\alpha\beta} \rho^\alpha \rho^\beta, \qquad \mathbf{M} = -M_{\alpha\beta} \rho^\alpha \rho^\beta \times \mathbf{n}.$$

These expressions for the components of stress resultants and stress couple tensors were used in [Koiter (1970)], [Donnell (1976)], [Libai and Simmonds (1998)], and many others.

Donnell analyzed the conditions under which a particular class of shallow shells that obey Assumption S can be used. The *theory of shallow shells* is usable for moderate deformations where the shell shape remains close to planar. For example, this is the case when the midsurface is a spherical cap which is a relatively small part of the complete sphere. This theory is attributed to Donnell, Marguerre, and Vlasov. In writing out æ, they neglect $\mathbf{B} \cdot \tilde{\mathbf{v}}$ in the expression for $\vartheta = \widetilde{\nabla} w$ as a small quantity of higher order. Thus, the strains in shallow shell theory are given by

$$2\epsilon = (\widetilde{\nabla}\tilde{\mathbf{v}}) \cdot \mathbf{A} + \mathbf{A} \cdot (\widetilde{\nabla}\tilde{\mathbf{v}})^T - 2w\mathbf{B}, \qquad æ = -\widetilde{\nabla}\widetilde{\nabla} w.$$

This simplification for æ was first introduced by the Russian engineer and mechanicist Vasilij Z. Vlasov [Vlasov (1949)].

In shallow shell theory, in the expressions for the stress resultants and stress couples, we use a "strange" plane geometry; we formally set the curvature components **B** to zero in æ but retain them in ϵ. The differential operations in this theory are produced in the curvilinear coordinate system. Note that Vlasov's initial equations were derived when the geometry of a shell was taken to be plane, but some nonzero curvatures were artificially included in ϵ. A particular case of shallow shell theory is the plate theory considered in § 7.8.

It is worth mentioning some aspects of the application of tensor analysis to shell theory. In the literature, one can find many versions of linear shell theory using various assumptions on the smallness of certain quantities. Some of these were derived in special coordinate systems like the ones constituted by the principal curvature lines. In a certain sense, for an isotropic material, such equations may lose their isotropic properties when we express them in arbitrary curvilinear coordinates. Such a change of the isotropic property was noted by [Zhilin (2006)] for an early version of the shell equations by Novozhilov. Tensor analysis allows us to avoid similar accidents.

7.5   Shell Energy

We saw the importance of strain and total energy in elasticity. Using these, we proved uniqueness of solution for boundary value problems of elasticity, and formulated variational principles of elasticity. In shell theory, the energy also plays an important role. Kirchhoff's hypotheses affect the form of the stresses and strains in a shell as a three-dimensional body, so we should derive the form of the strain energy density for the shell.

The strain energy density W of the shell as a three-dimensional body takes the form

$$W = \frac{1}{2}\boldsymbol{\sigma} \cdot\cdot\, \boldsymbol{\varepsilon} = \frac{1}{2}\sigma_j^i \varepsilon_i^j = \frac{1}{2}\sigma_\beta^\alpha \varepsilon_\alpha^\beta.$$

This follows from Kirchhoff's hypotheses that $\varepsilon_3^1 = 0$, $\varepsilon_3^2 = 0$, and $\sigma_3^3 = 0$.

Substituting (7.28) into W, we get

$$W = \frac{E}{2(1+\nu)}\left[\tilde{\varepsilon}\cdot\!\cdot\,\tilde{\varepsilon} + \frac{\nu}{1-\nu}\,\mathrm{tr}^2\,\tilde{\varepsilon}\right]$$

$$= \frac{E}{2(1+\nu)}\left[\varepsilon^\beta_\alpha\varepsilon^\alpha_\beta + \frac{\nu}{1-\nu}(\varepsilon^\alpha_\alpha)^2\right]. \tag{7.36}$$

Using (7.26), we transform this to

$$W = \frac{E}{2(1+\nu)}\left[\epsilon\cdot\!\cdot\,\epsilon + \frac{\nu}{1-\nu}\,\mathrm{tr}^2\,\epsilon\right] + \frac{Ez^2}{2(1+\nu)}\left[æ\cdot\!\cdot\,æ + \frac{\nu}{1-\nu}\,\mathrm{tr}^2\,æ\right]$$

$$+ \frac{Ez}{1+\nu}\left[\epsilon\cdot\!\cdot\,æ + \frac{\nu}{1-\nu}\,\mathrm{tr}\,\epsilon\,\mathrm{tr}\,æ\right]. \tag{7.37}$$

So, using Kirchhoff's hypotheses, we have derived the expression for W in terms of ϵ, $æ$, and z.

Consider the strain energy functional

$$E = \int_V W\,dV.$$

We recall that in curvilinear coordinates q^1, q^2, z the volume element is

$$dV = \sqrt{g}\,dq^1\,dq^2\,dz = \sqrt{a}G\,dq^1\,dq^2\,dz, \qquad G = \det(\mathbf{A} - z\mathbf{B}).$$

Under our assumptions on smallness of the displacements, W is a quantity of the second order of smallness because it is a quadratic form in the strains. Under Assumption S, we can neglect $z\mathbf{B}$ in comparison with $\mathbf{A}$, for which $\det\mathbf{A} = 1$, and change G to 1. Integrating (7.37) over z explicitly, we transform E to

$$E = \int_V W\sqrt{a}\,dq^1\,dq^2\,dz = \int_\Sigma U\sqrt{a}\,dq^1\,dq^2,$$

where $U = [\![W]\!]$ denotes the shell strain energy density per area:

$$U = \frac{Eh}{2(1+\nu)}\left[\epsilon\cdot\!\cdot\,\epsilon + \frac{\nu}{1-\nu}\,\mathrm{tr}^2\,\epsilon\right] + \frac{Eh^3}{24(1+\nu)}\left[æ\cdot\!\cdot\,æ + \frac{\nu}{1-\nu}\,\mathrm{tr}^2\,æ\right]. \tag{7.38}$$

U is a quadratic form in two tensors, ϵ and $æ$, defined on Σ. We see that U splits into two terms; the first depends only on ϵ, while the second depends only on $æ$. From a physical viewpoint, this decomposition means that the strain energy density splits into two parts. The first is due to the shell tangential deformation, i.e., the deformations in the tangent plane. The second is due to shell bending.

From (7.38) it follows that $U \geq 0$, and that $U = 0$ if and only if both $\varepsilon = \mathbf{0}$ and $\text{æ} = \mathbf{0}$. Indeed, we recall that $-1 < \nu \leq 1/2$ for an elastic material. Then

$$\epsilon \cdot\cdot \epsilon + \frac{\nu}{1 - \nu} \, \text{tr}^2 \, \epsilon > 0 \quad \text{when} \quad \epsilon \neq \mathbf{0}.$$

Similarly,

$$\text{æ} \cdot\cdot \text{æ} + \frac{\nu}{1 - \nu} \, \text{tr}^2 \, \text{æ} > 0 \quad \text{when} \quad \text{æ} \neq \mathbf{0}.$$

Exercise 7.12. Show that

$$\epsilon \cdot\cdot \epsilon + \frac{\nu}{1 - \nu} \, \text{tr}^2 \, \epsilon > 0 \quad \text{when} \quad \epsilon \neq \mathbf{0}.$$

Using the component notation for $\mathbf{T}$ and $\mathbf{M}$, we can represent U in another form:

$$U = \frac{1}{2} T_{\alpha\beta} \epsilon^{\alpha\beta} + \frac{1}{2} M_{\alpha\beta} \text{æ}^{\alpha\beta}. \tag{7.39}$$

From (7.39), with use of (7.32) and (7.35), it follows that

$$T_{\alpha\beta} = \frac{\partial U}{\partial \epsilon^{\alpha\beta}}, \qquad M_{\alpha\beta} = \frac{\partial U}{\partial \text{æ}^{\alpha\beta}}. \tag{7.40}$$

In component-free form these are

$$U = \frac{1}{2} \mathbf{T} \cdot\cdot \epsilon + \frac{1}{2} \mathbf{M} \cdot\cdot (\mathbf{n} \times \text{æ}) = \frac{1}{2} \mathbf{T} \cdot\cdot \epsilon + \frac{1}{2} (\mathbf{M} \times \mathbf{n}) \cdot\cdot \text{æ} \tag{7.41}$$

and

$$\mathbf{T} = \frac{\partial U}{\partial \epsilon}, \qquad \mathbf{M} \times \mathbf{n} = \frac{\partial U}{\partial \text{æ}}. \tag{7.42}$$

U inherits the positive definiteness of the energy density in three-dimensional linear elasticity, where the energy density $W = 0$ if and only if $\varepsilon = \mathbf{0}$. We recall that $\varepsilon = \mathbf{0}$ if and only if the displacement field is a small displacement of a rigid body, i.e., it is the sum of a parallel translation and a rotation of the body in space. We encounter a similar situation in shell theory. In Kirchhoff's theory, the expressions for ϵ and æ are cumbersome, so we will demonstrate that fact later only for a simpler case of the plate theory.

Exercise 7.13. Prove (7.41).

7.6   Boundary Conditions

To uniquely define a solution of the equilibrium problem for an elastic shell, we should supplement the equilibrium equations with boundary conditions. This is no elementary task: in the beginnings of shell theory, the question of boundary conditions was debated for a long time. Even the great Poisson proposed an erroneous number of boundary conditions.

The equilibrium equations are defined on the midsurface Σ, so the boundary conditions are given on its boundary contour ω.

First we will introduce the *kinematic boundary conditions*: on the boundary we assign the displacement $\mathbf{v}$ that defines the position of the boundary contour after deformation.   In the Kirchhoff–Love theory we should also assign the rotation of the normal $\mathbf{n}$ about the tangential to the boundary shell contour vector — that is, about the vector $\mathbf{n} \times \boldsymbol{\nu}$ (Fig. 7.6).

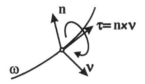

Fig. 7.4   The trihedron $\mathbf{n}, \boldsymbol{\nu}, \boldsymbol{\tau} = \mathbf{n} \times \boldsymbol{\nu}$ used to formulate the boundary conditions.

The rotation is defined by the formula $\boldsymbol{\vartheta} = \widetilde{\nabla}w + \mathbf{B} \cdot \tilde{\mathbf{v}}$, so the rotation about $\boldsymbol{\tau} = \mathbf{n} \times \boldsymbol{\nu}$ is

$$\vartheta_\nu \equiv \boldsymbol{\nu} \cdot \boldsymbol{\vartheta} = \boldsymbol{\nu} \cdot \widetilde{\nabla}w + \boldsymbol{\nu} \cdot \mathbf{B} \cdot \tilde{\mathbf{v}} = \frac{\partial w}{\partial \nu} + \boldsymbol{\nu} \cdot \mathbf{B} \cdot \tilde{\mathbf{v}},$$

where $\boldsymbol{\nu} \cdot \widetilde{\nabla}w = \partial w / \partial \nu$, the derivative with respect to the normal. Thus, in the Kirchhoff–Love theory, the *kinematic boundary conditions* consist of four scalar equations:

$$\mathbf{v}\big|_\omega = \mathbf{v}^0(s), \qquad \frac{\partial w}{\partial \nu} + \boldsymbol{\nu} \cdot \mathbf{B} \cdot \tilde{\mathbf{v}}\big|_\omega = \vartheta^0(s), \tag{7.43}$$

where $\mathbf{v}^0(s)$ and $\vartheta^0(s)$ are given on ω.

Some difficulties are involved in formulating static boundary conditions in the Kirchhoff–Love theory. At first glance, it seems we could assign the quantities $\boldsymbol{\nu} \cdot \mathbf{T}$ and $\boldsymbol{\nu} \cdot (\mathbf{M} \times \mathbf{n})$ on the boundary. But this provides five, not four, scalar conditions. Earlier we mentioned Poisson's error in constructing plate theory: he proposed too many static conditions on the boundary.

The derivation of the static boundary conditions can be based on the Lagrange variational principle for the shell, which states that in equilibrium the total energy functional takes its minimum value on the set of kinematically admissible displacements. The static boundary conditions become the natural boundary conditions for the problem of minimizing the total energy; they are dual to the kinematic conditions. For the shell, the proof of the Lagrange principle and the derivation of the static boundary conditions are cumbersome, so we refer to more specialized literature (e.g., [Green and Zerna (1954); Novozhilov, Chernykh, and Mikhailovskiy (1991)]) and merely sketch the procedure for deriving the boundary conditions. Later, this will be given in more detail for the technically simpler case of a plate.

A consequence of the Lagrange variational principle is the statement that the variation of the shell strain energy is equal to the work of external forces on the kinematically admissible displacements:

$$\delta E = \delta A.$$

It turns out that on the set of admissible displacements of the shell edge, the quantities $\mathbf{v} = \tilde{\mathbf{v}} + w\mathbf{n}$ and ϑ_ν, which constitute four scalar quantities, can be assigned independently. Hence the work of external forces on the shell edge takes the form

$$\delta A = \int_\omega \left(\boldsymbol{\varphi}^0(s) \cdot \delta \mathbf{v} - \ell^0(s)\delta\vartheta_\nu \right) ds,$$

where $\boldsymbol{\varphi}^0(s)$ and $\ell^0(s)$ are given functions on ω. It is seen that δA depends on $\delta\mathbf{v}$ and $\delta\vartheta_\nu$ linearly. With regard for (7.40) we get

$$\delta E = \int_\Sigma \delta U \sqrt{a}\, dq^1\, dq^2 = \int_\Sigma (T_{\alpha\beta}\, \delta\epsilon^{\alpha\beta} + M_{\alpha\beta}\, \delta\ae^{\alpha\beta})\sqrt{a}\, dq^1\, dq^2.$$

The further transformations of δE are done using integration by parts and the Gauss–Ostrogradsky theorem. We should represent $\delta\epsilon^{\alpha\beta}$ and $\delta\ae^{\alpha\beta}$ through $\delta\mathbf{v}$ and $\delta\boldsymbol{\vartheta}$ and eliminate their derivatives from the surface integrals. Unlike the case with three-dimensional elasticity, $\delta\ae^{\alpha\beta}$ contains the second derivatives of w and hence integration by parts should be done twice. For the boundary integrals we should also apply further transformations in such a way that they contain only the factors $\delta\mathbf{v}$ and $\delta\vartheta_\nu$; the techniques will be presented in the section for the plate. The equation $\delta E = \delta A$ holds for any admissible $\delta\mathbf{v}$ and $\delta\boldsymbol{\vartheta}$. Selecting only $\delta\mathbf{v}$ and $\delta\boldsymbol{\vartheta}$ satisfying the homogeneous kinematic conditions, we find that the minimum point $\mathbf{v}$ and $\boldsymbol{\vartheta}$ satisfies the equilibrium equations on Σ. Next, extending the set of $\delta\mathbf{v}$ and

$\delta\boldsymbol{\vartheta}$ to all the admissible $\delta\mathbf{v}$ and $\delta\boldsymbol{\vartheta}$, we derive the following set of static boundary conditions for $\mathbf{v}$ and $\boldsymbol{\vartheta}$. They are defined on ω:

$$\boldsymbol{\nu} \cdot \mathbf{T} \cdot \mathbf{A}\big|_\omega = \boldsymbol{\varphi}^0(s) \cdot \mathbf{A},$$

$$\boldsymbol{\nu} \cdot \mathbf{T} \cdot \mathbf{n}\big|_\omega - \frac{\partial}{\partial s}\left(\boldsymbol{\nu} \cdot (\mathbf{M} \times \mathbf{n}) \cdot \boldsymbol{\tau}\right)\big|_\omega = \boldsymbol{\varphi}^0(s) \cdot \mathbf{n}, \qquad (7.44)$$

$$\boldsymbol{\nu} \cdot (\mathbf{M} \times \mathbf{n}) \cdot \boldsymbol{\nu}\big|_\omega = \ell^0(s).$$

As with the kinematic boundary conditions, these are also four scalar equations. Here $\boldsymbol{\varphi}^0(s)$ and $\ell^0(s)$ are given functions on ω that represent the stress resultants and the bending couples on the shell edge.

The principal boundary conditions used in engineering are as follows.

Clamped (fixed) edge:

$$\mathbf{v}\big|_\omega = \mathbf{0}, \qquad \frac{\partial w}{\partial \nu}\bigg|_\omega = 0.$$

Simple support edge:

$$\mathbf{v}\big|_\omega = \mathbf{0}, \qquad \boldsymbol{\nu} \cdot (\mathbf{M} \times \mathbf{n}) \cdot \boldsymbol{\nu}\big|_\omega = 0.$$

Free edge:

$$\boldsymbol{\nu} \cdot \mathbf{T} \cdot \mathbf{A}\big|_\omega = \mathbf{0}, \qquad \boldsymbol{\nu} \cdot \mathbf{T} \cdot \mathbf{n}\big|_\omega - \frac{\partial}{\partial s}\left(\boldsymbol{\nu} \cdot (\mathbf{M} \times \mathbf{n}) \cdot \boldsymbol{\tau}\right)\big|_\omega = 0,$$

$$\boldsymbol{\nu} \cdot (\mathbf{M} \times \mathbf{n}) \cdot \boldsymbol{\nu}\big|_\omega = 0.$$

Mixed boundary conditions are also possible. For example, on a part of ω we might specify (7.43) and on the remainder (7.44). Other combinations of boundary conditions, four at each point of ω, define various boundary value problems. Admissible combinations of boundary conditions are defined by the variational setups of these problems.

7.7   A Few Remarks on the Kirchhoff–Love Theory

There exist various approaches in the Kirchhoff–Love theory. They use different simplifying assumptions that depend on the choice of constitutive relations for U, $\mathbf{T}$, and $\mathbf{M}$, as well as on the choice of deformation measures $\boldsymbol{\epsilon}$ and $\boldsymbol{\ae}$. In a linear theory, equation (7.38) for U is a common form of dependence of U on the deformation measures (see, e.g., [Donnell (1976); Koiter (1970); Novozhilov, Chernykh, and Mikhailovskiy (1991); Timoshenko (1985)]).

The versions may differ in their definitions of $\mathbf{T}$. We introduced $\mathbf{T}$ as

$$\mathbf{T} = [\![(\mathbf{A} - z\mathbf{B})^{-1} \cdot \boldsymbol{\sigma}]\!]$$

and used Assumption S to recast it in the form

$$\mathbf{T} = [\![\mathbf{A} \cdot \boldsymbol{\sigma}]\!].$$

In the Kirchhoff–Love literature, one can find both the initial and simplified definitions. Let us consider this difference in more detail. We introduce a special coordinate system on the shell surface, defined by the principal curvature curves on Σ. Now $\mathbf{B}$ takes the diagonal form

$$\mathbf{B} = \mathbf{e}_1\mathbf{e}_1/R_1 + \mathbf{e}_2\mathbf{e}_2/R_2,$$

where R_1 and R_2 are principal radii of curvature of Σ. Then

$$\mathbf{A} = \mathbf{e}_1\mathbf{e}_1 + \mathbf{e}_2\mathbf{e}_2$$

and

$$G = (1 - z/R_1)(1 - z/R_2).$$

The tensor $G(\mathbf{A} - z\mathbf{B})^{-1}$ takes the form

$$G(\mathbf{A} - z\mathbf{B})^{-1} = \left(1 - \frac{z}{R_2}\right)\mathbf{e}_1\mathbf{e}_1 + \left(1 - \frac{z}{R_1}\right)\mathbf{e}_2\mathbf{e}_2.$$

Taking

$$\mathbf{T} = [\![(\mathbf{A} - z\mathbf{B})^{-1} \cdot \boldsymbol{\sigma}]\!],$$

we get the following formulas for the components of $\mathbf{T}$:

$$T_{11} = \int_{-h/2}^{h/2}\left(1 - \frac{z}{R_2}\right)\sigma_{11}\,dz, \quad T_{22} = \int_{-h/2}^{h/2}\left(1 - \frac{z}{R_1}\right)\sigma_{22}\,dz,$$

$$T_{12} = T_{21} = \int_{-h/2}^{h/2}\sigma_{12}\,dz.$$

The alternative definition $\mathbf{T} = [\![\mathbf{A} \cdot \boldsymbol{\sigma}]\!]$ yields the expressions

$$T_{11} = \int_{-h/2}^{h/2}\sigma_{11}\,dz, \quad T_{22} = \int_{-h/2}^{h/2}\sigma_{22}\,dz, \quad T_{12} = T_{21} = \int_{-h/2}^{h/2}\sigma_{12}\,dz.$$

So the two definitions of $\mathbf{T}$ yield equilibrium equations that differ by quantities that are second-order small. The simplified form for $\mathbf{T}$ was used by a number of authors, e.g., [Goldenveizer (1976); Novozhilov, Chernykh, and Mikhailovskiy (1991); Timoshenko (1985)]. Koiter [Koiter (1970)] used (7.38) and (7.40) to relate the stress resultants and the bending couples with

U, but he presented more complex expressions to calculate them through the stresses in a three-dimensional body.

The components of $M_{\alpha\beta}$ are also commonly calculated using (7.34), and are related to U through (7.40). As is the case for $\mathbf{T}$, however, in the literature one can find other ways to introduce $\mathbf{M}$. In [Goldenveizer (1976)], for instance, the expressions for $M_{\alpha\beta}$ include terms that are quadratic in z.

Some works on shell theory feature a "mixed" formulation of the boundary value problems, introducing the stress resultants via the simplified formula $\mathbf{T} = [\![\mathbf{A} \cdot \boldsymbol{\sigma}]\!]$. The constitutive equations take the form (7.38), but the equilibrium equations include additional terms arising from the complete definition $\mathbf{T} = [\![(\mathbf{A} - z\mathbf{B})^{-1} \cdot \boldsymbol{\sigma}]\!]$. In this way, the couple stresses appear in all the equilibrium equations.

In [Koiter (1970); Novozhilov, Chernykh, and Mikhailovskiy (1991)] and certain other works, the above tensor $\boldsymbol{\epsilon}$ is also used. The second deformation measure $\boldsymbol{\mathit{æ}}$, which was derived by the simplifying replacement of $\boldsymbol{\vartheta}$ by $\widetilde{\nabla}w$, also has a few different forms (see the discussion in [Koiter (1970)]).

In the linear Kirchhoff–Love shell theory, the reader can encounter various forms of the equilibrium equations in displacements, as well as differences in the stress resultants and bending couples. Clearly, in different versions of shell theory the forms of the static boundary conditions can differ, and this is also reflected in the literature.

It is worth noting that in different references, the same equilibrium equations in stresses can lead to different equilibrium equations in displacements. For a thin shell with a sufficiently smooth midsurface and smooth loads, this difference, as a rule, is of the second order of smallness in comparison with the main terms in the equations.

7.8 Plate Theory

Resultant forces and moments in a plate; equilibrium equations

A particular case of a shell is a thin plate. The base surface Σ of a plate is part of a plane, so $\mathbf{B} = \mathbf{0}$. The relations for a plate follow from shell theory. The equilibrium equations are

$$\widetilde{\nabla} \cdot \mathbf{T} + \mathbf{q} = \mathbf{0}, \qquad \widetilde{\nabla} \cdot \mathbf{M} + \mathbf{T}_\times + \mathbf{m} = \mathbf{0}, \qquad (7.45)$$

where

$$\mathbf{T} = [\![\mathbf{A} \cdot \boldsymbol{\sigma}]\!], \qquad \mathbf{M} = -[\![\mathbf{A} \cdot z\boldsymbol{\sigma} \times \mathbf{n}]\!]. \tag{7.46}$$

In a Cartesian basis $\mathbf{i}_1, \mathbf{i}_2, \mathbf{i}_3 = \mathbf{n}$, we have

$$\boldsymbol{\sigma} = \sigma_{\alpha\beta}\mathbf{i}_\alpha\mathbf{i}_\beta + \sigma_{\alpha 3}(\mathbf{i}_\alpha\mathbf{n} + \mathbf{n}\mathbf{i}_\alpha) + \sigma_{33}\mathbf{n}\mathbf{n}, \quad \mathbf{A} = \mathbf{i}_\alpha\mathbf{i}_\alpha, \quad \widetilde{\nabla} = \mathbf{i}_\alpha \frac{\partial}{\partial x_\alpha}, \quad G = 1,$$

where x_1, x_2 are Cartesian coordinates in the plane. The stress resultants are

$$\mathbf{T} = T_{\alpha\beta}\mathbf{i}_\alpha\mathbf{i}_\beta + T_{\alpha n}\mathbf{i}_\alpha\mathbf{n}, \tag{7.47}$$

$$T_{\alpha\beta} = [\![\sigma_{\alpha\beta}]\!] \equiv \int_{-h/2}^{h/2} \sigma_{\alpha\beta}\, dz, \qquad T_{\alpha n} = [\![\sigma_{\alpha 3}]\!] \equiv \int_{-h/2}^{h/2} \sigma_{\alpha 3}\, dz.$$

Note that σ_{33} does not appear in the definition of the stress resultants or the transverse shear stress resultants. It follows that the matrix $T_{\alpha\beta}$ is symmetric: $T_{\alpha\beta} = T_{\beta\alpha}$. The quantity $\mathbf{T}_\times$ is calculated by the formula

$$\mathbf{T}_\times = T_{\alpha\beta}\mathbf{i}_\alpha \times \mathbf{i}_\beta + T_{\alpha n}\mathbf{i}_\alpha \times \mathbf{n} = T_{\alpha n}\mathbf{i}_\alpha \times \mathbf{n} = T_{2n}\mathbf{i}_1 - T_{1n}\mathbf{i}_2.$$

Next, we introduce the moments matrix

$$M_{\alpha\beta} = [\![z\sigma_{\alpha\beta}]\!] \equiv \int_{-h/2}^{h/2} z\sigma_{\alpha\beta}\, dz. \tag{7.48}$$

Here, $M_{\alpha\beta}$ is a symmetric matrix as well. With regard for the equality

$$\begin{aligned}\boldsymbol{\sigma} \times \mathbf{n} &= \sigma_{\alpha\beta}\mathbf{i}_\alpha\mathbf{i}_\beta \times \mathbf{n} \\ &= -\sigma_{11}\mathbf{i}_1\mathbf{i}_2 + \sigma_{12}\mathbf{i}_1\mathbf{i}_1 - \sigma_{21}\mathbf{i}_2\mathbf{i}_2 + \sigma_{22}\mathbf{i}_2\mathbf{i}_1,\end{aligned}$$

from (7.46) we obtain

$$\mathbf{M} = [\![z\sigma_{11}]\!]\mathbf{i}_1\mathbf{i}_2 - [\![z\sigma_{12}]\!]\mathbf{i}_1\mathbf{i}_1 + [\![z\sigma_{21}]\!]\mathbf{i}_2\mathbf{i}_2 - [\![z\sigma_{22}]\!]\mathbf{i}_2\mathbf{i}_1.$$

By (7.48), $\mathbf{M}$ takes the form

$$\begin{aligned}\mathbf{M} &= M_{11}\mathbf{i}_1\mathbf{i}_2 - M_{12}\mathbf{i}_1\mathbf{i}_1 + M_{21}\mathbf{i}_2\mathbf{i}_2 - M_{22}\mathbf{i}_2\mathbf{i}_1 \\ &= -M_{\alpha\beta}\mathbf{i}_\alpha\mathbf{i}_\beta \times \mathbf{n}.\end{aligned}$$

So the equilibrium equations in Cartesian coordinates are

$$\frac{\partial T_{\alpha\beta}}{\partial x_\alpha}\mathbf{i}_\beta + \frac{\partial T_{\alpha n}}{\partial x_\alpha}\mathbf{n} + \mathbf{q} = \mathbf{0}, \qquad -\frac{\partial M_{\alpha\beta}}{\partial x_\alpha}\mathbf{i}_\beta \times \mathbf{n} + \mathbf{T}_\times + \mathbf{m} = \mathbf{0}. \tag{7.49}$$

In component form, these are

$$\frac{\partial T_{11}}{\partial x_1} + \frac{\partial T_{21}}{\partial x_2} + q_1 = 0,$$

$$\frac{\partial T_{12}}{\partial x_1} + \frac{\partial T_{22}}{\partial x_2} + q_2 = 0,$$

$$\frac{\partial T_{1n}}{\partial x_1} + \frac{\partial T_{2n}}{\partial x_2} + q_n = 0, \tag{7.50}$$

and

$$\frac{\partial M_{11}}{\partial x_1} + \frac{\partial M_{21}}{\partial x_2} - T_{1n} + m_1 = 0,$$

$$\frac{\partial M_{12}}{\partial x_1} + \frac{\partial M_{22}}{\partial x_2} - T_{2n} + m_2 = 0, \tag{7.51}$$

where

$$q_\alpha = \mathbf{i}_\alpha \cdot \mathbf{q}, \qquad q_n = \mathbf{n} \cdot \mathbf{q}, \qquad m_\alpha = \mathbf{i}_\alpha \cdot (\mathbf{n} \times \mathbf{m}).$$

Note that $T_{12} = T_{21}$ and $M_{12} = M_{21}$; we keep the different notation for symmetry.

In applications of shell theory, $\mathbf{m}$ is often negligible in comparison with the external forces. Hence, for simplicity we put $\mathbf{m} = \mathbf{0}$. Now we exclude T_{1n} and T_{2n} from equations (7.51). For this, we differentiate the second equation of (7.51) with respect to x_2, the first from (7.51) with respect to x_1, and add the results. Using the third equation from (7.50), we obtain the well-known equation of plate theory

$$\frac{\partial^2 M_{11}}{\partial x_1{}^2} + 2\frac{\partial^2 M_{12}}{\partial x_1 \partial x_2} + \frac{\partial^2 M_{22}}{\partial x_2{}^2} + q_n = 0. \tag{7.52}$$

In index notation this reads

$$\frac{\partial^2 M_{\alpha\beta}}{\partial x_\alpha \partial x_\beta} + q_n = 0,$$

whereas in component-free notation it takes the form

$$\widetilde{\nabla} \cdot (\widetilde{\nabla} \cdot (\mathbf{M} \times \mathbf{n})) + \mathbf{q} \cdot \mathbf{n} = 0. \tag{7.53}$$

Indeed,

$$\mathbf{M} = -M_{\alpha\beta}\mathbf{i}_\alpha\mathbf{i}_\beta \times \mathbf{n}.$$

Using the identity

$$(\mathbf{i}_\beta \times \mathbf{n}) \times \mathbf{n} = -\mathbf{i}_\beta,$$

the proof of which is left to the reader, we get

$$\mathbf{M} \times \mathbf{n} = -M_{\alpha\beta}\mathbf{i}_\alpha(\mathbf{i}_\beta \times \mathbf{n}) \times \mathbf{n} = M_{\alpha\beta}\mathbf{i}_\alpha\mathbf{i}_\beta.$$

Next we obtain

$$\widetilde{\nabla} \cdot (\mathbf{M} \times \mathbf{n}) = \frac{\partial M_{\alpha\beta}}{\partial x_\alpha}\mathbf{i}_\beta, \qquad \widetilde{\nabla} \cdot (\widetilde{\nabla} \cdot (\mathbf{M} \times \mathbf{n})) = \frac{\partial^2 M_{\alpha\beta}}{\partial x_\alpha \partial x_\beta},$$

from which (7.53) follows.

In plate theory, Kirchhoff's kinematic hypotheses reduce to an assumption on the form of the displacement field, which is

$$\mathbf{u}(x_1, x_2, z) = \mathbf{v}(x_1, x_2) - z\widetilde{\nabla}w(x_1, x_2)$$
$$= \tilde{\mathbf{v}} + w\mathbf{n} - z\widetilde{\nabla}w, \qquad \tilde{\mathbf{v}} \cdot \mathbf{n} = 0. \qquad (7.54)$$

Let us find the expression for the strain tensor that corresponds to $\mathbf{u}$. We have

$$\nabla\mathbf{u} = \left(\widetilde{\nabla} + \mathbf{n}\frac{\partial}{\partial z}\right)(\tilde{\mathbf{v}} + w\mathbf{n} - z\widetilde{\nabla}w)$$
$$= \widetilde{\nabla}\tilde{\mathbf{v}} - z\widetilde{\nabla}\widetilde{\nabla}w + (\widetilde{\nabla}w)\mathbf{n} - \mathbf{n}\widetilde{\nabla}w,$$

and

$$(\nabla\mathbf{u})^T = (\widetilde{\nabla}\tilde{\mathbf{v}})^T - z\widetilde{\nabla}\widetilde{\nabla}w + \mathbf{n}\widetilde{\nabla}w - (\widetilde{\nabla}w)\mathbf{n}.$$

Thus,

$$2\varepsilon = \widetilde{\nabla}\tilde{\mathbf{v}} + (\widetilde{\nabla}\tilde{\mathbf{v}})^T - 2z\widetilde{\nabla}\widetilde{\nabla}w.$$

It follows that $\varepsilon \cdot \mathbf{n} = \mathbf{n} \cdot \varepsilon = \mathbf{0}$.

Using the notation

$$2\epsilon = \widetilde{\nabla}\tilde{\mathbf{v}} + (\widetilde{\nabla}\tilde{\mathbf{v}})^T, \qquad \text{æ} = -\widetilde{\nabla}\widetilde{\nabla}w,$$

we find that the plate strains split into two parts:

$$\varepsilon = \epsilon + z\text{æ}. \qquad (7.55)$$

Here ϵ represents the deformation in the plate plane, and æ is related to plate bending. Geometrically, æ describes an infinitesimal change of the plate curvatures due to bending.

Now let us use Kirchhoff's static hypothesis $\sigma_{33} = 0$. From Hooke's law it follows that

$$\sigma_{11} = \frac{E}{1 - \nu^2}(\varepsilon_{11} + \nu\varepsilon_{22}),$$

$$\sigma_{22} = \frac{E}{1 - \nu^2}(\varepsilon_{22} + \nu\varepsilon_{11}),$$

$$\sigma_{12} = \frac{E}{1 + \nu}\varepsilon_{12}. \tag{7.56}$$

Substituting (7.55) into these, we get

$$\sigma_{11} = \frac{E}{1 - \nu^2}(\epsilon_{11} + \nu\epsilon_{22}) + \frac{Ez}{1 - \nu^2}(\ae_{11} + \nu\ae_{22}),$$

$$\sigma_{22} = \frac{E}{1 - \nu^2}(\epsilon_{22} + \nu\epsilon_{11}) + \frac{Ez}{1 - \nu^2}(\ae_{22} + \nu\ae_{11}),$$

$$\sigma_{12} = \frac{E}{1 + \nu}\epsilon_{12} + \frac{Ez}{1 + \nu}\ae_{12}.$$

Now by (7.47) and (7.48), integrating the expressions for $\sigma_{\alpha\beta}$ with respect to z over thickness, we derive the expressions for the stress resultants and stress couples:

$$T_{11} = \frac{Eh}{1 - \nu^2}(\epsilon_{11} + \nu\epsilon_{22}),$$

$$T_{22} = \frac{Eh}{1 - \nu^2}(\epsilon_{22} + \nu\epsilon_{11}),$$

$$T_{12} = \frac{Eh}{1 + \nu}\epsilon_{12},$$

$$M_{11} = \frac{Eh^3}{12(1 - \nu^2)}(\ae_{11} + \nu\ae_{22}),$$

$$M_{22} = \frac{Eh^3}{12(1 - \nu^2)}(\ae_{22} + \nu\ae_{11}),$$

$$M_{12} = \frac{Eh^3}{12(1 + \nu)}\ae_{12}. \tag{7.57}$$

The constant

$$D = Eh^3/12(1 - \nu^2)$$

is the *bending stiffness*. Substituting the expressions for $M_{\alpha\beta}$ into (7.52) and taking into account that

$$\ae_{11} = -\frac{\partial^2 w}{\partial x_1^2}, \qquad \ae_{22} = -\frac{\partial^2 w}{\partial x_2^2}, \qquad \ae_{12} = -\frac{\partial^2 w}{\partial x_1 \partial x_2},$$

we obtain

$$D \left(\frac{\partial^4 w}{\partial x_1^4} + 2 \frac{\partial^4 w}{\partial x_1^2 \partial x_2^2} + \frac{\partial^4 w}{\partial x_2^4} \right) = q_n. \tag{7.58}$$

In component-free form this is

$$D \widetilde{\nabla}^4 w = q_n. \tag{7.59}$$

Because

$$\widetilde{\nabla}^4 = (\widetilde{\nabla}^2)^2 = \Delta^2,$$

we call (7.58) (or (7.59)) the *biharmonic equation*.

This equation has a long history in plate theory. The bending equation was published by Marie-Sophie Germain, who received a prize from the Paris Academy of Sciences in 1816. Later some of Germain's errors were corrected by Lagrange, hence the equation is now called *Germain's equation* or the *Germain–Lagrange equation*. The plate theory we have considered is attributed to Kirchhoff, who formulated correct boundary conditions for a plate. Moreover, Kirchhoff clearly introduced physically meaningful hypotheses regarding the distribution of displacements and stresses in the plate.

Note that the $T_{\alpha\beta}$ do not depend on w. Thus, the plate equilibrium equations split into two groups. The first consists of the two first equations of (7.50) with respect to $\tilde{\mathbf{v}}$, whereas the second consists of the equation (7.59) for variable w.

Let us write the equations for $\mathbf{T} \cdot \mathbf{A}$ in terms of the tangential displacement vector $\tilde{\mathbf{v}}$. In tensor form, equations (7.50) are

$$\widetilde{\nabla} \cdot (\mathbf{T} \cdot \mathbf{A}) + \mathbf{q} \cdot \mathbf{A} = \mathbf{0}. \tag{7.60}$$

The constitutive equation for $\mathbf{T} \cdot \mathbf{A}$ is

$$\mathbf{T} \cdot \mathbf{A} = \frac{Eh}{1 + \nu} \left[\epsilon + \frac{\nu}{1 - \nu} \mathbf{A} \operatorname{tr} \epsilon \right].$$

So we obtain

$$\begin{aligned}
\widetilde{\nabla} \cdot (\mathbf{T} \cdot \mathbf{A}) &= \frac{Eh}{1 + \nu} \left[\widetilde{\nabla} \cdot \epsilon + \frac{\nu}{1 - \nu} \widetilde{\nabla} \operatorname{tr} \epsilon \right] \\
&= \frac{Eh}{2(1 + \nu)} \left[\widetilde{\nabla} \cdot \widetilde{\nabla} \tilde{\mathbf{v}} + \widetilde{\nabla} \widetilde{\nabla} \cdot \tilde{\mathbf{v}} + \frac{2\nu}{1 - \nu} \widetilde{\nabla} \widetilde{\nabla} \cdot \tilde{\mathbf{v}} \right] \\
&= \frac{Eh}{2(1 + \nu)} \left[\widetilde{\nabla} \cdot \widetilde{\nabla} \tilde{\mathbf{v}} + \frac{1 + \nu}{1 - \nu} \widetilde{\nabla} \widetilde{\nabla} \cdot \tilde{\mathbf{v}} \right].
\end{aligned}$$

In terms of $\tilde{\mathbf{v}}$, equation (7.60) takes the form

$$\frac{Eh}{2(1+\nu)}\left[\tilde{\nabla}\cdot\tilde{\nabla}\tilde{\mathbf{v}}+\frac{1+\nu}{1-\nu}\tilde{\nabla}\tilde{\nabla}\cdot\tilde{\mathbf{v}}\right]+\mathbf{q}\cdot\mathbf{A}=\mathbf{0}. \qquad (7.61)$$

There may be some concern over equations (7.51), which contain the transverse shear stress resultants T_{1n} and T_{2n}. It may seem that by the kinematic hypothesis $\mathbf{n}\cdot\boldsymbol{\varepsilon}=\mathbf{0}$ and the last equation from (7.47), both T_{1n} and T_{2n} should vanish. But Kirchhoff's kinematic hypothesis restricts only the form of deformation. So in Kirchhoff's plate theory, T_{1n} and T_{2n} are not defined by the strains; they are the reactions due to the strain constraints. Note that the equations (7.51) are used to determine T_{1n} and T_{2n}.

Boundary conditions

In plate theory as a particular case of shell theory, on the boundary contour ω we can assign the kinematic conditions

$$\tilde{\mathbf{v}}=\mathbf{v}^0(s), \qquad w=w^0(s), \qquad \frac{\partial w}{\partial\nu}=\vartheta^0(s), \qquad (7.62)$$

or the static conditions

$$\boldsymbol{\nu}\cdot\mathbf{T}\cdot\mathbf{A}=\boldsymbol{\varphi}^0(s)\cdot\mathbf{A}, \qquad \boldsymbol{\nu}\cdot\mathbf{T}\cdot\mathbf{n}-\frac{\partial}{\partial s}\left(\boldsymbol{\nu}\cdot(\mathbf{M}\times\mathbf{n})\cdot\boldsymbol{\tau}\right)=\varphi_n^0(s),$$

$$\boldsymbol{\nu}\cdot(\mathbf{M}\times\mathbf{n})\cdot\boldsymbol{\nu}=\ell^0(s). \qquad (7.63)$$

For a mixed boundary value problem, these sets of conditions can be given on different portions of ω. Alternatively, at a point we can include conditions from (7.62) and the dual conditions from (7.63). At each point, four scalar conditions must be appointed.

We split the boundary load into tangential and normal parts:

$$\boldsymbol{\varphi}^0=\boldsymbol{\varphi}^0\cdot\mathbf{A}+\varphi_n^0\mathbf{n}, \qquad \varphi_n^0=\boldsymbol{\varphi}^0\cdot\mathbf{n}.$$

From equilibrium equations (7.51), we find the transverse shear stress resultants in terms of the stress couples. As in our earlier treatment of shell theory, we set $m_1=m_2=0$. We have

$$T_{1n}=\frac{\partial M_{11}}{\partial x_1}+\frac{\partial M_{21}}{\partial x_2}, \qquad T_{2n}=\frac{\partial M_{12}}{\partial x_1}+\frac{\partial M_{22}}{\partial x_2}.$$

These can be written as the single vector equation

$$\mathbf{T}\cdot\mathbf{n}=\tilde{\nabla}\cdot(\mathbf{M}\times\mathbf{n}).$$

There are various types of mixed boundary conditions. We restrict ourselves to the two defined by (7.62) and (7.63). Note that $\mathbf{T}\cdot\mathbf{A}$ depends only on $\tilde{\mathbf{v}}$, whereas $\mathbf{M}$ is defined through w. Thus, like the equilibrium equations, the boundary conditions in plate theory split into two sets. One is for normal displacement w:

$$w\big|_{\omega_1} = w^0(s), \qquad \frac{\partial w}{\partial \nu}\bigg|_{\omega_1} = \vartheta^0(s), \qquad (7.64)$$

$$\boldsymbol{\nu} \cdot (\tilde{\nabla} \cdot (\mathbf{M} \times \mathbf{n}))\big|_{\omega_2} - \frac{\partial}{\partial s}\,(\boldsymbol{\nu} \cdot (\mathbf{M} \times \mathbf{n}) \cdot \boldsymbol{\tau})\big|_{\omega_2} = \varphi_n^0(s),$$

$$\boldsymbol{\nu} \cdot (\mathbf{M} \times \mathbf{n}) \cdot \boldsymbol{\nu}\big|_{\omega_2} = \ell^0(s).$$

The other is for tangential displacements $\tilde{\mathbf{v}}$:

$$\tilde{\mathbf{v}}\big|_{\omega_3} = \mathbf{v}^0(s), \qquad \boldsymbol{\nu} \cdot \mathbf{T} \cdot \mathbf{A}\big|_{\omega_4} = \boldsymbol{\varphi}^0(s) \cdot \mathbf{A}. \qquad (7.65)$$

We have supposed the plate contour to be partitioned in such a way that $\omega_2 = \omega \setminus \omega_1$ and $\omega_4 = \omega \setminus \omega_3$.

In Appendix A we list some common homogeneous boundary conditions in terms of w for the plate.

Strain energy for the plate

Let us consider the strain energy density for a plate in more detail. It is

$$U = \frac{Eh}{2(1+\nu)}\left[\boldsymbol{\epsilon} \cdot\cdot\, \boldsymbol{\epsilon} + \frac{\nu}{1-\nu}\,\mathrm{tr}^2\,\boldsymbol{\epsilon}\right] + \frac{Eh^3}{24(1+\nu)}\left[\text{æ} \cdot\cdot\, \text{æ} + \frac{\nu}{1-\nu}\,\mathrm{tr}^2\,\text{æ}\right]. \tag{7.66}$$

A plate is a particular case of a shell considered above, but with $\mathbf{B} = \mathbf{0}$. Hence we can use the expression for shell strain density

$$U = \frac{1}{2}T_{\alpha\beta}\epsilon^{\alpha\beta} + \frac{1}{2}M_{\alpha\beta}\text{æ}^{\alpha\beta} = \frac{1}{2}\mathbf{T} \cdot\cdot\, \boldsymbol{\epsilon} + \frac{1}{2}\mathbf{M} \cdot\cdot\, (\mathbf{n} \times \text{æ}).$$

Let us rewrite U in terms of the displacements:

$$U = \frac{Eh}{2(1+\nu)}\left[\frac{\partial u_\alpha}{\partial x_\beta}\frac{\partial u_\alpha}{\partial x_\beta} + \frac{\nu}{1-\nu}\left(\frac{\partial u_\alpha}{\partial x_\alpha}\right)^2\right]$$

$$+ \frac{Eh^3}{24(1+\nu)}\left[\frac{\partial^2 w}{\partial x_\alpha \partial x_\beta}\frac{\partial^2 w}{\partial x_\alpha \partial x_\beta} + \frac{\nu}{1-\nu}\left(\frac{\partial^2 w}{\partial x_\alpha \partial x_\alpha}\right)^2\right].$$

Rigid motions

Now we consider the consequences of the equation $U = 0$. Because U is positive definite, it is zero if and only if both $\boldsymbol{\epsilon}$ and æ are zero. But $\boldsymbol{\epsilon} = \mathbf{0}$ implies that the plate can move in its plane as a rigid body only. In other words, it can translate and rotate about a normal to its midplane. The corresponding displacement field takes the form

$$\tilde{\mathbf{v}} = \mathbf{v}_0 + \omega_0 \mathbf{n} \times (\boldsymbol{\rho} - \boldsymbol{\rho}_0),$$

where $\mathbf{v}_0$ is an arbitrary fixed vector in the midplane, i.e., $\mathbf{v}_0 \cdot \mathbf{n} = 0$, the angle ω_0 is an arbitrary but fixed rotation angle about $\mathbf{n}$, the vector $\boldsymbol{\rho}$ is the position vector of a point on the midplane, and $\boldsymbol{\rho}_0$ is an arbitrary but fixed vector in the same plane. This formula is a particular case of (6.41) for the plane deformation.

The equation

$$\mathbf{æ} \equiv -\widetilde{\nabla}\widetilde{\nabla}w = \mathbf{0}$$

implies that w is a linear function in x_1 and x_2:

$$w = w_0 + w_1 x_1 + w_2 x_2,$$

where w_0, w_1, and w_2 are arbitrary fixed scalars. Physically, w corresponds to a translation of the whole plate in the normal direction and a rotation with respect to the axes $\mathbf{i}_1, \mathbf{i}_2$. Thus, as for three-dimensional elasticity, $U = 0$ only for those displacements that represent small displacements of the plate as a rigid body.

Lagrange variational principle in plate theory

In shell theory, the Lagrange variational principle of elasticity changes somewhat because of Kirchhoff's hypotheses. We will formulate Lagrange's variational principle for the plate. Recall that in three-dimensional elasticity, the total energy functional takes the form

$$E(\mathbf{u}) = \int_V W(\varepsilon)\,dV - \int_V \rho \mathbf{f} \cdot \mathbf{u}\,dV - \int_S \mathbf{t}^0 \cdot \mathbf{u}\,dS.$$

We found that for a thin shell, the integral over V is represented through the integral over the midplane Σ. The same is true for the plate:

$$\int_V W(\varepsilon)\,dV = \int_\Sigma U\,d\Sigma,$$

where U is defined by (7.66). Let us consider the remaining terms in E, which represent the work of external forces. In terms of the plate theory, they take the form

$$\int_\Sigma \mathbf{q} \cdot \mathbf{v}\,d\Sigma + \int_{\omega_4} \boldsymbol{\varphi}^0(s) \cdot \tilde{\mathbf{v}}\,dS + \int_{\omega_2} \left(\varphi_n^0 w - \ell^0 \frac{\partial w}{\partial \nu} \right) dS.$$

So the total energy functional for the plate splits into two parts:

$$E = E_{\tilde{\mathbf{v}}}(\tilde{\mathbf{v}}) + E_w(w),$$

where

$$E_{\tilde{\mathbf{v}}}(\tilde{\mathbf{v}}) = \frac{Eh}{2(1+\nu)} \int_{\Sigma} \left[\boldsymbol{\epsilon} \cdot\cdot \boldsymbol{\epsilon} + \frac{\nu}{1-\nu} \operatorname{tr}^2 \boldsymbol{\epsilon} \right] d\Sigma$$
$$- \int_{\Sigma} \mathbf{q} \cdot \tilde{\mathbf{v}}\, d\Sigma - \int_{\omega_4} \boldsymbol{\varphi}^0(s) \cdot \tilde{\mathbf{v}}\, dS$$

is due to tangential displacements, and

$$E_w(w) = \frac{Eh^3}{24(1+\nu)} \int_{\Sigma} \left[\boldsymbol{\ae} \cdot\cdot \boldsymbol{\ae} + \frac{\nu}{1-\nu} \operatorname{tr}^2 \boldsymbol{\ae} \right] d\Sigma$$
$$- \int_{\Sigma} w\mathbf{q} \cdot \mathbf{n}\, d\Sigma - \int_{\omega_2} \varphi_n^0 w\, dS + \int_{\omega_2} \ell^0 \frac{\partial w}{\partial \nu}\, dS$$

accounts for bending.

The arguments $\tilde{\mathbf{v}}$ and w of $E_{\tilde{\mathbf{v}}}$ and E_w are independent, so the problem of minimizing the total energy naturally splits into two minimization problems. Moreover, we can formulate two variational principles: one for tangential deformations, and one for plate bending.

Lagrange's variational principle for tangential deformation

Theorem 7.1. *On the set of admissible tangential displacement fields, a stationary point $\tilde{\mathbf{v}}$ of $E_{\tilde{\mathbf{v}}}$ satisfies the plate equilibrium equations* (7.60) *on Σ, or, equivalently* (7.61) *and the boundary condition*

$$\boldsymbol{\nu} \cdot \mathbf{T} \cdot \mathbf{A}|_{\omega_4} = \boldsymbol{\varphi}^0(s) \cdot \mathbf{A},$$

and conversely. The stationary point, if it exists, minimizes $E_{\tilde{\mathbf{v}}}$.

The set of admissible tangential displacements $\tilde{\mathbf{v}}$ consists of vector functions twice differentiable in Σ and satisfying the kinematic boundary condition on ω_3 from (7.65). The proof mimics that of Lagrange's principle in three-dimensional elasticity (Exercise 7.18).

Note that we represented the work of external forces using the general definition of work and common sense. The fact that the equilibrium equations and the kinematic boundary conditions really define a stationary point of the energy functional shows that the work functional was written correctly.

Lagrange's variational principle for plate bending

Theorem 7.2. *On the set of admissible deflection fields, a stationary point w of E_w satisfies the plate equilibrium equations (7.53) or (7.59) in Σ and the static boundary conditions from (7.64) on ω_2, and conversely. The stationary point, if it exists, minimizes E_w.*

Note that the set of admissible deflections w consists of the functions that possess continuous derivatives in Σ up to order four and satisfy two kinematic boundary conditions on ω_1 from (7.64).

Proof. Let us define the variation of E_w. We have

$$\delta \frac{Eh^3}{24(1+\nu)} \left[\text{æ} \cdot\cdot \text{æ} + \frac{\nu}{1-\nu} \text{tr}^2 \text{æ} \right] = \frac{Eh^3}{12(1+\nu)} \left[\text{æ} + \frac{\nu}{1-\nu} \text{tr}\,\text{æ} \right] \cdot\cdot \delta\text{æ}$$

$$= M_{\alpha\beta}\,\delta\text{æ}^{\alpha\beta}$$

$$= -M_{\alpha\beta}\frac{\partial^2\,\delta w}{\partial x_\alpha x_\beta}.$$

Using this and the Gauss–Ostrogradsky theorem, we get

$$-\int_\Sigma M_{\alpha\beta}\frac{\partial^2\,\delta w}{\partial x_\alpha x_\beta}\,d\Sigma = \int_\Sigma \frac{\partial M_{\alpha\beta}}{\partial x_\alpha}\frac{\partial\,\delta w}{\partial x_\beta}\,d\Sigma - \int_\omega \nu_\alpha M_{\alpha\beta}\frac{\partial\,\delta w}{\partial x_\beta}\,ds$$

$$= -\int_\Sigma \frac{\partial^2 M_{\alpha\beta}}{\partial x_\alpha \partial x_\beta}\,\delta w\,d\Sigma + \int_\omega \nu_\beta \frac{\partial M_{\alpha\beta}}{\partial x_\alpha}\,\delta w\,ds$$

$$- \int_\omega \nu_\alpha M_{\alpha\beta}\frac{\partial\,\delta w}{\partial x_\beta}\,ds$$

$$= -\int_\Sigma \widetilde{\nabla}\cdot(\widetilde{\nabla}\cdot(\mathbf{M}\times\mathbf{n}))\,\delta w\,d\Sigma$$

$$+ \int_\omega \boldsymbol{\nu}\cdot(\widetilde{\nabla}\cdot(\mathbf{M}\times\mathbf{n}))\,\delta w\,ds$$

$$- \int_\omega \boldsymbol{\nu}\cdot(\mathbf{M}\times\mathbf{n})\cdot\widetilde{\nabla}\delta w\,ds, \qquad (7.67)$$

where $\nu_\alpha = \boldsymbol{\nu}\cdot\mathbf{i}_\alpha$ and s is the length parameter over ω.

Let us consider the last integral in (7.67). On ω the expression $\widetilde{\nabla}\delta w$ can be represented as

$$\widetilde{\nabla}\delta w = \boldsymbol{\nu}\frac{\partial\,\delta w}{\partial\nu} + \boldsymbol{\tau}\frac{\partial\,\delta w}{\partial\tau},$$

where $\boldsymbol{\tau} = \mathbf{n}\times\boldsymbol{\nu}$ is the vector tangent to ω. Using the contour length parameter s over ω, we get

$$\frac{\partial\,\delta w}{\partial\tau} = \frac{\partial\,\delta w}{\partial s}.$$

Note that on ω the expressions $\partial\,\delta w/\partial\nu$ and δw are independent. The expressions $\partial\,\delta w/\partial s$ and δw are dependent because the value $\partial\,\delta w/\partial s$ is uniquely defined when δw is known on ω, whereas $\partial\,\delta w/\partial\nu$ is not defined by the values of δw on ω. To exclude the derivative $\partial\,\delta w/\partial s$, we integrate by parts in the integral over the closed contour ω:

$$\int_\omega f\frac{\partial g}{\partial s}\,ds = -\int_\omega g\frac{\partial f}{\partial s}\,ds,$$

where f and g are arbitrary functions of s. Applying this to the last integral in (7.67), we get

$$-\int_\omega \boldsymbol{\nu}\cdot(\mathbf{M}\times\mathbf{n})\cdot\tilde{\nabla}\delta w\,ds = -\int_\omega \left[\boldsymbol{\nu}\cdot(\mathbf{M}\times\mathbf{n})\cdot\boldsymbol{\nu}\frac{\partial\,\delta w}{\partial\nu}\right.$$
$$\left.-\frac{\partial}{\partial s}\left(\boldsymbol{\nu}\cdot(\mathbf{M}\times\mathbf{n})\cdot\boldsymbol{\tau}\right)\delta w\right]ds.$$

On ω_1 we have

$$\delta w = 0 = \frac{\partial\,\delta w}{\partial\nu},$$

hence

$$\delta E_w = -\int_\Sigma \left[\tilde{\nabla}\cdot(\tilde{\nabla}\cdot(\mathbf{M}\times\mathbf{n})) + \mathbf{q}\cdot\mathbf{n}\right]\delta w\,d\Sigma$$
$$+\int_{\omega_2}\left[\boldsymbol{\nu}\cdot(\tilde{\nabla}\cdot(\mathbf{M}\times\mathbf{n}))\right.$$
$$\left.-\frac{\partial}{\partial s}(\boldsymbol{\nu}\cdot(\mathbf{M}\times\mathbf{n})\cdot\boldsymbol{\tau}) - \varphi_n^0\right]\delta w\,ds$$
$$+\int_{\omega_2}\left[\boldsymbol{\nu}\cdot(\mathbf{M}\times\mathbf{n})\cdot\boldsymbol{\nu} - \ell^0\right]\frac{\partial\,\delta w}{\partial\nu}\,ds. \qquad (7.68)$$

Using a standard procedure in the calculus of variations (see, e.g., [Lebedev and Cloud (2003)]), we can show that the equation $\delta E_w = 0$, which holds for all admissible δw, implies both the equilibrium equation for the plate

$$\tilde{\nabla}\cdot(\tilde{\nabla}\cdot(\mathbf{M}\times\mathbf{n})) + \mathbf{q}\cdot\mathbf{n} = 0 \quad \text{in } \Sigma$$

and the static boundary conditions

$$\boldsymbol{\nu}\cdot(\tilde{\nabla}\cdot(\mathbf{M}\times\mathbf{n})) - \frac{\partial}{\partial s}(\boldsymbol{\nu}\cdot(\mathbf{M}\times\mathbf{n})\cdot\boldsymbol{\tau}) - \varphi_n^0 = 0$$

and

$$\boldsymbol{\nu}\cdot(\mathbf{M}\times\mathbf{n})\cdot\boldsymbol{\nu} - \ell^0 = 0 \quad \text{on } \omega_2.$$

This derivation also indicates that we have introduced the work for the transverse load correctly: a stationary point of E_w satisfies (7.53) on Σ and the boundary conditions (7.64) on ω_2.

Now we can prove the converse statement: a solution w of the boundary value problem (7.53), (7.64) is a stationary point of E_w. Multiplying (7.53) by an admissible δw, integrating appropriately by parts and therefore doing the above transformations in the reverse order, we arrive at the equation $\delta E_w = 0$. This completes the proof.

In a manner similar to the proof of minimality in the Lagrange principle in elasticity, we can show minimality of a stationary point in the bending problem as well. The reader is encouraged to produce a complete proof. Note that in the proof for three-dimensional elasticity, the only thing that mattered was the structure of the total energy functional: it is a sum of quadratic and linear functionals, and its quadratic portion is positive definite. $\qquad\square$

Note that by considering a plate as a three-dimensional body, we obtain a system of simultaneous equations in displacements of the second order, accompanied by three boundary conditions at each point of the boundary. Transformation of the problem to a two-dimensional plate problem brings us again to three differential equations in displacements, but the equation for w is of fourth order; it is supplemented with two conditions at each point of the plate edge, and the other two equations are of second order. The increase in order of the system is the cost of reducing a three-dimensional problem to a two-dimensional one.

In shell theory there are other variational principles. Their importance in this theory is even more than in elasticity. They are the basis for numerical methods and investigations of qualitative questions of the theory. Moreover, variational principles are used to construct various versions of shell theory. For example, we can introduce the type of distribution of the stresses and displacements along the thickness, and derive the equilibrium equations, find admissible sets of boundary conditions, and justify well-posedness of the corresponding boundary value problems.

Uniqueness of solution

For an equilibrium problem in plate theory, a solution is unique in the same meaning as for a problem in three-dimensional elasticity. That is, if some part of the boundary contour is clamped, then the solution is truly unique.

When the plate can move as a rigid body, the solution is uniquely defined by the equations and boundary conditions up to a small rigid motion. We will briefly outline a proof.

Suppose to the contrary the existence of two solutions $\mathbf{v}_1$ and $\mathbf{v}_2$ to the boundary value problem (7.60), (7.65), (7.53), (7.64). The difference $\mathbf{v} = \mathbf{v}_2 - \mathbf{v}_1$ is a solution of this homogeneous boundary value problem. It is easy to understand that it should be a stationary point of the total energy functional

$$\int_\Sigma U \, d\Sigma.$$

This implies that

$$\delta \int_\Sigma U \, d\Sigma = 0.$$

But U is a homogeneous quadratic form, so putting $\delta \mathbf{v} = \mathbf{v}$ in δU we get $\delta U = 2U(\mathbf{v})$ and thus

$$2 \int_\Sigma U \, d\Sigma = 0.$$

But earlier we considered the question of when the strain energy functional takes zero value: the equation

$$\int_\Sigma U \, d\Sigma = 0$$

implies that $\mathbf{v}$ is a small rigid displacement of the plate. So the uniqueness of a solution for an equilibrium problem is demonstrated: a solution is uniquely defined up to a rigid displacement if there are no kinematic boundary constraints.

Exercise 7.14. Derive (7.50) from (7.49).

Exercise 7.15. Show that $\mathrm{tr}\,\boldsymbol{\varepsilon} = \widetilde{\nabla} \cdot \tilde{\mathbf{v}} - z\widetilde{\nabla} \cdot \widetilde{\nabla} w$.

Exercise 7.16. Let w and its first and second derivatives be small. Demonstrate that æ is the curvature tensor of the bent surface of the plate. Use the results of Exercise 5.33.

Exercise 7.17. Derive (7.58).

Exercise 7.18. Demonstrate the variational principle for tangential displacements in the theory of plates.

Exercise 7.19. Show that at a point of plate equilibrium, E_w takes its minimum value.

Remarks

Kirchhoff's plate theory was extended to shell theory. In shell theory, Kirchhoff's hypotheses are called the Kirchhoff–Love hypotheses. In the Kirchhoff–Love shell theory, the equations are written on the midsurface. The strain energy density for the shell also splits into two parts, for tangential and bending deformations. However, unlike the splitting of the plate energy, these terms are dependent. Moreover, all three equilibrium equations in displacements contain all components of the displacement vector. Technically, the Kirchhoff–Love shell theory is more complicated than plate theory, but qualitatively they are quite similar. In shell theory, we have Lagrange's variational principle and uniqueness of solution to a boundary value problem. In both cases, we can also prove existence of a solution, but in shell theory this requires more advanced mathematical tools.

7.9 On Non-Classical Theories of Plates and Shells

Reissner's approach to plate and shell theory

We recall that Kirchhoff's hypotheses yield only an approximation to the real deformations of thin-walled bodies. In the mechanics of plates and shells, there are other approaches to the representation of three-dimensional deformation, where shear deformation, normal extension, and other factors are taken into account. In particular, the Reissner and Mindlin approaches allow us to construct more precise two-dimensional models for the solution of a three-dimensional elastic problem for a shell. The main features of the theory will be demonstrated using Reissner's plate equations.

Unlike (7.54), in Reissner's approach we take the displacement field to be of the form

$$\mathbf{u}(x_1, x_2, z) = \mathbf{v}(x_1, x_2) - z\boldsymbol{\vartheta}(x_1, x_2)$$
$$= \tilde{\mathbf{v}} + w\mathbf{n} - z\boldsymbol{\vartheta}, \qquad \boldsymbol{\vartheta} \cdot \mathbf{n} = 0. \qquad (7.69)$$

The components of $\boldsymbol{\vartheta}$ can be interpreted as the rotation angles of the shell cross section. In the general case, unlike in the Kirchhoff theory, $\boldsymbol{\vartheta} \neq \tilde{\nabla}w$ and $\boldsymbol{\vartheta}$ is considered to be independent of $\mathbf{v}$. Thus, in Reissner's theory there are five unknown variables: the components v_1, v_2, w of $\mathbf{v}$ and the components ϑ_1, ϑ_2 of $\boldsymbol{\vartheta}$. In Reissner's theory, the transverse shear stress resultants are on equal footing with the stress resultants and the bending and twisting stress couples. Unlike what happens in Kirchhoff's model, they are independent and so it is necessary to formulate additional constitutive equations along the lines of (7.47). The equilibrium equations for Reissner's plate take the same form as in Kirchhoff's theory, i.e., (7.45):

$$\widetilde{\nabla} \cdot \mathbf{T} + \mathbf{q} = \mathbf{0}, \qquad \widetilde{\nabla} \cdot \mathbf{M} + \mathbf{T}_\times + \mathbf{m} = \mathbf{0}, \qquad (7.70)$$

where $\mathbf{T}$ and $\mathbf{M}$ are the stress resultant tensor and the stress couple tensor, respectively, and $\mathbf{q}$ and $\mathbf{m}$ represent the forces and couples distributed over Σ. Note that $\mathbf{m} \cdot \mathbf{n} = 0$. In the component representation, they coincide with (7.50) and (7.51).

The dynamic equations of the theory are

$$\widetilde{\nabla} \cdot \mathbf{T} + \mathbf{q} = \rho \ddot{\mathbf{v}} + \rho \mathbf{\Theta}_1 \cdot \ddot{\boldsymbol{\vartheta}},$$
$$\widetilde{\nabla} \cdot \mathbf{M} + \mathbf{T}_\times + \mathbf{m} = \rho \mathbf{\Theta}_1^{\mathrm{T}} \cdot \ddot{\mathbf{v}} + \rho \mathbf{\Theta}_2 \cdot \ddot{\boldsymbol{\vartheta}}, \qquad (7.71)$$

where ρ is the surface shell density, $\mathbf{\Theta}_1$ and $\mathbf{\Theta}_2$ are the inertia tensors, and

$$\mathbf{\Theta}_2^{T} = \mathbf{\Theta}_2.$$

The boundary conditions for the plate are

$$\mathbf{v}\big|_{\omega_1} = \mathbf{v}^0(s), \qquad \boldsymbol{\vartheta}\big|_{\omega_3} = \boldsymbol{\vartheta}^0(s), \qquad (7.72)$$
$$\boldsymbol{\nu} \cdot \mathbf{T}\big|_{\omega_2} = \boldsymbol{\varphi}(s), \qquad \boldsymbol{\nu} \cdot \mathbf{M}\big|_{\omega_4} = \boldsymbol{\ell}(s), \qquad (7.73)$$

where $\mathbf{v}^0(s)$ and $\boldsymbol{\vartheta}^0(s)$ are given vector functions of the length parameter s such that $\boldsymbol{\vartheta}^0 \cdot \mathbf{n} = 0$. They define the displacements and the rotation on some part of the boundary contour, respectively. A given $\boldsymbol{\varphi}(s)$ and $\boldsymbol{\ell}(s)$ would define the stress resultants and the stress couples acting on the rest of the plate edge, $\boldsymbol{\ell} \cdot \mathbf{n} = 0$.

In Reissner's plate theory, the constitutive equations are

$$\mathbf{T} \cdot \mathbf{A} = \mathbf{C} \cdots \boldsymbol{\mu}, \qquad \mathbf{T} \cdot \mathbf{n} = \boldsymbol{\Gamma} \cdot \boldsymbol{\gamma}, \qquad \mathbf{M} = \mathbf{D} \cdots \boldsymbol{\kappa}. \qquad (7.74)$$

Here $\mathbf{T} \cdot \mathbf{A}$ is the in-plane stress resultant tensor, $\mathbf{T} \cdot \mathbf{n}$ represents the transverse shear stress resultants, and $\mathbf{M}$ is the stress couple tensor. The strain measures are denoted as follows: $\boldsymbol{\mu}$ is the in-plane strain tensor, $\boldsymbol{\gamma}$ is the vector of transverse shear strains, and $\boldsymbol{\kappa}$ is the tensor of out-of-plane strains. Their definitions are

$$\boldsymbol{\mu} = \frac{1}{2}\left(\widetilde{\nabla}\tilde{\mathbf{v}} + (\widetilde{\nabla}\tilde{\mathbf{v}})^T\right), \qquad \boldsymbol{\gamma} = \widetilde{\nabla}w - \mathbf{n} \times \boldsymbol{\vartheta}, \qquad \boldsymbol{\kappa} = \widetilde{\nabla}\boldsymbol{\vartheta}.$$

The remaining notation is for the fourth-order tensors $\mathbf{C}$ and $\mathbf{D}$ and the second-order tensor $\boldsymbol{\Gamma}$ that describe the effective stiffness properties of the plate; they depend on the material properties of the plate and on the cross-section geometry. In the case of an isotropic plate having properties symmetric about the midplane, the effective stiffness tensors take the form

$$\mathbf{C} = C_{11}\mathbf{A}\mathbf{A} + C_{22}(\mathbf{A}_2\mathbf{A}_2 + \mathbf{A}_4\mathbf{A}_4),$$

$$\mathbf{D} = D_{22}(\mathbf{A}_2\mathbf{A}_2 + \mathbf{A}_4\mathbf{A}_4) + D_{33}\mathbf{A}_3\mathbf{A}_3,$$

$$\mathbf{\Gamma} = \Gamma\mathbf{A},$$

where

$$\mathbf{A} = \mathbf{i}_1\mathbf{i}_1 + \mathbf{i}_2\mathbf{i}_2, \qquad\qquad \mathbf{A}_2 = \mathbf{i}_1\mathbf{i}_1 - \mathbf{i}_2\mathbf{i}_2,$$

$$\mathbf{A}_3 = \mathbf{i}_1\mathbf{i}_2 - \mathbf{i}_2\mathbf{i}_1, \qquad\qquad \mathbf{A}_4 = \mathbf{i}_1\mathbf{i}_2 + \mathbf{i}_2\mathbf{i}_1,$$

and $\mathbf{i}_1$ and $\mathbf{i}_2$ are the unit basis vectors with $\mathbf{i}_1 \cdot \mathbf{i}_2 = 0$ on the midplane. Denote $\mathbf{A}_1 = \mathbf{A}$. The following orthogonality relation can be obtained:

$$\frac{1}{2}\mathbf{A}_i \cdot\cdot\ \mathbf{A}_j = \delta_{ij} \qquad (i, j = 1, 2, 3, 4).$$

For an isotropic homogeneous plate, the components of the stiffness tensors are defined as follows:

$$C_{11} = \frac{Eh}{2(1-\nu)}, \qquad\qquad C_{22} = \frac{Eh}{2(1+\nu)} = \mu h,$$

$$D_{33} = \frac{Eh^3}{24(1-\nu)}, \qquad\qquad D_{22} = \frac{Eh^3}{24(1+\nu)} = \frac{\mu h^3}{12}.$$

The classical bending stiffness for a plate is

$$D = D_{33} + D_{22} = \frac{Eh^3}{12(1-\nu^2)}. \tag{7.75}$$

The surface density and the inertia tensors are

$$\rho = \rho_0 h, \qquad \mathbf{\Theta}_1 = \mathbf{0}, \qquad \mathbf{\Theta}_2 = \Theta\mathbf{A}, \qquad \Theta = \frac{\rho_0 h^3}{12}, \tag{7.76}$$

where ρ_0 is the plate material density.

The transverse shear stiffness is given by

$$\Gamma = k\mu h, \tag{7.77}$$

where k is a *shear correction factor* first introduced by Timoshenko in the theory of beams[2]. For the value of k, Mindlin[3] proposed $k = \pi^2/12$ while

---

[2]Timoshenko, S. P. On the correction for shear of the differential equation for transverse vibrations of prismatic bars. *Phil. Mag. Ser.* 6, **41**, 744–746, 1921.

[3]Mindlin, R.D. Influence of rotatory inertia and shear on flexural motions of isotropic, elastic plates. *Trans. ASME J. Appl. Mech.*, **18**, 31–38, 1951.

Reissner[4] proposed a similar value of $k = 5/6$. The literature mentions other values of k; for example, for plates that are strongly nonhomogeneous in thickness, k can significantly differ from the above values[5].

The strain energy density for Reissner's plate is given by

$$U = \frac{1}{2}\boldsymbol{\mu} \cdots \mathbf{C} \cdots \boldsymbol{\mu} + \frac{1}{2}\boldsymbol{\kappa} \cdots \mathbf{D} \cdots \boldsymbol{\kappa} + \frac{1}{2}\boldsymbol{\gamma} \cdot \boldsymbol{\Gamma} \cdot \boldsymbol{\gamma}.$$

It is seen that

$$\mathbf{T} \cdot \mathbf{A} = \frac{\partial U}{\partial \boldsymbol{\mu}}, \qquad \mathbf{T} \cdot \mathbf{n} = \frac{\partial U}{\partial \boldsymbol{\gamma}}, \qquad \mathbf{M} = \frac{\partial U}{\partial \boldsymbol{\kappa}}.$$

The five scalar equations of (7.50) and (7.51) contain five scalar unknowns v_1, v_2, w, ϑ_1, and ϑ_2. In Reissner's model, equation (7.59) for w changes to the following:

$$D\widetilde{\nabla}^4 w = q_n - \frac{D}{\Gamma}\Delta q_n. \tag{7.78}$$

Because of the additional term $(D/\Gamma)\Delta q_n$, for strongly nonhomogeneous loads the results obtained from Reissner's theory can differ significantly from those obtained from Kirchhoff's model.

For advanced readers, we propose

Exercise 7.20. Derive (7.78).

Reissner's theory is usually used for non-thin plates, for dynamic problems, and for cases of anisotropic materials having small shear stiffness. A more detailed presentation of the theory can be found in [Wang, Reddy, and Lee (2000); Zhilin (2006)]; see also Altenbach[6] and Grigolyuk and Selezov[7].

Plate and shell theories of higher order

The approximation of the displacement field (7.69) is linear in z. In the literature on shell theory, we encounter displacement approximations in z of

---

[4]Reissner, E. On the theory of bending of elastic plates. *J. Math. Physics*, **23**, 184–194, 1944.

[5]Altenbach, H., and Eremeyev, V.A. Direct approach based analysis of plates composed of functionally graded materials. *Arch. Appl. Mech.*, **78**, 775–794, 2008.

[6]Altenbach, H. An alternative determination of transverse shear stiffnesses for sandwich and laminated plates. *Int. J. Solids Struct.*, **37**, 3503–3520, 2000. Altenbach, H. On the determination of transverse shear stiffnesses of orthotropic plates. *ZAMP*, **51**, 629–649, 2000.

[7]Grigolyuk, E.I., and Selezov, I.T. *Nonclassical theories of vibration of beams, plates and shells* (in Russian). In: Itogi nauki i tekhniki, Mekhanika tverdogo deformiruemogo tela, vol 5, VINITI, Moskva, 1973.

higher order. The natural idea of expanding the displacement in a series of powers in z belongs to Cauchy and Poisson. We could say that Germain's equation for plate deflection represents the zero-order theory, as here the deflection field does not depend on z. Equation (7.69) can be considered as a basis of the first-order theory. Some versions of higher-order theory were developed in a number of works; certain difficulties arise, however, such as how one should define the boundary conditions and physically interpret moments of higher order.

As an example, we will consider the *third-order plate theory*. In [Wang, Reddy, and Lee (2000)] there is a review of the literature on this version of the theory. The displacement field is approximated by the expressions

$$\mathbf{u}(x_1, x_2, z) = \mathbf{v}(x_1, x_2) + z\boldsymbol{\phi}(x_1, x_2) + z^3\boldsymbol{\psi}(x_1, x_2)$$
$$= \tilde{\mathbf{v}} + w\mathbf{n} + z\boldsymbol{\phi} + z^3\boldsymbol{\psi}, \qquad (7.79)$$

where $\tilde{\mathbf{v}} \cdot \mathbf{n} = 0$, and $\boldsymbol{\phi}$ and $\boldsymbol{\psi}$ are functions such that $\boldsymbol{\phi} \cdot \mathbf{n} = 0$ and $\boldsymbol{\psi} \cdot \mathbf{n} = 0$. In particular, Reddy proposed the following expression for $\boldsymbol{\psi}$:

$$\boldsymbol{\psi} = -\alpha\left(\boldsymbol{\phi} + \tilde{\nabla}w\right),$$

where $\alpha = 4/(3h^2)$ and h is the plate thickness. The strain tensor becomes

$$\varepsilon = \tilde{\varepsilon} + \frac{1}{2}\left(\boldsymbol{\gamma}\mathbf{n} + \mathbf{n}\boldsymbol{\gamma}\right), \qquad (7.80)$$

where

$$\tilde{\varepsilon} = \epsilon + \frac{z}{2}\left(\tilde{\nabla}\boldsymbol{\phi} + (\tilde{\nabla}\boldsymbol{\phi})^T\right) + \frac{z^3}{2}\left(\tilde{\nabla}\boldsymbol{\psi} + (\tilde{\nabla}\boldsymbol{\psi})^T\right)$$
$$= \epsilon + \frac{z}{2}\left(\tilde{\nabla}\boldsymbol{\phi} + (\tilde{\nabla}\boldsymbol{\phi})^T\right) - \alpha\frac{z^3}{2}\left(\tilde{\nabla}\boldsymbol{\phi} + (\tilde{\nabla}\boldsymbol{\phi})^T + 2\tilde{\nabla}\tilde{\nabla}w\right),$$
$$\boldsymbol{\gamma} = \boldsymbol{\phi} + 3z^2\boldsymbol{\psi} = \boldsymbol{\phi} - 3z^2\alpha\boldsymbol{\phi} - 3z^2\alpha\tilde{\nabla}w,$$
$$2\epsilon = \tilde{\nabla}\tilde{\mathbf{v}} + (\tilde{\nabla}\tilde{\mathbf{v}})^T.$$

In the third-order theory, we introduce the stress resultants $T_{\alpha\beta}$, the transverse shear stress resultants $T_{\alpha n}$, and the stress couples $M_{\alpha\beta}$. But we introduce some additional quantities such as the higher-order stress resultants $P_{\alpha\beta}$ and R_α:

$$T_{\alpha\beta} = [\![\sigma_{\alpha\beta}]\!] \equiv \int_{-h/2}^{h/2} \sigma_{\alpha\beta}\, dz, \qquad T_{\alpha n} = [\![\sigma_{\alpha n}]\!] \equiv \int_{-h/2}^{h/2} \sigma_{\alpha 3}\, dz,$$

$$M_{\alpha\beta} = [\![z\sigma_{\alpha\beta}]\!] \equiv \int_{-h/2}^{h/2} z\sigma_{\alpha\beta}\, dz, \tag{7.81}$$

$$P_{\alpha\beta} = [\![z^3\sigma_{\alpha\beta}]\!] \equiv \int_{-h/2}^{h/2} z^3\sigma_{\alpha\beta}\, dz, \quad R_\alpha = [\![z^2\sigma_{\alpha n}]\!] \equiv \int_{-h/2}^{h/2} z^2\sigma_{\alpha 3}\, dz.$$

Exercise 7.21. Using (7.80) and (7.56), express (7.81) in terms of the components of ϕ and w.

The equilibrium equations split into two systems. The first is for *the plane state*; it is of the form we had in Kirchhoff's theory:

$$\frac{\partial T_{11}}{\partial x_1} + \frac{\partial T_{21}}{\partial x_2} + q_1 = 0, \qquad \frac{\partial T_{12}}{\partial x_1} + \frac{\partial T_{22}}{\partial x_2} + q_2 = 0. \tag{7.82}$$

Equations (7.82) are supplemented with the boundary conditions

$$\tilde{\mathbf{v}}\big|_{\omega_1} = \tilde{\mathbf{v}}^0(s), \qquad \boldsymbol{\nu} \cdot \mathbf{T} \cdot \mathbf{A}\big|_{\omega_2} = \boldsymbol{\varphi}^0(s), \tag{7.83}$$

where

$$\mathbf{n} \cdot \tilde{\mathbf{v}}^0 = 0, \qquad \mathbf{n} \cdot \boldsymbol{\varphi}^0 = 0, \qquad \omega_1 \cup \omega_2 = \omega, \qquad \omega_1 \cap \omega_2 = \emptyset.$$

The constitutive equations are

$$T_{11} = \frac{Eh}{1-\nu^2}(\epsilon_{11} + \nu\epsilon_{22}),$$

$$T_{22} = \frac{Eh}{1-\nu^2}(\epsilon_{22} + \nu\epsilon_{11}),$$

$$T_{12} = \frac{Eh}{1+\nu}\epsilon_{12}.$$

The *boundary value problem for plate bending* is presented by the equilibrium equations

$$\frac{\partial T_{1n}}{\partial x_1} + \frac{\partial T_{2n}}{\partial x_2} + \alpha\left(\frac{\partial^2 P_{11}}{\partial x_1^2} + 2\frac{\partial^2 P_{12}}{\partial x_1 \partial x_2} + \frac{\partial^2 P_{22}}{\partial x_2^2}\right) + q_n = 0,$$

$$\frac{\partial \widehat{M}_{11}}{\partial x_1} + \frac{\partial \widehat{M}_{21}}{\partial x_2} - \widehat{T}_{1n} = 0,$$

$$\frac{\partial \widehat{M}_{12}}{\partial x_1} + \frac{\partial \widehat{M}_{22}}{\partial x_2} - \widehat{T}_{2n} = 0, \tag{7.84}$$

where

$$\widehat{M}_{\alpha\beta} = M_{\alpha\beta} - \alpha P_{\alpha\beta}, \qquad \widehat{T}_{\alpha n} = T_{\alpha n} - 3\alpha R_\alpha.$$

The kinematic boundary conditions for the bending problem consist of four scalar equations:

$$w\big|_\omega = w^0(s), \qquad \frac{\partial w}{\partial n}\bigg|_\omega = w^0_n(s), \qquad \boldsymbol{\phi}\big|_\omega = \boldsymbol{\phi}^0(s),$$

where $\mathbf{n} \cdot \boldsymbol{\phi}^0 = 0$. The static boundary conditions are expressed in terms of stress resultants, stress couples, and moments of higher order; the interested reader should consult the original sources [Wang, Reddy, and Lee (2000)] and Kienzler[8], Levinson[9], and Reddy[10].

Note that the third- and higher-order shell and plate theories show why variational formulations are needed. These are the only way to introduce correct static boundary conditions.

Micropolar shells or 6th-parametrical shell theory

As an example of a non-classical shell theory, we will sketch the theory of micropolar shells. Its roots originate in the work of the Cosserat brothers, Eugèn and François [Cosserat and Cosserat (1909)]. The micropolar theory of shells is presented in the works of Eremeyev and Zubov [Eremeyev and Zubov (2008)], Zhilin [Zhilin (2006)], and others.

In this theory, the dynamical equations and shell kinematics coincide with those of a 6-parametric shell theory, a nonlinear version of which is presented in the books by Libai and Simmonds [Libai and Simmonds (1998)] and Chróścielewski et al [Chróścielewski, Makowski, and Pietraszkiewicz (2004)]. There exists a micropolar plate theory by Eringen [Eringen (1999)]. It is based on integration over the thickness of the plate when its material is considered as a Cosserat continuum. This differs from 6-parametrical shell-plate theory, as it contains more than 6 unknown scalar functions.

The kinematics of the shell surface are described by six scalar quantities, three of which are the components of the displacement vector $\mathbf{v}$ and the rest of which are the components of the microrotation vector $\boldsymbol{\vartheta}$. So a shell particle has six degrees of freedom described by the components of $\mathbf{v}$

---

[8]Kienzler, R. On consistent plate theories. *Arch. Appl. Mech.*, **72**, 229–247, 2002.

[9]Levinson, M. An accurate, simple theory of the statics and dynamics of elastic plates. *Mech. Res. Comm.*, **7**, 343–350, 1980.

[10]Reddy, J.N. A simple higher-order theory for laminated composite plates. *Trans. ASME J. Appl. Mech.*, **51**, 745–752, 1984.

and $\boldsymbol{\vartheta}$. The vector $\boldsymbol{\vartheta}$ satisfies $\boldsymbol{\vartheta} \cdot \mathbf{n} \neq 0$; the vectors $\mathbf{v}$ and $\boldsymbol{\vartheta}$ are mutually independent. This is analogous to a two-dimensional version of the Cosserat medium with couple stresses and the rotation interaction of the particles.

For a micropolar shell, we can assign the couple load acting on the shell surface. Here the order of the equilibrium equations is 12, so we should supplement them with 6 conditions on the shell edge. On the part of the edge that is free of geometrical constraints, we can assign the distribution of forces and couples. This extends the range of applicability of the theory in comparison with the theories considered above: we can, say, assign boundary conditions that describe a shell clamped to a rigid body.

The micropolar theory is used, in particular, to describe branching shells and thin-walled bodies with complex intrinsic structure that include multilayered or composite plates and shells, shells with inner partition, with stringers and those similar to honeycomb or made of highly porous materials.

We present the micropolar shell equations for small deformation. The dynamical equations are

$$\widetilde{\nabla} \cdot \mathbf{T} + \mathbf{q} = \rho \ddot{\mathbf{v}} + \rho \boldsymbol{\Theta}_1 \cdot \ddot{\boldsymbol{\vartheta}},$$

$$\widetilde{\nabla} \cdot \mathbf{M} + \mathbf{T}_\times + \mathbf{m} = \rho \boldsymbol{\Theta}_1^{\mathrm{T}} \cdot \ddot{\mathbf{v}} + \rho \boldsymbol{\Theta}_2 \cdot \ddot{\boldsymbol{\vartheta}}, \tag{7.85}$$

where $\mathbf{T}$ and $\mathbf{M}$ are surface stress and couple stress tensors analogous to those in Kirchhoff's theory, ρ is the shell surface density, $\boldsymbol{\Theta}_1$ and $\boldsymbol{\Theta}_2$ are the inertia tensors, $\boldsymbol{\Theta}_2$ is symmetric, $\boldsymbol{\Theta}_2^T = \boldsymbol{\Theta}_2$, and $\mathbf{q}$ and $\mathbf{m}$ are distributed surface forces and couples, respectively. Besides, in this theory, we can assign a twisting couple (or drilling moment) on the shell surface. The equilibrium equations take the form

$$\widetilde{\nabla} \cdot \mathbf{T} + \mathbf{q} = \mathbf{0}, \qquad \widetilde{\nabla} \cdot \mathbf{M} + \mathbf{T}_\times + \mathbf{m} = \mathbf{0}. \tag{7.86}$$

The conditions at a boundary point can be kinematic

$$\mathbf{v}\big|_{\omega_1} = \mathbf{v}^0(s), \qquad \boldsymbol{\vartheta}\big|_{\omega_3} = \boldsymbol{\vartheta}^0(s), \tag{7.87}$$

or static

$$\boldsymbol{\nu} \cdot \mathbf{T}\big|_{\omega_2} = \boldsymbol{\varphi}(s), \qquad \boldsymbol{\nu} \cdot \mathbf{M}\big|_{\omega_4} = \boldsymbol{\ell}(s), \tag{7.88}$$

where $\mathbf{v}^0(s)$ and $\boldsymbol{\vartheta}^0(s)$ are given vector functions of the length parameter s; they define the displacements and microrotation of the shell edge. The functions $\boldsymbol{\varphi}(s)$ and $\boldsymbol{\ell}(s)$ determine the surface stresses and stress couples on the edge. Here

$$\omega = \omega_1 \cup \omega_2 = \omega_3 \cup \omega_4, \qquad \omega_2 = \omega \setminus \omega_1, \qquad \omega_4 = \omega \setminus \omega_3.$$

The form of the equilibrium and dynamical equations for a micropolar shell does not differ from (7.70) and (7.71). In general, in this theory $\mathbf{m} \cdot \mathbf{n} \neq 0$ and $\boldsymbol{\ell} \cdot \mathbf{n} \neq 0$. In particular, $\mathbf{M} \cdot \mathbf{n} \neq \mathbf{0}$.

The strain measures take the form

$$\boldsymbol{\epsilon} = \widetilde{\nabla}\mathbf{v} + \mathbf{A} \times \boldsymbol{\vartheta}, \qquad \varkappa = \widetilde{\nabla}\boldsymbol{\vartheta}, \qquad (7.89)$$

where $\boldsymbol{\epsilon}$ and $\varkappa$ are non-symmetric strain and bending strain tensors respectively.

The constitutive equations for an elastic shell are represented though the strain energy density $U = U(\boldsymbol{\epsilon}, \varkappa)$:

$$\mathbf{T} = \frac{\partial U}{\partial \boldsymbol{\epsilon}}, \qquad \mathbf{M} = \frac{\partial U}{\partial \varkappa}. \qquad (7.90)$$

For an isotropic shell, U is a quadratic form in its variables:

$$\begin{aligned}
2U &= \alpha_1 \operatorname{tr}^2 \boldsymbol{\epsilon} + \alpha_2 \operatorname{tr} \tilde{\boldsymbol{\epsilon}}^2 + \alpha_3 \operatorname{tr}\left(\tilde{\boldsymbol{\epsilon}} \cdot \tilde{\boldsymbol{\epsilon}}^T\right) + \alpha_4 \mathbf{n} \cdot \boldsymbol{\epsilon}^T \cdot \boldsymbol{\epsilon} \cdot \mathbf{n} \\
&\quad + \beta_1 \operatorname{tr}^2 \varkappa + \beta_2 \operatorname{tr} \tilde{\varkappa}^2 + \beta_3 \operatorname{tr}\left(\tilde{\varkappa} \cdot \tilde{\varkappa}^T\right) + \beta_4 \mathbf{n} \cdot \varkappa^T \cdot \varkappa \cdot \mathbf{n}, \\
\tilde{\boldsymbol{\epsilon}} &= \boldsymbol{\epsilon} \cdot \mathbf{A}, \quad \tilde{\varkappa} = \varkappa \cdot \mathbf{A},
\end{aligned} \qquad (7.91)$$

where α_k and β_k $(k = 1, 2, 3, 4)$ are elastic moduli. Substituting (7.91) into (7.90), we get

$$\begin{aligned}
\mathbf{T} &= \alpha_1 \mathbf{A} \operatorname{tr} \boldsymbol{\epsilon} + \alpha_2 \tilde{\boldsymbol{\epsilon}}^T + \alpha_3 \tilde{\boldsymbol{\epsilon}} + \alpha_4 (\boldsymbol{\epsilon} \cdot \mathbf{n})\mathbf{n}, \\
\mathbf{M} &= \beta_1 \mathbf{A} \operatorname{tr} \varkappa + \beta_2 \tilde{\varkappa}^T + \beta_3 \tilde{\varkappa} + \beta_4 (\varkappa \cdot \mathbf{n})\mathbf{n}.
\end{aligned} \qquad (7.92)$$

Equations (7.86)–(7.88) and (7.92) constitute a linear boundary value problem with respect to the fields of displacement and microrotation; they describe the equilibrium of a micropolar shell in the case of small deformation. For dynamic problems, equations (7.86) change to (7.85). Under some additional assumptions, Reissner's shell theory is a consequence of the micropolar theory.

As in the other shell theories, we can formulate some variational principles. *Lagrange's variational principle* for an elastic micropolar shell starts with formulation of the total energy functional

$$E(\mathbf{v}, \boldsymbol{\vartheta}) = \int_{\Sigma} U(\boldsymbol{\epsilon}, \varkappa)\, d\Sigma - A(\mathbf{v}, \boldsymbol{\vartheta}), \qquad (7.93)$$

where the potential of external loads $A(\mathbf{v}, \boldsymbol{\vartheta})$ is

$$A(\mathbf{v}, \boldsymbol{\vartheta}) = \int_{\Sigma} (\mathbf{q} \cdot \mathbf{v} + \mathbf{m} \cdot \boldsymbol{\vartheta})\, d\Sigma + \int_{\omega_2} \boldsymbol{\varphi} \cdot \mathbf{v}\, ds + \int_{\omega_4} \boldsymbol{\ell} \cdot \boldsymbol{\vartheta}\, ds.$$

The functional $E(\mathbf{v}, \boldsymbol{\vartheta})$ is considered on the set of twice continuously differentiable fields of displacements and microrotations that satisfy (7.87). The pair $(\mathbf{v}, \boldsymbol{\vartheta})$ that satisfies (7.86) and (7.88) is a stationary point of $E(\mathbf{v}, \boldsymbol{\vartheta})$. Lagrange's stationary principle is minimal: on the equilibrium solution, the functional (7.93) attains its minimum.

Rayleigh's variational principle now takes the following form.

On the set of functions with boundary conditions $\mathbf{v}\big|_{\omega_1} = \mathbf{0}$, $\boldsymbol{\vartheta}\big|_{\omega_3} = \mathbf{0}$ *that obey the constraint*

$$K(\mathbf{v}^\circ, \boldsymbol{\vartheta}^\circ) \equiv \int_\Sigma \rho \left(\frac{1}{2} \mathbf{v}^\circ \cdot \mathbf{v}^\circ + \mathbf{v}^\circ \cdot \boldsymbol{\Theta}_1 \cdot \boldsymbol{\vartheta}^\circ + \frac{1}{2} \boldsymbol{\vartheta}^\circ \cdot \boldsymbol{\Theta}_2 \cdot \boldsymbol{\vartheta}^\circ \right) d\Sigma = 1,$$

the eigenoscillation modes of the shell are stationary points of the strain energy functional

$$E(\mathbf{v}^\circ, \boldsymbol{\vartheta}^\circ) = \int_\Sigma U(\boldsymbol{\epsilon}^\circ, \boldsymbol{\varkappa}^\circ) \, d\Sigma, \tag{7.94}$$

where $\boldsymbol{\epsilon}^\circ = \widetilde{\nabla} \mathbf{v}^\circ + \mathbf{A} \times \boldsymbol{\vartheta}^\circ$ *and* $\boldsymbol{\varkappa}^\circ = \widetilde{\nabla} \boldsymbol{\vartheta}^\circ$.

Rayleigh's principle also includes the reverse statement: on the set of functions satisfying the above restrictions, the stationary points of E are the modes of eigenoscillation. Here $\mathbf{v}^\circ$ and $\boldsymbol{\vartheta}^\circ$ are the amplitudes of the oscillations of the displacements and microrotations as the solutions of the eigenoscillation problem are sought in the form $\mathbf{v} = \mathbf{v}^\circ e^{i\omega t}$, $\boldsymbol{\vartheta} = \boldsymbol{\vartheta}^\circ e^{i\omega t}$. Rayleigh's quotient takes the form

$$R(\mathbf{v}^\circ, \boldsymbol{\vartheta}^\circ) = \frac{E(\mathbf{v}^\circ, \boldsymbol{\vartheta}^\circ)}{K(\mathbf{v}^\circ, \boldsymbol{\vartheta}^\circ)}.$$

Now the lowest eigenfrequency of the shell, which is the minimal frequency, is equal to the minimum value of the functional R.

Finally, we would like to repeat that, contrary to the situation with linear elasticity, shell theory is still under development; investigators continue to seek better models and numerical methods for the solution of practical problems. This is a consequence of the evident fact that it is impossible to find an accurate approximation to the results of a three-dimensional elasticity problem under all the circumstances when one solves the same problem using one of the two-dimensional models of shell theory.

Appendix A

Formulary

For convenience we list the main formulas obtained in each chapter. The symbol ∀ denotes the universal quantifier (read as "for all" or "for every").

Chapter 1

Dot product

$$\mathbf{a} \cdot \mathbf{b} = |\mathbf{a}||\mathbf{b}| \cos \theta = a_1 b_1 + a_2 b_2 + a_3 b_3$$

Cross product

$$\mathbf{a} \times \mathbf{b} = \begin{vmatrix} \mathbf{i}_1 & \mathbf{i}_2 & \mathbf{i}_3 \\ a_1 & a_2 & a_3 \\ b_1 & b_2 & b_3 \end{vmatrix}$$

Scalar triple product

$$\mathbf{a} \cdot (\mathbf{b} \times \mathbf{c}) = \begin{vmatrix} a_1 & a_2 & a_3 \\ b_1 & b_2 & b_3 \\ c_1 & c_2 & c_3 \end{vmatrix}$$

Chapter 2

Reciprocal (dual) basis

Kronecker delta symbol

$$\delta_j^i = \begin{cases} 1 & j = i \\ 0 & j \neq i \end{cases}$$

Definition of reciprocal basis

$$\mathbf{e}_j \cdot \mathbf{e}^i = \delta_j^i$$

Components of a vector

$$\mathbf{x} = x^i \mathbf{e}_i = x_i \mathbf{e}^i \qquad \begin{aligned} x^i &= \mathbf{x} \cdot \mathbf{e}^i \\ x_i &= \mathbf{x} \cdot \mathbf{e}_i \end{aligned}$$

Relations between dual bases

$$\mathbf{e}^i = \frac{1}{V}(\mathbf{e}_j \times \mathbf{e}_k) \qquad\qquad \mathbf{e}_i = \frac{1}{V'}(\mathbf{e}^j \times \mathbf{e}^k)$$

where

$$(i, j, k) = (1, 2, 3) \text{ or } (2, 3, 1) \text{ or } (3, 1, 2)$$

and

$$\begin{aligned} V &= \mathbf{e}_1 \cdot (\mathbf{e}_2 \times \mathbf{e}_3) \\ V' &= \mathbf{e}^1 \cdot (\mathbf{e}^2 \times \mathbf{e}^3) \end{aligned} \qquad\qquad V' = 1/V$$

Metric coefficients

$$\begin{aligned} g^{jq} &= \mathbf{e}^j \cdot \mathbf{e}^q \\ g_{ip} &= \mathbf{e}_i \cdot \mathbf{e}_p \end{aligned} \qquad\qquad g_{ij}\, g^{jk} = \delta_i^k$$

In Cartesian frames,

$$g_{ij} = \delta_i^j \qquad\qquad g^{ij} = \delta_j^i$$

Dot products in mixed and unmixed bases

$$\mathbf{a} \cdot \mathbf{b} = a^i b^j g_{ij} = a_i b_j g^{ij} = a^i b_i = a_i b^i$$

Raising and lowering of indices

$$x_j = x^i \, g_{ij} \qquad\qquad x^i = x_j \, g^{ij}$$

Frame transformation

Equations of transformation

$$\mathbf{e}_i = A_i^j \, \tilde{\mathbf{e}}_j \qquad\qquad A_i^j = \mathbf{e}_i \cdot \tilde{\mathbf{e}}^j$$

$$\tilde{\mathbf{e}}_i = \tilde{A}_i^j \, \mathbf{e}_j \qquad\qquad \tilde{A}_i^j = \tilde{\mathbf{e}}_i \cdot \mathbf{e}^j$$

where

$$\tilde{A}_i^j \, A_j^k = A_i^j \, \tilde{A}_j^k = \delta_i^k$$

Vector components and transformation laws

$$\mathbf{x} = x^i \, \mathbf{e}_i = x_i \, \mathbf{e}^i = \tilde{x}^i \, \tilde{\mathbf{e}}_i = \tilde{x}_i \, \tilde{\mathbf{e}}^i$$

and

$$\tilde{x}^i = A_j^i \, x^j \qquad\qquad x^i = \tilde{A}_j^i \, \tilde{x}^j$$

$$\tilde{x}_i = \tilde{A}_i^j \, x_j \qquad\qquad x_i = A_i^j \, \tilde{x}_j$$

Miscellaneous

Permutation (Levi–Civita) symbol

$$\epsilon_{ijk} = \mathbf{e}_i \cdot (\mathbf{e}_j \times \mathbf{e}_k) = \begin{cases} +V & (i,j,k) \text{ an even permutation of } (1,2,3) \\ -V & (i,j,k) \text{ an odd permutation of } (1,2,3) \\ 0 & \text{two or more indices equal} \end{cases}$$

$$\epsilon^{ijk} = \mathbf{e}^i \cdot (\mathbf{e}^j \times \mathbf{e}^k) = \begin{cases} +V' & (i,j,k) \text{ an even permutation of } (1,2,3) \\ -V' & (i,j,k) \text{ an odd permutation of } (1,2,3) \\ 0 & \text{two or more indices equal} \end{cases}$$

Useful identities

$$\epsilon_{ijk}\,\epsilon^{pqr} = \begin{vmatrix} \delta_i^p & \delta_i^q & \delta_i^r \\ \delta_j^p & \delta_j^q & \delta_j^r \\ \delta_k^p & \delta_k^q & \delta_k^r \end{vmatrix} \qquad\qquad \epsilon_{ijk}\,\epsilon^{pqk} = \delta_i^p\,\delta_j^q - \delta_i^q\,\delta_j^p$$

Determinant of Gram matrix

$$V^2 = \det[g_{ij}]$$

Cross product

$$\mathbf{a} \times \mathbf{b} = \mathbf{e}^i\,\epsilon_{ijk}\,a^j\,b^k = \mathbf{e}_i\,\epsilon^{ijk}\,a_j\,b_k$$

Chapter 3

Dyad product

Properties

$$(\lambda\mathbf{a}) \otimes \mathbf{b} = \mathbf{a} \otimes (\lambda\mathbf{b}) = \lambda(\mathbf{a} \otimes \mathbf{b})$$
$$(\mathbf{a} + \mathbf{b}) \otimes \mathbf{c} = \mathbf{a} \otimes \mathbf{c} + \mathbf{b} \otimes \mathbf{c}$$
$$\mathbf{a} \otimes (\mathbf{b} + \mathbf{c}) = \mathbf{a} \otimes \mathbf{b} + \mathbf{a} \otimes \mathbf{c}$$

Dot products of dyad with vector

$$\mathbf{ab} \cdot \mathbf{c} = (\mathbf{b} \cdot \mathbf{c})\mathbf{a}$$
$$\mathbf{c} \cdot (\mathbf{ab}) = (\mathbf{c} \cdot \mathbf{a})\mathbf{b}$$

Tensors from operator viewpoint

Equality of tensors

$$\mathbf{A} = \mathbf{B} \iff \forall \mathbf{x},\ \mathbf{A} \cdot \mathbf{x} = \mathbf{B} \cdot \mathbf{x}$$

Components

$$a^{ij} = \mathbf{e}^i \cdot \mathbf{A} \cdot \mathbf{e}^j$$
$$a_{ij} = \mathbf{e}_i \cdot \mathbf{A} \cdot \mathbf{e}_j$$
$$a^i_{.j} = \mathbf{e}^i \cdot \mathbf{A} \cdot \mathbf{e}_j$$
$$a_i^{.j} = \mathbf{e}_i \cdot \mathbf{A} \cdot \mathbf{e}^j$$

Definition of sum $\mathbf{A} + \mathbf{B}$

$$\forall \mathbf{x}, \ (\mathbf{A} + \mathbf{B}) \cdot \mathbf{x} = \mathbf{A} \cdot \mathbf{x} + \mathbf{B} \cdot \mathbf{x}$$

Definition of scalar multiple $c\mathbf{A}$

$$\forall \mathbf{x}, \ (c\mathbf{A}) \cdot \mathbf{x} = c(\mathbf{A} \cdot \mathbf{x})$$

Definition of dot product $\mathbf{A} \cdot \mathbf{B}$

$$(\mathbf{A} \cdot \mathbf{B}) \cdot \mathbf{x} = \mathbf{A} \cdot (\mathbf{B} \cdot \mathbf{x})$$

Definition of pre-multiplication $\mathbf{y} \cdot \mathbf{A}$

$$\forall \mathbf{x}, \ (\mathbf{y} \cdot \mathbf{A}) \cdot \mathbf{x} = \mathbf{y} \cdot (\mathbf{A} \cdot \mathbf{x})$$

Definition of unit tensor $\mathbf{E}$

$$\forall \mathbf{x}, \ \mathbf{E} \cdot \mathbf{x} = \mathbf{x} \cdot \mathbf{E} = \mathbf{x}$$

Unit tensor components

$$\mathbf{E} = \mathbf{e}^i \mathbf{e}_i = \mathbf{e}_j \mathbf{e}^j = g_{ij} \mathbf{e}^i \mathbf{e}^j = g^{ij} \mathbf{e}_i \mathbf{e}_j$$

Inverse tensor

$$\mathbf{A} \cdot \mathbf{A}^{-1} = \mathbf{E}$$

$$(\mathbf{A} \cdot \mathbf{B})^{-1} = \mathbf{B}^{-1} \cdot \mathbf{A}^{-1}$$

Nonsingular tensor $\mathbf{A}$

$$\mathbf{A} \cdot \mathbf{x} = 0 \implies \mathbf{x} = 0$$

Determinant of a tensor

$$\det \mathbf{A} = \left| a_i^{\cdot j} \right| = \left| a_{\cdot m}^k \right| = \frac{1}{g} \left| a_{st} \right| = g \left| a^{pq} \right|$$
$$= \frac{1}{6} \, \epsilon_{ijk} \, \epsilon^{mnp} \, a_m^{\cdot \, i} \, a_n^{\cdot j} \, a_p^{\cdot k}$$

Dyadic components under transformation

Transformation to reciprocal basis

$$a_{km} = a^{ij} \, g_{ki} \, g_{jm}$$

More general transformation

$$\mathbf{e}_i = A_i^j \, \tilde{\mathbf{e}}_j \implies \tilde{a}^{ij} = a^{km} \, A_k^i \, A_m^j$$

$$\tilde{\mathbf{e}}_i = \tilde{A}_i^j \, \mathbf{e}_j \implies a^{ij} = \tilde{a}^{km} \, \tilde{A}_k^i \, \tilde{A}_m^j$$

$$A_j^k \, \tilde{A}_k^i = \delta_j^i$$

$$\mathbf{A} = \tilde{a}^{ij} \, \tilde{\mathbf{e}}_i \, \tilde{\mathbf{e}}_j = \tilde{a}_{kl} \, \tilde{\mathbf{e}}^k \, \tilde{\mathbf{e}}^l = \tilde{a}_i^{\cdot \, j} \, \tilde{\mathbf{e}}^i \, \tilde{\mathbf{e}}_j = \tilde{a}_{\cdot \, l}^k \, \tilde{\mathbf{e}}_k \, \tilde{\mathbf{e}}^l$$
$$= a^{ij} \, \mathbf{e}_i \, \mathbf{e}_j = a_{kl} \, \mathbf{e}^k \, \mathbf{e}^l = a_i^{\cdot \, j} \, \mathbf{e}^i \, \mathbf{e}_j = a_{\cdot \, l}^k \, \mathbf{e}_k \, \mathbf{e}^l$$

where

$$\tilde{a}^{ij} = A_k^i \, A_l^j \, a^{kl} \qquad\qquad a^{ij} = \tilde{A}_k^i \, \tilde{A}_l^j \, \tilde{a}^{kl}$$
$$\tilde{a}_{ij} = \tilde{A}_i^k \, \tilde{A}_j^l \, a_{kl} \qquad\qquad a_{ij} = A_i^k \, A_j^l \, \tilde{a}_{kl}$$
$$\tilde{a}_{\cdot \, j}^i = A_k^i \, \tilde{A}_j^l \, a_{\cdot \, l}^k \qquad\qquad a_{\cdot \, j}^i = \tilde{A}_k^i \, A_j^l \, \tilde{a}_{\cdot \, l}^k$$
$$\tilde{a}_i^{\cdot \, j} = \tilde{A}_i^k \, A_l^j \, a_k^{\cdot \, l} \qquad\qquad a_i^{\cdot \, j} = A_i^k \, \tilde{A}_l^j \, \tilde{a}_k^{\cdot \, l}$$

More dyadic operations

Dot product

$$\mathbf{ab} \cdot \mathbf{cd} = (\mathbf{b} \cdot \mathbf{c})\mathbf{ad}$$

$$\mathbf{A} \cdot (\lambda \mathbf{a} + \mu \mathbf{b}) = \lambda \mathbf{A} \cdot \mathbf{a} + \mu \mathbf{A} \cdot \mathbf{b}$$
$$(\lambda \mathbf{A} + \mu \mathbf{B}) \cdot \mathbf{a} = \lambda \mathbf{A} \cdot \mathbf{a} + \mu \mathbf{B} \cdot \mathbf{a}$$

Double dot product

$$\mathbf{ab} \cdot\cdot \mathbf{cd} = (\mathbf{b} \cdot \mathbf{c})(\mathbf{a} \cdot \mathbf{d})$$

Scalar product of second-order tensors

$$\mathbf{ab} \bullet \mathbf{cd} = (\mathbf{a} \cdot \mathbf{c})(\mathbf{b} \cdot \mathbf{d})$$

Second-order tensor topics

Transpose

$$\mathbf{A}^T = a^{ji}\,\mathbf{e}_i\,\mathbf{e}_j = a_{ji}\,\mathbf{e}^i\,\mathbf{e}^j = a^j_{.i}\,\mathbf{e}^i\,\mathbf{e}_j = a_j^{.i}\,\mathbf{e}_i\,\mathbf{e}^j$$

$$\mathbf{A} \cdot \mathbf{x} = \mathbf{x} \cdot \mathbf{A}^T$$

$$(\mathbf{A}^T)^T = \mathbf{A}$$

$$(\mathbf{A} \cdot \mathbf{B})^T = \mathbf{B}^T \cdot \mathbf{A}^T$$

$$\mathbf{a} \cdot \mathbf{C}^T \cdot \mathbf{b} = \mathbf{b} \cdot \mathbf{C} \cdot \mathbf{a}$$

$$\det \mathbf{A}^{-1} = (\det \mathbf{A})^{-1} \qquad (\mathbf{A} \cdot \mathbf{B})^{-1} = \mathbf{B}^{-1} \cdot \mathbf{A}^{-1}$$
$$(\mathbf{A}^T)^{-1} = (\mathbf{A}^{-1})^T \qquad (\mathbf{A}^{-1})^{-1} = \mathbf{A}$$

Tensors raised to powers

$$\mathbf{A}^0 = \mathbf{E} \qquad\qquad \mathbf{A}^n = \mathbf{A} \cdot \mathbf{A}^{n-1} \text{ for } n = 1, 2, 3, \ldots$$

$$\mathbf{A}^{-n} = \mathbf{A}^{-n+1} \cdot \mathbf{A}^{-1} \text{ for } n = 2, 3, 4, \ldots$$

$$e^{\mathbf{A}} = \mathbf{E} + \frac{\mathbf{A}}{1!} + \frac{\mathbf{A}^2}{2!} + \frac{\mathbf{A}^3}{3!} + \cdots$$

Symmetric and antisymmetric tensors

symmetric: $\qquad\qquad \mathbf{A} = \mathbf{A}^T; \quad \forall \mathbf{x}, \ \mathbf{A} \cdot \mathbf{x} = \mathbf{x} \cdot \mathbf{A}$

antisymmetric: $\qquad\qquad \mathbf{A} = -\mathbf{A}^T$

$$\mathbf{A} = \frac{1}{2}\left(\mathbf{A} + \mathbf{A}^T\right) + \frac{1}{2}\left(\mathbf{A} - \mathbf{A}^T\right)$$

Eigenpair

$$\mathbf{A} \cdot \mathbf{x} = \lambda \mathbf{x}$$

Viète formulas for invariants

$$I_1(\mathbf{A}) = \lambda_1 + \lambda_2 + \lambda_3 \qquad\qquad = \operatorname{tr} \mathbf{A}$$

$$I_2(\mathbf{A}) = \lambda_1\lambda_2 + \lambda_1\lambda_3 + \lambda_2\lambda_3 \qquad = \frac{1}{2}[\operatorname{tr}^2 \mathbf{A} - \operatorname{tr} \mathbf{A}^2]$$

$$I_3(\mathbf{A}) = \lambda_1\lambda_2\lambda_3 \qquad\qquad\qquad = \det \mathbf{A}$$

Orthogonal tensor

$$\mathbf{Q} \cdot \mathbf{Q}^T = \mathbf{Q}^T \cdot \mathbf{Q} = \mathbf{E}$$

Polar decompositions

$\mathbf{A} = \mathbf{S} \cdot \mathbf{Q} \qquad\qquad \mathbf{Q}$ orthogonal

$\mathbf{A} = \mathbf{Q} \cdot \mathbf{S}' \qquad\qquad \mathbf{S}, \mathbf{S}'$ positive definite and symmetric

Chapter 4

Vector fields

Some rules for differentiating vector functions

$$\frac{d(\mathbf{e}_1(t) + \mathbf{e}_2(t))}{dt} = \frac{d\mathbf{e}_1(t)}{dt} + \frac{d\mathbf{e}_2(t)}{dt}$$

$$\frac{d(c\,\mathbf{e}(t))}{dt} = c\frac{d\mathbf{e}(t)}{dt}$$

$$\frac{d(f(t)\mathbf{e}(t))}{dt} = \frac{df(t)}{dt}\,\mathbf{e}(t) + f(t)\frac{d\mathbf{e}(t)}{dt}$$

$$\frac{d}{dt}(\mathbf{e}_1(t) \cdot \mathbf{e}_2(t)) = \mathbf{e}_1'(t) \cdot \mathbf{e}_2(t) + \mathbf{e}_1(t) \cdot \mathbf{e}_2'(t)$$

$$\frac{d}{dt}(\mathbf{e}_1(t) \times \mathbf{e}_2(t)) = \mathbf{e}_1'(t) \times \mathbf{e}_2(t) + \mathbf{e}_1(t) \times \mathbf{e}_2'(t)$$

and

$$\frac{d}{dt}[\mathbf{e}_1(t)\,,\mathbf{e}_2(t)\,,\mathbf{e}_3(t)] = [\mathbf{e}_1'(t)\,,\mathbf{e}_2(t)\,,\mathbf{e}_3(t)]$$
$$+ [\mathbf{e}_1(t)\,,\mathbf{e}_2'(t)\,,\mathbf{e}_3(t)]$$
$$+ [\mathbf{e}_1(t)\,,\mathbf{e}_2(t)\,,\mathbf{e}_3'(t)]$$

where

$$[\mathbf{e}_1(t)\,,\mathbf{e}_2(t)\,,\mathbf{e}_3(t)] = (\mathbf{e}_1(t) \times \mathbf{e}_2(t)) \cdot \mathbf{e}_3(t)$$

Tangent vectors to coordinate lines

$$\mathbf{r}_i = \frac{\partial \mathbf{r}}{\partial q^i} \qquad (i = 1, 2, 3)$$

Jacobian

$$\sqrt{g} = \mathbf{r}_1 \cdot (\mathbf{r}_2 \times \mathbf{r}_3) = \left|\frac{\partial x^i}{\partial q^j}\right|$$

Pointwise definition of reciprocal basis

$$\mathbf{r}^i \cdot \mathbf{r}_j = \delta_j^i$$

Definition of metric coefficients

$$g_{ij} = \mathbf{r}_i \cdot \mathbf{r}_j$$
$$g^{ij} = \mathbf{r}^i \cdot \mathbf{r}^j$$
$$g_i^j = \mathbf{r}_i \cdot \mathbf{r}^j = \delta_i^j$$

Transformation laws

$$\mathbf{r}_i = A_i^j \, \tilde{\mathbf{r}}_j \qquad\qquad A_i^j = \frac{\partial \tilde{q}^j}{\partial q^i}$$

$$\tilde{\mathbf{r}}_i = \tilde{A}_i^j \, \mathbf{r}_j \qquad\qquad \tilde{A}_i^j = \frac{\partial q^j}{\partial \tilde{q}^i}$$

$$\tilde{f}^i = A_j^i \, f^j \qquad \tilde{f}_i = \tilde{A}_i^j \, f_j \qquad f^i = \tilde{A}_j^i \, \tilde{f}^j \qquad f_i = A_i^j \, \tilde{f}_j$$

Differentials and the nabla operator

Metric forms

$$(ds)^2 = d\mathbf{r} \cdot d\mathbf{r} = g_{ij} \, dq^i \, dq^j$$

Nabla operator

$$\nabla = \mathbf{r}^i \frac{\partial}{\partial q^i}$$

Gradient of a vector function

$$d\mathbf{f} = d\mathbf{r} \cdot \nabla \mathbf{f} = \nabla \mathbf{f}^T \cdot d\mathbf{r}$$

Divergence of vector

$$\operatorname{div} \mathbf{f} = \nabla \cdot \mathbf{f} = \mathbf{r}^i \cdot \frac{\partial \mathbf{f}}{\partial q^i}$$

Rotation and curl of vector

$$\operatorname{rot} \mathbf{f} = \nabla \times \mathbf{f} = \mathbf{r}^i \times \frac{\partial \mathbf{f}}{\partial q^i} \qquad\qquad \boldsymbol{\omega} = \frac{1}{2} \operatorname{rot} \mathbf{f}$$

Divergence and rotation of second-order tensor

$$\nabla \cdot \mathbf{A} = \mathbf{r}^i \cdot \frac{\partial}{\partial q^i} \mathbf{A} \qquad\qquad \nabla \times \mathbf{A} = \mathbf{r}^i \times \frac{\partial}{\partial q^i} \mathbf{A}$$

Differentiation of a vector function

Christoffel coefficients of the second kind

$$\frac{\partial \mathbf{r}_i}{\partial q^j} = \Gamma_{ij}^k \, \mathbf{r}_k \qquad\qquad \frac{\partial \mathbf{r}^j}{\partial q^i} = -\Gamma_{it}^j \, \mathbf{r}^t$$

$$\Gamma_{ij}^k = \Gamma_{ji}^k$$

Christoffel coefficients of the first kind

$$\frac{1}{2} \left(\frac{\partial g_{it}}{\partial q^j} + \frac{\partial g_{tj}}{\partial q^i} - \frac{\partial g_{ji}}{\partial q^t} \right) = \Gamma_{ijt}$$

$$\Gamma_{ijk} = \Gamma_{jik}$$

Covariant differentiation

$$\frac{\partial \mathbf{f}}{\partial q^i} = \mathbf{r}^k \, \nabla_i f_k = \mathbf{r}_j \nabla_i f^j$$

$$\nabla_k f_i = \frac{\partial f_i}{\partial q^k} - \Gamma_{ki}^j \, f_j \qquad\qquad \nabla_k f^i = \frac{\partial f^i}{\partial q^k} + \Gamma_{kt}^i \, f^t$$

$$\nabla \mathbf{f} = \mathbf{r}^i \, \mathbf{r}^j \, \nabla_i f_j = \mathbf{r}^i \, \mathbf{r}_j \, \nabla_i f^j$$

Covariant differentiation of second-order tensor

$$\frac{\partial}{\partial q^k} \mathbf{A} = \nabla_k a^{ij} \, \mathbf{r}_i \, \mathbf{r}_j = \nabla_k a_{ij} \, \mathbf{r}^i \, \mathbf{r}^j = \nabla_k a_i^{\cdot j} \, \mathbf{r}^i \, \mathbf{r}_j = \nabla_k a_{\cdot j}^i \, \mathbf{r}_i \, \mathbf{r}^j$$

$$\nabla_k a^{ij} = \frac{\partial a^{ij}}{\partial q^k} + \Gamma^i_{ks} \, a^{sj} + \Gamma^j_{ks} \, a^{is} \qquad \nabla_k a_{ij} = \frac{\partial a_{ij}}{\partial q^k} - \Gamma^s_{ki} \, a_{sj} - \Gamma^s_{kj} \, a_{is}$$

$$\nabla_k a_i^{\cdot j} = \frac{\partial a_i^{\cdot j}}{\partial q^k} - \Gamma^s_{ki} \, a_s^{\cdot j} + \Gamma^j_{ks} \, a_i^{\cdot s} \qquad \nabla_k a_{\cdot j}^i = \frac{\partial a_{\cdot j}^i}{\partial q^k} + \Gamma^i_{ks} \, a_{\cdot j}^s - \Gamma^s_{kj} \, a_{\cdot s}^i$$

Differential operations

$$\nabla \times \mathbf{f} = \mathbf{r}_k \, \epsilon^{ijk} \, \frac{\partial f_j}{\partial q^i}$$

$$\Gamma^i_{in} = \frac{1}{\sqrt{g}} \frac{\partial \sqrt{g}}{\partial q^n}$$

$$\nabla \cdot \mathbf{f} = \frac{1}{\sqrt{g}} \frac{\partial}{\partial q^i} \left(\sqrt{g} f^i \right)$$

$$\nabla \times \mathbf{A} = \epsilon^{kin} \, \mathbf{r}_n \, \frac{\partial}{\partial q^k} \left(\mathbf{r}^j \, a_{ij} \right)$$

$$\nabla \cdot \mathbf{A} = \frac{1}{\sqrt{g}} \frac{\partial}{\partial q^i} \left(\sqrt{g} \, a^{ij} \, \mathbf{r}_j \right)$$

$$\nabla^2 f = \nabla \cdot \nabla f = g^{ij} \left(\frac{\partial^2 f}{\partial q^i \partial q^j} - \Gamma^k_{ij} \frac{\partial f}{\partial q^k} \right)$$

$$\nabla^2 f = \nabla^j \nabla_j f \qquad \nabla^j = g^{ij} \nabla_i$$

$$\nabla^2 \mathbf{f} = \mathbf{r}_j \, \nabla^i \nabla_i f^j$$

$$\nabla \nabla \cdot \mathbf{f} = \mathbf{r}^i \, \nabla_i \nabla_j f^j$$

$$\nabla \times \nabla \times \mathbf{f} = \nabla \nabla \cdot \mathbf{f} - \nabla^2 \mathbf{f}$$

Orthogonal coordinate systems

Lamé coefficients

$$(H_i)^2 = g_{ii}$$
$$\mathbf{r}^i = \mathbf{r}_i/(H_i)^2 \qquad (i = 1, 2, 3)$$
$$\hat{\mathbf{r}}_i = \mathbf{r}_i/H_i$$

Differentiation in the orthogonal basis

$$\nabla = \frac{\hat{\mathbf{r}}_i}{H_i} \frac{\partial}{\partial q^i}$$

$$\nabla \mathbf{f} = \hat{\mathbf{r}}_i \hat{\mathbf{r}}_j \left(\frac{1}{H_i} \frac{\partial f_j}{\partial q^i} - \frac{f_i}{H_i H_j} \frac{\partial H_i}{\partial q^j} + \delta_{ij} \frac{f_k}{H_k} \frac{1}{H_i} \frac{\partial H_i}{\partial q^k} \right)$$

$$\nabla \cdot \mathbf{f} = \frac{1}{H_1 H_2 H_3} \left(\frac{\partial}{\partial q^1} (H_2 H_3 f_1) + \frac{\partial}{\partial q^2} (H_3 H_1 f_2) + \frac{\partial}{\partial q^3} (H_1 H_2 f_3) \right)$$

$$\nabla \times \mathbf{f} = \frac{1}{2} \frac{\hat{\mathbf{r}}_i \times \hat{\mathbf{r}}_j}{H_i H_j} \left(\frac{\partial}{\partial q^i} (H_j f_j) - \frac{\partial}{\partial q^j} (H_i f_i) \right)$$

$$\nabla^2 f = \frac{1}{H_1 H_2 H_3} \left[\frac{\partial}{\partial q^1} \left(\frac{H_2 H_3}{H_1} \frac{\partial f}{\partial q^1} \right) \right.$$
$$+ \frac{\partial}{\partial q^2} \left(\frac{H_3 H_1}{H_2} \frac{\partial f}{\partial q^2} \right)$$
$$\left. + \frac{\partial}{\partial q^3} \left(\frac{H_1 H_2}{H_3} \frac{\partial f}{\partial q^3} \right) \right]$$

Integration formulas

Transformation of multiple integral

$$\int_V f(x_1, x_2, x_3) \, dx_1 \, dx_2 \, dx_3 = \int_V f(q^1, q^2, q^3) J \, dq^1 \, dq^2 \, dq^3$$

$$J = \sqrt{g} = \left| \frac{\partial x_i}{\partial q^j} \right|$$

Integration by parts

$$\int_V \frac{\partial f}{\partial x_k} g \, dx_1 \, dx_2 \, dx_3 = - \int_V \frac{\partial g}{\partial x_k} f \, dx_1 \, dx_2 \, dx_3 + \int_S f g \, n_k \, dS$$

Miscellaneous results

$$\int_V \nabla f \, dV = \int_S f \mathbf{n} \, dS \qquad\qquad \int_V \nabla \cdot \mathbf{f} \, dV = \int_S \mathbf{n} \cdot \mathbf{f} \, dS$$

$$\int_V \nabla \mathbf{f} \, dV = \int_S \mathbf{n} \mathbf{f} \, dS \qquad\qquad \int_V \nabla \times \mathbf{f} \, dV = \int_S \mathbf{n} \times \mathbf{f} \, dS$$

$$\int_V \nabla \mathbf{A} \, dV = \int_S \mathbf{n} \mathbf{A} \, dS$$

$$\int_V \nabla \cdot \mathbf{A} \, dV = \int_S \mathbf{n} \cdot \mathbf{A} \, dS$$

$$\int_V \nabla \times \mathbf{A} \, dV = \int_S \mathbf{n} \times \mathbf{A} \, dS$$

$$\oint_\Gamma \mathbf{f} \cdot d\mathbf{r} = \int_S (\mathbf{n} \times \nabla) \cdot \mathbf{f} \, dS$$

$$\oint_\Gamma d\mathbf{r} \cdot \mathbf{A} = \int_S (\mathbf{n} \times \nabla) \cdot \mathbf{A} \, dS$$

$$\oint_\Gamma \mathbf{A} \cdot d\mathbf{r} = \int_S (\mathbf{n} \times \nabla) \cdot \mathbf{A}^T \, dS$$

Chapter 5

Elementary theory of curves

Parametrization

$$\mathbf{r} = \mathbf{r}(t) \qquad \text{or} \qquad \mathbf{r} = \mathbf{r}(s)$$

Length

$$s = \int_a^b |\mathbf{r}'(t)| \, dt$$

Unit tangent

$$\boldsymbol{\tau}(s) = \mathbf{r}'(s)$$

Equation of tangent line

$$\mathbf{r} = \mathbf{r}(t_0) + \lambda \mathbf{r}'(t_0)$$

$$\frac{x - x(t_0)}{x'(t_0)} = \frac{y - y(t_0)}{y'(t_0)} = \frac{z - z(t_0)}{z'(t_0)}$$

Curvature

$$k_1 = |\mathbf{r}''(s)| \qquad\qquad k_1^2 = \frac{(\mathbf{r}'(t) \times \mathbf{r}''(t))^2}{(\mathbf{r}'^2(t))^3}$$

Radius of curvature

$$R = 1/k_1$$

Principal normal, binormal

$$\boldsymbol{\nu} = \frac{\mathbf{r}''(s)}{k_1} \qquad\qquad \boldsymbol{\beta} = \boldsymbol{\tau} \times \boldsymbol{\nu}$$

Osculating plane

$$[\mathbf{r} - \mathbf{r}(s_0)] \cdot \boldsymbol{\beta}(s_0) = 0$$

$$\begin{vmatrix} x - x(t_0) & y - y(t_0) & z - z(t_0) \\ x'(t_0) & y'(t_0) & z'(t_0) \\ x''(t_0) & y''(t_0) & z''(t_0) \end{vmatrix} = 0$$

Torsion

$$k_2 = -\frac{(\mathbf{r}'(s) \times \mathbf{r}''(s)) \cdot \mathbf{r}'''(s)}{k_1^2} \qquad\qquad k_2 = -\frac{(\mathbf{r}'(t) \times \mathbf{r}''(t)) \cdot \mathbf{r}'''(t)}{(\mathbf{r}'(t) \times \mathbf{r}''(t))^2}$$

Serret–Frenet equations

$$\boldsymbol{\tau}' = k_1 \boldsymbol{\nu}$$
$$\boldsymbol{\nu}' = -k_1 \boldsymbol{\tau} - k_2 \boldsymbol{\beta}$$
$$\boldsymbol{\beta}' = k_2 \boldsymbol{\nu}$$

Theory of surfaces

Parametrization

$$\mathbf{r} = \mathbf{r}(u^1, u^2)$$

Tangent vectors, unit normal

$$\mathbf{r}_i = \frac{\partial \mathbf{r}}{\partial u^i} \qquad (i = 1, 2)$$

$$\mathbf{n} = \frac{\mathbf{r}_1 \times \mathbf{r}_2}{|\mathbf{r}_1 \times \mathbf{r}_2|}$$

First fundamental form

$$(ds)^2 = g_{ij}\, du^i\, du^j = E(du^1)^2 + 2F\, du^1\, du^2 + G(du^2)^2$$

$$E = \mathbf{r}_1 \cdot \mathbf{r}_1 \qquad\qquad F = \mathbf{r}_1 \cdot \mathbf{r}_2 \qquad\qquad G = \mathbf{r}_2 \cdot \mathbf{r}_2$$

Orthogonality of curves

$$E\, du^1\, d\tilde{u}^1 + F(du^1\, d\tilde{u}^2 + du^2\, d\tilde{u}^1) + G\, du^2\, d\tilde{u}^2 = 0$$

Area

$$S = \int_A \sqrt{EG - F^2}\, du^1\, du^2$$

Second fundamental form

$$d^2\mathbf{r} \cdot \mathbf{n} = L(du^1)^2 + 2M\,du^1\,du^2 + N(du^2)^2 = -d\mathbf{r} \cdot d\mathbf{n}$$

$$L = \frac{\partial^2 \mathbf{r}}{(\partial u^1)^2} \cdot \mathbf{n} \qquad M = \frac{\partial^2 \mathbf{r}}{\partial u^1 \partial u^2} \cdot \mathbf{n} \qquad N = \frac{\partial^2 \mathbf{r}}{(\partial u^2)^2} \cdot \mathbf{n}$$

Normal curvature, mean curvature, Gaussian curvature

$$k_0 = k_1 \cos \vartheta \qquad \vartheta = \text{angle between } \boldsymbol{\nu} \text{ and } \mathbf{n}$$

$$H = \frac{1}{2}(k_{\min} + k_{\max}) = \frac{1}{2}\frac{LG - 2MF + NE}{EG - F^2}$$

$$K = k_{\min} k_{\max} = \frac{LN - M^2}{EG - F^2}$$

Surface given by $z = f(x, y)$ in Cartesian coordinates

Subscripts x, y denote partial derivatives with respect to x, y respectively.

$$E = 1 + f_x{}^2 \qquad F = f_x f_y \qquad G = 1 + f_y{}^2$$

$$EG - F^2 = 1 + f_x{}^2 + f_y{}^2 \qquad S = \int_D \sqrt{1 + f_x{}^2 + f_y{}^2}\, dx\, dy$$

$$\mathbf{n} = \frac{-f_x \mathbf{i}_1 - f_y \mathbf{i}_2 + \mathbf{i}_3}{\sqrt{1 + f_x{}^2 + f_y{}^2}}$$

$$L = \mathbf{r}_{xx} \cdot \mathbf{n} = \frac{f_{xx}}{\sqrt{1 + f_x{}^2 + f_y{}^2}}$$

$$M = \mathbf{r}_{xy} \cdot \mathbf{n} = \frac{f_{xy}}{\sqrt{1 + f_x{}^2 + f_y{}^2}}$$

$$N = \mathbf{r}_{yy} \cdot \mathbf{n} = \frac{f_{yy}}{\sqrt{1 + f_x{}^2 + f_y{}^2}}$$

$$K = \frac{f_{xx}f_{yy} - f_{xy}^2}{\left(1 + f_x^2 + f_y^2\right)^2}$$

Surface of revolution about z-axis

$$x = \phi(u) \qquad z = \psi(u)$$

$$(ds)^2 = \left(\phi'^2 + \psi'^2\right) du^2 + \phi^2\, dv^2$$

$$-d\mathbf{n} \cdot d\mathbf{r} = \frac{\psi''\phi' - \phi''\psi'}{\sqrt{\phi'^2 + \psi'^2}}\, du^2 + \frac{\psi'\phi}{\sqrt{\phi'^2 + \psi'^2}}\, dv^2$$

Surface gradient operator and Gauss–Ostrogradsky theorems

Surface gradient operator

$$\widetilde{\nabla} = \mathbf{r}^i \frac{\partial}{\partial u^i} \qquad (i = 1, 2)$$

Surface analogues of the Gauss–Ostrogradsky (divergence) theorem

$$\int_S \left(\widetilde{\nabla} \cdot \mathbf{X} + 2H\mathbf{n} \cdot \mathbf{X}\right) dS = \oint_\Gamma \boldsymbol{\nu} \cdot \mathbf{X}\, ds$$

$$\int_S \left(\widetilde{\nabla}\mathbf{X} + 2H\mathbf{n}\mathbf{X}\right) dS = \oint_\Gamma \boldsymbol{\nu}\mathbf{X}\, ds$$

$$\int_S \left(\widetilde{\nabla} \times \mathbf{X} + 2H\mathbf{n} \times \mathbf{X}\right) dS = \oint_\Gamma \boldsymbol{\nu} \times \mathbf{X}\, ds$$

$$\int_S \widetilde{\nabla} \times (\mathbf{n}\mathbf{X})\, dS = \oint_\Gamma \boldsymbol{\tau}\mathbf{X}\, ds$$

Chapter 6

Stress tensor

Relation between stress tensor and stress vector

$$\mathbf{t} = \mathbf{n} \cdot \boldsymbol{\sigma}$$

Equilibrium and motion equations

$$\nabla \cdot \boldsymbol{\sigma} + \rho \mathbf{f} = \mathbf{0} \qquad\qquad \nabla \cdot \boldsymbol{\sigma} + \rho \mathbf{f} = \rho \frac{d^2 \mathbf{u}}{dt^2}$$

Strain tensor

$$\varepsilon = \frac{1}{2}\left(\nabla \mathbf{u} + (\nabla \mathbf{u})^T\right)$$

Compatibility deformation conditions

$$\nabla \times (\nabla \times \varepsilon)^T = \mathbf{0}$$

Hooke's law

$$\boldsymbol{\sigma} = \mathbf{C} \cdot\cdot \, \varepsilon \qquad \begin{aligned} c_{ijmn} &= c_{jimn} = c_{ijnm} \\ c_{ijmn} &= c_{mnij} \end{aligned}$$

Isotropic material

$$\boldsymbol{\sigma} = \lambda \mathbf{E}\, \mathrm{tr}\,\varepsilon + 2\mu\varepsilon$$

Recalculation of elastic moduli for isotropic body

Moduli	λ, μ	k, μ	μ, ν	E, ν	E, μ
λ	λ	$k - \frac{2}{3}\mu$	$\frac{2\mu\nu}{1-2\nu}$	$\frac{\nu E}{(1+\nu)(1-2\nu)}$	$\frac{(E-2\mu)\mu}{3\mu-E}$
$\mu = G$	μ	μ	μ	$\frac{E}{2(1+\nu)}$	μ
k	$\lambda + \frac{2}{3}\mu$	k	$\frac{2\mu(1+\nu)}{3(1-2\nu)}$	$\frac{E}{3(1-2\nu)}$	$\frac{E\mu}{3(3\mu-E)}$
E	$\frac{\mu(3\lambda+2\mu)}{\lambda+\mu}$	$\frac{9k\mu}{3k+\mu}$	$2\mu(1+\nu)$	E	E
ν	$\frac{\lambda}{2(\lambda+\mu)}$	$\frac{3k-2\mu}{6k+2\mu}$	ν	ν	$\frac{1}{2}\frac{E}{\mu} - 1$

Principal equations in Cartesian coordinates

Equations of motion

$$\frac{\partial \sigma_{ij}}{\partial x_i} + \rho f_j = \rho \frac{\partial^2 u_j}{\partial t^2}$$

$$\frac{\partial \sigma_{11}}{\partial x_1} + \frac{\partial \sigma_{21}}{\partial x_2} + \frac{\partial \sigma_{31}}{\partial x_3} + \rho f_1 = \rho \frac{\partial^2 u_1}{\partial t^2}$$

$$\frac{\partial \sigma_{12}}{\partial x_1} + \frac{\partial \sigma_{22}}{\partial x_2} + \frac{\partial \sigma_{32}}{\partial x_3} + \rho f_2 = \rho \frac{\partial^2 u_2}{\partial t^2}$$

$$\frac{\partial \sigma_{13}}{\partial x_1} + \frac{\partial \sigma_{23}}{\partial x_2} + \frac{\partial \sigma_{33}}{\partial x_3} + \rho f_3 = \rho \frac{\partial^2 u_3}{\partial t^2}$$

Strains

$$\varepsilon_{11} = \frac{\partial u_1}{\partial x_1} \qquad\qquad \varepsilon_{12} = \frac{1}{2}\left(\frac{\partial u_1}{\partial x_2} + \frac{\partial u_2}{\partial x_1}\right)$$

$$\varepsilon_{22} = \frac{\partial u_2}{\partial x_2} \qquad\qquad \varepsilon_{13} = \frac{1}{2}\left(\frac{\partial u_1}{\partial x_3} + \frac{\partial u_3}{\partial x_1}\right)$$

$$\varepsilon_{33} = \frac{\partial u_3}{\partial x_3} \qquad\qquad \varepsilon_{23} = \frac{1}{2}\left(\frac{\partial u_2}{\partial x_3} + \frac{\partial u_3}{\partial x_2}\right)$$

Principal equations in a curvilinear coordinate system

Equations of motion

$$\frac{\partial}{\partial q^i}\left(\sqrt{g}\,\sigma^{ij}\right) + \Gamma^j_{mn}\sigma^{mn} + \rho\sqrt{g}f^j = \rho\sqrt{g}\frac{\partial^2 u^j}{\partial t^2}$$

Strains

$$\boldsymbol{\varepsilon} = \varepsilon_{st}\mathbf{r}^s\mathbf{r}^t \qquad\qquad \varepsilon_{st} = \frac{1}{2}\left(\frac{\partial u_s}{\partial q_t} + \frac{\partial u_t}{\partial q_s}\right) - \Gamma^r_{st}u_r$$

Principal equations in cylindrical coordinates

Equations of motion

$$\frac{\partial \sigma_{rr}}{\partial r} + \frac{\sigma_{rr} - \sigma_{\phi\phi}}{r} + \frac{1}{r}\frac{\partial \sigma_{r\phi}}{\partial \phi} + \frac{\partial \sigma_{zr}}{\partial z} + \rho f_r = \rho\frac{\partial^2 u_r}{\partial t^2}$$

$$\frac{\partial \sigma_{r\phi}}{\partial r} + 2\frac{\sigma_{r\phi}}{r} + \frac{1}{r}\frac{\partial \sigma_{\phi\phi}}{\partial \phi} + \frac{\partial \sigma_{z\phi}}{\partial z} + \rho f_\phi = \rho\frac{\partial^2 u_\phi}{\partial t^2}$$

$$\frac{\partial \sigma_{rz}}{\partial r} + \frac{\sigma_{rz}}{r} + \frac{1}{r}\frac{\partial \sigma_{z\phi}}{\partial \phi} + \frac{\partial \sigma_{zz}}{\partial z} + \rho f_z = \rho\frac{\partial^2 u_z}{\partial t^2}$$

Strains

$$\varepsilon_{rr} = \frac{\partial u_r}{\partial r} \qquad\qquad \varepsilon_{r\phi} = \frac{1}{2}\left(\frac{\partial u_\phi}{\partial r} + \frac{1}{r}\frac{\partial u_r}{\partial \phi} - \frac{u_\phi}{r}\right)$$

$$\varepsilon_{\phi\phi} = \frac{1}{r}\frac{\partial u_\phi}{\partial \phi} + \frac{u_r}{r} \qquad\qquad \varepsilon_{\phi z} = \frac{1}{2}\left(\frac{1}{r}\frac{\partial u_z}{\partial \phi} + \frac{\partial u_\phi}{\partial z}\right)$$

$$\varepsilon_{zz} = \frac{\partial u_z}{\partial z} \qquad\qquad \varepsilon_{rz} = \frac{1}{2}\left(\frac{\partial u_r}{\partial z} + \frac{\partial u_z}{\partial r}\right)$$

Equilibrium equations in displacements

$$\mu\left(\Delta u_r - \frac{u_r}{r^2} - \frac{2}{r^2}\frac{\partial u_\phi}{\partial \phi}\right)$$
$$+ (\lambda + \mu)\frac{\partial}{\partial r}\left[\frac{1}{r}\frac{\partial}{\partial r}(ru_r) + \frac{1}{r}\frac{\partial u_\phi}{\partial \phi} + \frac{\partial u_z}{\partial z}\right] + \rho f_r = 0$$

$$\mu\left(\Delta u_\phi - \frac{u_\phi}{r^2} + \frac{2}{r^2}\frac{\partial u_r}{\partial \phi}\right)$$
$$+ (\lambda + \mu)\frac{1}{r}\frac{\partial}{\partial \phi}\left[\frac{1}{r}\frac{\partial}{\partial r}(ru_r) + \frac{1}{r}\frac{\partial u_\phi}{\partial \phi} + \frac{\partial u_z}{\partial z}\right] + \rho f_\phi = 0$$

$$\mu\Delta u_z + (\lambda + \mu)\frac{\partial}{\partial r}\left[\frac{1}{r}\frac{\partial}{\partial r}(ru_r) + \frac{1}{r}\frac{\partial u_\phi}{\partial \phi} + \frac{\partial u_z}{\partial z}\right] + \rho f_z = 0$$

where

$$\Delta = \frac{\partial^2}{\partial r^2} + \frac{1}{r}\frac{\partial}{\partial r} + \frac{1}{r^2}\frac{\partial^2}{\partial \phi^2} + \frac{\partial^2}{\partial z^2}$$

Principal equations in spherical coordinates

Equations of motion

$$\frac{\partial \sigma_{rr}}{\partial r} + \frac{1}{r}\frac{\partial \sigma_{r\theta}}{\partial \theta} + \frac{1}{r\sin\theta}\frac{\partial \sigma_{r\phi}}{\partial \phi}$$
$$+ \frac{1}{r}\left(2\sigma_{rr} - \sigma_{\theta\theta} - \sigma_{\phi\phi} + \sigma_{r\theta}\cot\theta\right) + \rho f_r = \rho\frac{\partial^2 u_r}{\partial t^2}$$
$$\frac{\partial \sigma_{r\phi}}{\partial r} + \frac{1}{r}\frac{\partial \sigma_{\theta\phi}}{\partial \theta} + \frac{1}{r\sin\theta}\frac{\partial \sigma_{\phi\phi}}{\partial \phi}$$
$$+ \frac{1}{r}\left(3\sigma_{r\phi} + 2\sigma_{\theta\phi}\cot\theta\right) + \rho f_\phi = \rho\frac{\partial^2 u_\phi}{\partial t^2}$$
$$\frac{\partial \sigma_{r\theta}}{\partial r} + \frac{1}{r}\frac{\partial \sigma_{\theta\theta}}{\partial \theta} + \frac{1}{r\sin\theta}\frac{\partial \sigma_{\theta\phi}}{\partial \phi}$$
$$+ \frac{1}{r}\left[(\sigma_{\theta\theta} - \sigma_{\phi\phi})\cot\theta + 3\sigma_{r\theta}\right] + \rho f_\theta = \rho\frac{\partial^2 u_\theta}{\partial t^2}$$

Strains

$$\varepsilon_{rr} = \frac{\partial u_r}{\partial r}$$
$$\varepsilon_{\theta\theta} = \frac{1}{r}\frac{\partial u_\theta}{\partial \theta} + \frac{u_r}{r}$$
$$\varepsilon_{\phi\phi} = \frac{1}{r\sin\theta}\frac{\partial u_\phi}{\partial \phi} + \frac{u_\theta}{r}\cot\theta + \frac{u_r}{r}$$
$$\varepsilon_{\theta\phi} = \frac{1}{2r}\left(\frac{\partial u_\phi}{\partial \theta} + \frac{1}{\sin\theta}\frac{\partial u_\theta}{\partial \phi} - u_\phi\cot\theta\right)$$
$$\varepsilon_{r\phi} = \frac{1}{2}\left(\frac{1}{r\sin\theta}\frac{\partial u_r}{\partial \phi} + \frac{\partial u_\phi}{\partial r} - \frac{u_\phi}{r}\right)$$
$$\varepsilon_{r\theta} = \frac{1}{2}\left(\frac{\partial u_\theta}{\partial r} + \frac{1}{r}\frac{\partial u_r}{\partial \theta} - \frac{u_\theta}{r}\right)$$

Equilibrium equations in displacements

$$\mu\left\{\Delta u_r - \frac{2}{r^2}\left[u_r + \frac{1}{\sin\theta}\frac{\partial}{\partial\theta}(u_\theta\sin\theta) + \frac{1}{\sin\theta}\frac{\partial u_\phi}{\partial\phi}\right]\right\}$$
$$+ (\lambda+\mu)\frac{\partial}{\partial r}\left[\frac{1}{r^2}\frac{\partial}{\partial r}(r^2 u_r) + \frac{1}{r\sin\theta}\frac{\partial}{\partial\theta}(u_\theta\sin\theta) + \frac{1}{r\sin\theta}\frac{\partial u_\phi}{\partial\phi}\right] + \rho f_r = 0$$

$$\mu\left\{\Delta u_\theta - \frac{2}{r^2}\left[\frac{\partial u_r}{\partial\theta} - \frac{1}{2\sin^2\theta}u_\theta - \frac{\cos\theta}{\sin^2\theta}\frac{\partial u_\phi}{\partial\phi}\right]\right\}$$
$$+ \frac{\lambda+\mu}{r}\frac{\partial}{\partial\theta}\left[\frac{1}{r^2}\frac{\partial}{\partial r}(r^2 u_r) + \frac{1}{r\sin\theta}\frac{\partial}{\partial\theta}(u_\theta\sin\theta) + \frac{1}{r\sin\theta}\frac{\partial u_\phi}{\partial\phi}\right] + \rho f_\theta = 0$$

$$\mu\left\{\Delta u_\phi + \frac{2}{r^2\sin\theta}\left[\frac{\partial u_r}{\partial\phi} + \cot\frac{\partial u_\theta}{\partial\phi} - \frac{u_\phi}{2\sin\theta}\right]\right\}$$
$$+ \frac{\lambda+\mu}{r\sin\theta}\frac{\partial}{\partial\phi}\left[\frac{1}{r^2}\frac{\partial}{\partial r}(r^2 u_r) + \frac{1}{r\sin\theta}\frac{\partial}{\partial\theta}(u_\theta\sin\theta) + \frac{1}{r\sin\theta}\frac{\partial u_\phi}{\partial\phi}\right] + \rho f_\phi = 0$$

where

$$\Delta = \frac{1}{r^2}\frac{\partial}{\partial r}\left(r^2\frac{\partial}{\partial r}\right) + \frac{1}{r^2\sin\theta}\frac{\partial}{\partial\theta}\left(\sin\theta\frac{\partial}{\partial\theta}\right) + \frac{1}{r^2\sin\theta}\frac{\partial^2}{\partial\phi^2}$$

Chapter 7

Formulas of Surface Theory

Position vector of the middle surface and base vectors

$$\boldsymbol{\rho} = \boldsymbol{\rho}(q^1, q^2)$$
$$\boldsymbol{\rho}_1 = \frac{\partial\boldsymbol{\rho}}{\partial q^1}$$
$$\boldsymbol{\rho}_2 = \frac{\partial\boldsymbol{\rho}}{\partial q^2}$$
$$\boldsymbol{\rho}_\alpha \cdot \boldsymbol{\rho}^\beta = \delta_\alpha^\beta \quad (\alpha, \beta = 1, 2)$$

Normal to mid-surface

$$\mathbf{n} = \frac{\boldsymbol{\rho}_1 \times \boldsymbol{\rho}_2}{|\boldsymbol{\rho}_1 \times \boldsymbol{\rho}_2|}$$

Metric tensor

$$\mathbf{A} = \boldsymbol{\rho}^\alpha \boldsymbol{\rho}_\alpha = \mathbf{E} - \mathbf{nn}$$

Nabla operator

$$\widetilde{\nabla} = \boldsymbol{\rho}^\alpha \frac{\partial}{\partial q^\alpha}$$

Curvature tensor

$$\mathbf{B} = b_{\alpha\beta} \boldsymbol{\rho}^\alpha \boldsymbol{\rho}^\beta = -\widetilde{\nabla}\mathbf{n}$$

Derivatives of $\boldsymbol{\rho}_\alpha$ and $\boldsymbol{\rho}^\alpha$

$$\frac{\partial \boldsymbol{\rho}_\alpha}{\partial q^\beta} = \Gamma^\gamma_{\alpha\beta} \boldsymbol{\rho}_\gamma + b_{\alpha\beta}\mathbf{n} \qquad\qquad \frac{\partial \boldsymbol{\rho}^\alpha}{\partial q^\beta} = -\Gamma^\alpha_{\beta\gamma} \boldsymbol{\rho}^\gamma + b^\alpha_\beta \mathbf{n}$$

Gradient of vector $\mathbf{v}$

$$\widetilde{\nabla}\mathbf{v} = (\widetilde{\nabla}\tilde{\mathbf{v}}) \cdot \mathbf{A} - w\mathbf{B} + (\widetilde{\nabla}w + \mathbf{B} \cdot \tilde{\mathbf{v}})\mathbf{n}$$

$$\mathbf{v} = \tilde{\mathbf{v}} + w\mathbf{n}$$
$$\tilde{\mathbf{v}} = v_1(q^1, q^2)\boldsymbol{\rho}^1 + v_2(q^1, q^2)\boldsymbol{\rho}^2$$
$$w = v_3 = v^3 = \mathbf{v} \cdot \mathbf{n}$$

Divergence of second-order tensor $\mathbf{T}$

$$\widetilde{\nabla} \cdot \mathbf{T} = \boldsymbol{\rho}^\gamma \frac{\partial}{\partial q^\gamma} \cdot \left(T^{\alpha\beta} \boldsymbol{\rho}_\alpha \boldsymbol{\rho}_\beta + T^{3\beta} \mathbf{n}\boldsymbol{\rho}_\beta + T^{\alpha 3} \boldsymbol{\rho}_\alpha \mathbf{n} + T^{33}\mathbf{nn} \right)$$

$$= \frac{\partial T^{\alpha\beta}}{\partial q^\alpha} \boldsymbol{\rho}_\beta + T^{\alpha\beta} \boldsymbol{\rho}^\gamma \cdot \frac{\partial \boldsymbol{\rho}_\alpha}{\partial q^\gamma} \boldsymbol{\rho}_\beta + T^{\alpha\beta} \frac{\partial \boldsymbol{\rho}_\beta}{\partial q^\alpha} + T^{3\beta} \boldsymbol{\rho}^\gamma \cdot \frac{\partial \mathbf{n}}{\partial q^\gamma} \boldsymbol{\rho}_\beta$$

$$+ T^{33} \boldsymbol{\rho}^\gamma \cdot \frac{\partial \mathbf{n}}{\partial q^\gamma} \mathbf{n} + \frac{\partial T^{\alpha 3}}{\partial q^\alpha} \mathbf{n} + T^{\alpha 3} \boldsymbol{\rho}^\gamma \cdot \frac{\partial \boldsymbol{\rho}_\alpha}{\partial q^\gamma} \mathbf{n} + T^{\alpha 3} \frac{\partial \mathbf{n}}{\partial q^\alpha}$$

$$= \frac{\partial T^{\alpha\beta}}{\partial q^\alpha} \boldsymbol{\rho}_\beta + T^{\alpha\beta} \Gamma^\gamma_{\alpha\gamma} \boldsymbol{\rho}_\beta + T^{\alpha\beta} \Gamma^\gamma_{\beta\alpha} \boldsymbol{\rho}_\gamma + T^{\alpha\beta} b_{\alpha\beta}\mathbf{n} - T^{3\beta} b^\gamma_\gamma \boldsymbol{\rho}_\beta$$

$$- T^{33} b^\gamma_\gamma \mathbf{n} + \frac{\partial T^{\alpha 3}}{\partial q^\alpha} \mathbf{n} + T^{\alpha 3} \Gamma^\gamma_{\alpha\gamma} \mathbf{n} - T^{\alpha 3} b^\beta_\alpha \boldsymbol{\rho}_\beta$$

Kinematics in a Neighborhood of a Shell Mid-Surface

Position vector

$$\mathbf{r} = \mathbf{r}(q^1, q^2, z) = \boldsymbol{\rho}(q^1, q^2) + z\mathbf{n}$$

Basis vectors

$$\mathbf{r}_\alpha = \frac{\partial \mathbf{r}}{\partial q^\alpha} = \boldsymbol{\rho}_\alpha + z\frac{\partial \mathbf{n}}{\partial q^\alpha} = (\mathbf{A} - z\mathbf{B}) \cdot \boldsymbol{\rho}_\alpha$$

$$\mathbf{r}^\alpha = (\mathbf{A} - z\mathbf{B})^{-1} \cdot \boldsymbol{\rho}^\alpha$$

$$\mathbf{r}_3 = \mathbf{r}^3 = \mathbf{n}$$

Spatial nabla operator

$$\nabla = \mathbf{r}^\alpha \frac{\partial}{\partial q^\alpha} + \mathbf{n}\frac{\partial}{\partial z} = (\mathbf{A} - z\mathbf{B})^{-1} \cdot \widetilde{\nabla} + \mathbf{n}\frac{\partial}{\partial z}$$

Shell Equilibrium Equations

Stress resultant tensor and couple stress tensor

$$\mathbf{T} = [\![(\mathbf{A} - z\mathbf{B})^{-1} \cdot \boldsymbol{\sigma}]\!]$$

$$\mathbf{M} = -[\![(\mathbf{A} - z\mathbf{B})^{-1} \cdot z\boldsymbol{\sigma} \times \mathbf{n}]\!]$$

where

$$[\![f]\!] = \int_{-h/2}^{h/2} Gf\,dz$$

Equilibrium equations

$$\widetilde{\nabla} \cdot \mathbf{T} + \mathbf{q} = 0 \qquad\qquad \mathbf{T}_\times = \boldsymbol{\rho}_\alpha \times [\![\boldsymbol{\sigma}^\alpha]\!]$$

$$\widetilde{\nabla} \cdot \mathbf{M} + \mathbf{T}_\times + \mathbf{m} = 0 \qquad\qquad \boldsymbol{\sigma}^\alpha = \mathbf{r}^\alpha \cdot \boldsymbol{\sigma}$$

Strain measures

$$\boldsymbol{\epsilon} = \frac{1}{2}\left[(\widetilde{\nabla}\tilde{\mathbf{v}}) \cdot \mathbf{A} + \mathbf{A} \cdot (\widetilde{\nabla}\tilde{\mathbf{v}})^T\right] - w\mathbf{B}$$

$$\text{æ} = -\frac{1}{2}\left((\widetilde{\nabla}\tilde{\boldsymbol{\vartheta}}) \cdot \mathbf{A} + \mathbf{A} \cdot (\widetilde{\nabla}\tilde{\boldsymbol{\vartheta}})^T\right)$$

Constitutive equations for Kirchhoff–Love shells

$$\mathbf{T} \cdot \mathbf{A} = \frac{Eh}{1+\nu} \left[\epsilon + \frac{\nu}{1-\nu} \mathbf{A} \operatorname{tr} \epsilon \right]$$

$$\mathbf{M} = -\frac{Eh^3}{12(1+\nu)} \left[\text{æ} + \frac{\nu}{1-\nu} \mathbf{A} \operatorname{tr} \text{æ} \right] \times \mathbf{n}$$

Component representation

$$\mathbf{T} \cdot \mathbf{A} = T_{\alpha\beta} \boldsymbol{\rho}^\alpha \boldsymbol{\rho}^\beta \qquad\qquad \epsilon = \epsilon_{\alpha\beta} \boldsymbol{\rho}^\alpha \boldsymbol{\rho}^\beta$$

$$\mathbf{M} = -M_{\alpha\beta} \boldsymbol{\rho}^\alpha \boldsymbol{\rho}^\beta \times \mathbf{n} \qquad\qquad \text{æ} = \text{æ}_{\alpha\beta} \boldsymbol{\rho}^\alpha \boldsymbol{\rho}^\beta$$

$$T_{\alpha\beta} = [\![\sigma_{\alpha\beta}]\!] = \int_{-h/2}^{h/2} \sigma_{\alpha\beta} \, dz$$

$$M_{\alpha\beta} = [\![z\sigma_{\alpha\beta}]\!] = \int_{-h/2}^{h/2} z\sigma_{\alpha\beta} \, dz$$

$$T_{\alpha\beta} = \frac{Eh}{1+\nu} \left[\epsilon_{\alpha\beta} + \frac{\nu}{1-\nu} a_{\alpha\beta} \epsilon_\gamma^\gamma \right]$$

$$M_{\alpha\beta} = \frac{Eh}{1+\nu} \left[\text{æ}_{\alpha\beta} + \frac{\nu}{1-\nu} a_{\alpha\beta} \text{æ}_\gamma^\gamma \right]$$

Shell surface energy density

$$U = \frac{Eh}{2(1+\nu)} \left[\epsilon \cdot\cdot \epsilon + \frac{\nu}{1-\nu} \operatorname{tr}^2 \epsilon \right] + \frac{Eh^3}{24(1+\nu)} \left[\text{æ} \cdot\cdot \text{æ} + \frac{\nu}{1-\nu} \operatorname{tr}^2 \text{æ} \right]$$

Common principal boundary conditions

(1) Clamped (fixed) edge

$$\mathbf{v}\big|_\omega = \mathbf{0} \qquad\qquad \frac{\partial w}{\partial \nu}\bigg|_\omega = 0$$

(2) Simple support edge

$$\mathbf{v}\big|_\omega = \mathbf{0} \qquad\qquad \boldsymbol{\nu} \cdot (\mathbf{M} \times \mathbf{n}) \cdot \boldsymbol{\nu}\big|_\omega = 0$$

(3) Free edge

$$\boldsymbol{\nu} \cdot \mathbf{T} \cdot \mathbf{A}\big|_{\omega} = \mathbf{0}$$

$$\boldsymbol{\nu} \cdot \mathbf{T} \cdot \mathbf{n}\big|_{\omega} - \frac{\partial}{\partial s} \left(\boldsymbol{\nu} \cdot (\mathbf{M} \times \mathbf{n}) \cdot \boldsymbol{\tau}\right)\big|_{\omega} = 0$$

$$\boldsymbol{\nu} \cdot (\mathbf{M} \times \mathbf{n}) \cdot \boldsymbol{\nu}\big|_{\omega} = 0$$

Deflection of a plate

Resultant moments

$$M_{11} = \frac{Eh^3}{12(1 - \nu^2)}(\text{æ}_{11} + \nu\text{æ}_{22}) \qquad \text{æ}_{11} = -\frac{\partial^2 w}{\partial x_1^2}$$

$$M_{22} = \frac{Eh^3}{12(1 - \nu^2)}(\text{æ}_{22} + \nu\text{æ}_{11}) \qquad \text{æ}_{22} = -\frac{\partial^2 w}{\partial x_2^2}$$

$$M_{12} = \frac{Eh^3}{12(1 + \nu)}\text{æ}_{12} \qquad \text{æ}_{12} = -\frac{\partial^2 w}{\partial x_1 \partial x_2}$$

Bending stiffness

$$D = \frac{Eh^3}{12(1 - \nu^2)}$$

Equilibrium equation

$$D\widetilde{\nabla}^4 w = q_n$$

Equilibrium equation in Cartesian coordinates

$$D\left(\frac{\partial^4 w}{\partial x_1^4} + 2\frac{\partial^4 w}{\partial x_1^2 \partial x_2^2} + \frac{\partial^4 w}{\partial x_2^4}\right) = q_n$$

Appendix B

Hints and Answers

Chapter 1 Exercises

Exercise 1.1 The given equation states that $\mathbf{a}, \mathbf{b}, \mathbf{c}$ are not coplanar; because $\mathbf{a}$ has a nonzero component perpendicular to the plane of $\mathbf{b}$ and $\mathbf{c}$, it cannot be in this plane.

Chapter 1 Problems

Problem 1.1 (a) $(5, 7, 4)$; (b) $(7, 7, -2)$; (c) $(17, 11, -4)$.

Problem 1.2 (a) $(0, 1, 0)$; (b) $(1, -10, 2)$; (c) $(12, 8, 6)$.

Problem 1.3 (a) $\mathbf{x} = -\frac{1}{2}\mathbf{a} + 2\mathbf{b}$; (b) $\mathbf{x} = 2\mathbf{b} - 2\mathbf{a}$; (c) $\mathbf{x} = \frac{1}{12}(\mathbf{c} - \mathbf{a}) + \frac{4}{3}\mathbf{b}$.

Problem 1.4 (a) 2, $(0, 1, -1)$; (b) 11, $(-7, 5, -1)$; (c) 4, $(1, -1, 0)$; (d) -2, $(-2, -3, -1)$; (e) -1, $(-1, -1, -1)$; (f) 1, $(0, 0, -5)$; (g) 1, $(5, -3, -1)$; (h) -1, $(-4, -2, -7)$; (i) -8, $(-16, 24, -14)$.

Problem 1.9 (a) 1; (b) -1; (c) 0.

Problem 1.10 (a) -1; (b) -4; (c) 5; (d) 5; (e) -179; (f) 0; (g) 0; (h) 13; (i) -2.

Problem 1.11 (a) -1; (b) 4; (c) 3 ; (d) -7; (e) -826; (f) -18; (g) 0; (h) 105; (i) 0.

Chapter 2 Exercises

Exercise 2.1 Suppose that $\mathbf{e}^i$ and $\mathbf{f}^i$ are two vectors such that $x^i = \mathbf{x} \cdot \mathbf{e}^i$ and $x^i = \mathbf{x} \cdot \mathbf{f}^i$ for all $\mathbf{x}$. Then $\mathbf{x} \cdot (\mathbf{e}^i - \mathbf{f}^i) = 0$ for all $\mathbf{x}$, from which it follows that $\mathbf{e}^i - \mathbf{f}^i = 0$. So $\mathbf{f}^i = \mathbf{e}^i$.

Exercise 2.2 We must show that the equation $\alpha_1 \mathbf{e}^1 + \alpha_2 \mathbf{e}^2 + \alpha_3 \mathbf{e}^3 = \mathbf{0}$ implies $\alpha_1 = \alpha_2 = \alpha_3 = 0$. To get $\alpha_1 = 0$, for example, we simply dot-multiply the equation by $\mathbf{e}_1$ and use $\mathbf{e}_i \cdot \mathbf{e}^j = \delta_i^j$.

Exercise 2.3 We have

$$\mathbf{e}^2 \times \mathbf{e}^3 = \frac{1}{V^2}(\mathbf{e}_3 \times \mathbf{e}_1) \times (\mathbf{e}_1 \times \mathbf{e}_2)$$

$$= \frac{1}{V^2}\{[\mathbf{e}_2 \cdot (\mathbf{e}_3 \times \mathbf{e}_1)]\mathbf{e}_1 - [\mathbf{e}_1 \cdot (\mathbf{e}_3 \times \mathbf{e}_1)]\mathbf{e}_2\}$$

$$= \frac{1}{V^2}\mathbf{e}_1[\mathbf{e}_2 \cdot (\mathbf{e}_3 \times \mathbf{e}_1)]$$

by the vector triple product identity, hence

$$V' = \frac{1}{V^2}\mathbf{e}^1 \cdot \mathbf{e}_1[\mathbf{e}_2 \cdot (\mathbf{e}_3 \times \mathbf{e}_1)] = \frac{1}{V^2}[\mathbf{e}_1 \cdot (\mathbf{e}_2 \times \mathbf{e}_3)] = \frac{1}{V}.$$

Exercise 2.4 (b) First use common properties of determinants to establish the identity

$$[\mathbf{a} \cdot (\mathbf{b} \times \mathbf{c})][\mathbf{u} \cdot (\mathbf{v} \times \mathbf{w})] = \begin{vmatrix} \mathbf{a} \cdot \mathbf{u} & \mathbf{a} \cdot \mathbf{v} & \mathbf{a} \cdot \mathbf{w} \\ \mathbf{b} \cdot \mathbf{u} & \mathbf{b} \cdot \mathbf{v} & \mathbf{b} \cdot \mathbf{w} \\ \mathbf{c} \cdot \mathbf{u} & \mathbf{c} \cdot \mathbf{v} & \mathbf{c} \cdot \mathbf{w} \end{vmatrix}. \tag{*}$$

Write

$$[\mathbf{a} \cdot (\mathbf{b} \times \mathbf{c})][\mathbf{u} \cdot (\mathbf{v} \times \mathbf{w})] = \begin{vmatrix} a_1 & a_2 & a_3 \\ b_1 & b_2 & b_3 \\ c_1 & c_2 & c_3 \end{vmatrix} \begin{vmatrix} u_1 & u_2 & u_3 \\ v_1 & v_2 & v_3 \\ w_1 & w_2 & w_3 \end{vmatrix}$$

in a Cartesian frame. Then

$$[\mathbf{a} \cdot (\mathbf{b} \times \mathbf{c})][\mathbf{u} \cdot (\mathbf{v} \times \mathbf{w})] = \begin{vmatrix} a_1 & a_2 & a_3 \\ b_1 & b_2 & b_3 \\ c_1 & c_2 & c_3 \end{vmatrix} \begin{vmatrix} u_1 & v_1 & w_1 \\ u_2 & v_2 & w_2 \\ u_3 & v_3 & w_3 \end{vmatrix}$$

$$= \begin{vmatrix} a_1u_1 + a_2u_2 + a_3u_3 & a_1v_1 + a_2v_2 + a_3v_3 & a_1w_1 + a_2w_2 + a_3w_3 \\ b_1u_1 + b_2u_2 + b_3u_3 & b_1v_1 + b_2v_2 + b_3v_3 & b_1w_1 + b_2w_2 + b_3w_3 \\ c_1u_1 + c_2u_2 + c_3u_3 & c_1v_1 + c_2v_2 + c_3v_3 & c_1w_1 + c_2w_2 + c_3w_3 \end{vmatrix}.$$

(Note that our intermediate steps occurred in a Cartesian frame, but the final result is still frame independent.) Finally, write

$$V^2 = [\mathbf{e}_1 \cdot (\mathbf{e}_2 \times \mathbf{e}_3)]^2 = \begin{vmatrix} \mathbf{e}_1 \cdot \mathbf{e}_1 & \mathbf{e}_1 \cdot \mathbf{e}_2 & \mathbf{e}_1 \cdot \mathbf{e}_3 \\ \mathbf{e}_2 \cdot \mathbf{e}_1 & \mathbf{e}_2 \cdot \mathbf{e}_2 & \mathbf{e}_2 \cdot \mathbf{e}_3 \\ \mathbf{e}_3 \cdot \mathbf{e}_1 & \mathbf{e}_3 \cdot \mathbf{e}_2 & \mathbf{e}_3 \cdot \mathbf{e}_3 \end{vmatrix} = g.$$

Exercise 2.5

(a) $|\mathbf{x}|^2 = x_k x^k = g_{ij} x^i x^j = g^{mn} x_m x_n.$
(b) $\mathbf{x} \cdot \mathbf{y} = x^k y_k = x_k y^k = g_{ij} x^i y^j = g^{mn} x_m y_n.$

Exercise 2.6 The transformation matrix is of the form

$$\begin{pmatrix} \cos\theta & \sin\theta & 0 \\ -\sin\theta & \cos\theta & 0 \\ 0 & 0 & 1 \end{pmatrix},$$

where θ is the angle of rotation.

Exercise 2.7 Equate two different expressions for $\mathbf{x}$ as is done in the text.

Exercise 2.8

$$\epsilon^{ijk} = \begin{cases} +V', & (i,j,k) \text{ an even permutation of } (1,2,3), \\ -V', & (i,j,k) \text{ an odd permutation of } (1,2,3), \\ 0, & \text{two or more indices equal.} \end{cases}$$

(Recall that $V' = 1/V$.) Then use the determinantal identity established in Exercise 2.4: put $\mathbf{a} = \mathbf{e}_i$, $\mathbf{b} = \mathbf{e}_j$, $\mathbf{c} = \mathbf{e}_k$, $\mathbf{u} = \mathbf{e}^p$, $\mathbf{v} = \mathbf{e}^q$, $\mathbf{w} = \mathbf{e}^r$. To prove the vector triple product identity we write

$$\mathbf{a} \times (\mathbf{b} \times \mathbf{c}) = \mathbf{e}^i \epsilon_{ijk} a^j \epsilon^{kpq} b_p c_q$$

$$= \mathbf{e}^i (\delta^p_i \delta^q_j - \delta^q_i \delta^p_j) a^j b_p c_q$$

$$= \mathbf{e}^i b_i (a^q c_q) - \mathbf{e}^i c_i (a^p b_p).$$

Exercise 2.9 Using the scalar triple product identity and (2.13) we have

$$(\mathbf{a} \times \mathbf{b}) \cdot (\mathbf{c} \times \mathbf{d}) = \mathbf{a} \cdot [\mathbf{b} \times (\mathbf{c} \times \mathbf{d})]$$

$$= \mathbf{a} \cdot [(\mathbf{b} \cdot \mathbf{d})\mathbf{c} - (\mathbf{b} \cdot \mathbf{c})\mathbf{d}],$$

and the result follows.

Chapter 2 Problems

Problem 2.1 (a) $\mathbf{e}^1 = (2\mathbf{i}_1 + 3\mathbf{i}_2 - 2\mathbf{i}_3)/9$, $\mathbf{e}^2 = (-3\mathbf{i}_1 + 3\mathbf{i}_2 + \mathbf{i}_3)/9$, $\mathbf{e}^3 = (5\mathbf{i}_1 - 6\mathbf{i}_2 + 4\mathbf{i}_3)/9$; (b) $\mathbf{e}^1 = -\mathbf{i}_1 + \mathbf{i}_3$, $\mathbf{e}^2 = (-5\mathbf{i}_1 - 3\mathbf{i}_2 + 7\mathbf{i}_3)/13$, $\mathbf{e}^3 = (12\mathbf{i}_1 + 2\mathbf{i}_2 - 9\mathbf{i}_3)/13$; (c) $\mathbf{e}^1 = (\mathbf{i}_1 + \mathbf{i}_2)/2$, $\mathbf{e}^2 = (\mathbf{i}_1 - \mathbf{i}_2)/2$, $\mathbf{e}^3 = 1/3\mathbf{i}_3$; (d) $\mathbf{e}^1 = \cos\phi\mathbf{i}_1 + \sin\phi\mathbf{i}_2$, $\mathbf{e}^2 = -\sin\phi\mathbf{i}_1 + \cos\phi\mathbf{i}_2$, $\mathbf{e}^3 = \mathbf{i}_3$.

Problem 2.2 Using the formulas $\tilde{\mathbf{e}}^1 = (\tilde{\mathbf{e}}_2 \times \tilde{\mathbf{e}}_3)/V$, etc., first find

$$\tilde{\mathbf{e}}^1 = (1,1,0), \qquad \tilde{\mathbf{e}}^2 = (-1,0,-1), \qquad \tilde{\mathbf{e}}^3 = (-1,-2,2).$$

Then, with regard for the formula $A_i^j = \mathbf{e}_i \cdot \tilde{\mathbf{e}}^j$, arrive at $\begin{pmatrix} 3 & -1 & -6 \\ 2 & -3 & 2 \\ 1 & -2 & 1 \end{pmatrix}$.

Problem 2.3 $\frac{1}{3} \begin{pmatrix} 5 & 14 & 2 \\ -10 & -10 & -1 \\ 2 & -1 & -1 \end{pmatrix}$.

Problem 2.4 (a) a^k; (b) $a_i a^i$; (c) 3; (d) δ_i^k; (e) 3; (f) 3.

Problem 2.7 (a) 0; (b) -6; (c) ϵ_{inm}; (d) 0; (e) 0; (f) $2\delta_k^n$.

Problem 2.8 $(\mathbf{a} \times \mathbf{b}) \times \mathbf{c} = \mathbf{b}(\mathbf{a} \cdot \mathbf{c}) - \mathbf{a}(\mathbf{b} \cdot \mathbf{c})$.

Chapter 3 Exercises

Exercise 3.1 (b)

$$\begin{pmatrix} 1 & 0 & 0 \\ 0 & 0 & 0 \\ 0 & 0 & 0 \end{pmatrix}, \qquad \begin{pmatrix} 0 & 0 & 0 \\ 0 & 1 & 0 \\ 0 & 0 & 0 \end{pmatrix}, \qquad \begin{pmatrix} 0 & 0 & 0 \\ 0 & 0 & 0 \\ 1 & 0 & 0 \end{pmatrix}.$$

Exercise 3.2 $\mathbf{E}$ can be written in any of the forms

$$\mathbf{E} = e^{ij}\mathbf{e}_i\mathbf{e}_j = e_{ij}\mathbf{e}^i\mathbf{e}^j = e_{\cdot j}^i\mathbf{e}_i\mathbf{e}^j = e_i^{\cdot j}\mathbf{e}^i\mathbf{e}_j.$$

Let us choose the first form as an example. Equation (3.4) with $\mathbf{x} = \mathbf{e}^k$ implies $e^{ij}\mathbf{e}_i\mathbf{e}_j \cdot \mathbf{e}^k = \mathbf{e}^k$. Pre-dotting this with $\mathbf{e}^m$ yields $e^{mk} = g^{mk}$.

Exercise 3.3 $(\mathbf{A} \cdot \mathbf{B}) \cdot (\mathbf{B}^{-1} \cdot \mathbf{A}^{-1}) = \mathbf{A} \cdot (\mathbf{B} \cdot \mathbf{B}^{-1}) \cdot \mathbf{A}^{-1} = \mathbf{A} \cdot \mathbf{E} \cdot \mathbf{A}^{-1} = \mathbf{A} \cdot \mathbf{A}^{-1} = \mathbf{E}$.

Exercise 3.4 (a) $\tilde{B} = A^T B A$. (b) $\tilde{B} = ABA^T$.

Exercise 3.5 We are given $\mathbf{e}_i = A_i^j \tilde{\mathbf{e}}_j$. Now it is necessary to understand what the needed tensor does. We denote it by $\mathbf{A}$. It is the map $\mathbf{y} = \mathbf{A} \cdot \mathbf{x}$ that must take $\mathbf{x} = \mathbf{e}_i$ into $\mathbf{y} = A_i^j \tilde{\mathbf{e}}_j$ for each $i = 1, 2, 3$. Let us write out these equations:

$$\mathbf{A} \cdot \mathbf{e}_i = A_i^j \tilde{\mathbf{e}}_j.$$

The decisive step is to suppose that $\mathbf{A}$ should be written in a "mixed basis" as $\mathbf{A} = a^m{}_{.n} \tilde{\mathbf{e}}_m \mathbf{e}^n$. We have

$$a^m{}_{.n} \tilde{\mathbf{e}}_m \mathbf{e}^n \cdot \mathbf{e}_i = A_i^j \tilde{\mathbf{e}}_j$$

and it follows that

$$a^m{}_{.i} \tilde{\mathbf{e}}_m = A_i^j \tilde{\mathbf{e}}_j.$$

Thus (because the indices are dummy)

$$a^j{}_{.i} = A_i^j$$

and so the tensor is $\mathbf{A} = a^m{}_{.n} \tilde{\mathbf{e}}_m \mathbf{e}^n$ where $a^m{}_{.n} = A_n^m$.

Exercise 3.7

(a) $a^{ij} \mathbf{e}_i \mathbf{e}_j \cdots \mathbf{e}^k \mathbf{e}_k = a^{ij} \delta_j^k g_{ik} = a_k{}^{.k}$.

(b) $a_{ij} \mathbf{e}^i \mathbf{e}^j \cdots b^{kn} \mathbf{e}_k \mathbf{e}_n = a_{ij} b^{kn} \delta_k^j \delta_n^i = a_{ik} b^{ki}$. Having obtained this we may use the metric tensor to raise and lower indices and thereby obtain other forms; for example, $a_{ik} b^{ki} = a^p{}_{.k} g_{pi} b^{ki} = a^p{}_{.k} b^k{}_{.p}$.

Exercise 3.10

(a) Write

$$\mathbf{A} = a^{ij} \mathbf{e}_i \mathbf{e}_j, \qquad\qquad \mathbf{B} = b^{kn} \mathbf{e}_k \mathbf{e}_n,$$
$$\mathbf{A}^T = a^{ji} \mathbf{e}_i \mathbf{e}_j, \qquad\qquad \mathbf{B}^T = b^{nk} \mathbf{e}_k \mathbf{e}_n,$$

and show that both sides can be written as $a^{jp} b_p{}^{.k} \mathbf{e}_k \mathbf{e}_j$.

(b) $\mathbf{b} \cdot \mathbf{C} \cdot \mathbf{a} = \mathbf{b} \cdot (\mathbf{C} \cdot \mathbf{a}) = \mathbf{b} \cdot (\mathbf{a} \cdot \mathbf{C}^T) = (\mathbf{a} \cdot \mathbf{C}^T) \cdot \mathbf{b} = \mathbf{a} \cdot \mathbf{C}^T \cdot \mathbf{b}$.

(c) See [Lurie (2005)].

(d) The relation $(\mathbf{A}^T)^{-1} = (\mathbf{A}^{-1})^T$ follows from the two relations

$$\mathbf{A}^T \cdot (\mathbf{A}^{-1})^T = (\mathbf{A}^{-1} \cdot \mathbf{A})^T = \mathbf{E}^T = \mathbf{E},$$
$$(\mathbf{A}^{-1})^T \cdot \mathbf{A}^T = (\mathbf{A} \cdot \mathbf{A}^{-1})^T = \mathbf{E}^T = \mathbf{E}.$$

Exercise 3.11

$$\begin{pmatrix} a^{11} & a^{12} & a^{13} \\ a^{12} & a^{22} & a^{23} \\ a^{13} & a^{23} & a^{33} \end{pmatrix}, \qquad \begin{pmatrix} 0 & a^{12} & a^{13} \\ -a^{12} & 0 & a^{23} \\ -a^{13} & -a^{23} & 0 \end{pmatrix}.$$

Exercise 3.12 With $\mathbf{A} = a^{ij}\mathbf{e}_i\mathbf{e}_j$ and $\mathbf{B} = b^{kn}\mathbf{e}_k\mathbf{e}_n$ we have

$$\mathbf{A} \cdot\cdot \mathbf{B} = a^{ij}b^{kn}g_{jk}g_{in}. \tag{*}$$

Since $\mathbf{A}$ is symmetric and $\mathbf{B}$ is antisymmetric, we may also write $\mathbf{A} \cdot\cdot \mathbf{B} = -a^{ji}b^{nk}g_{jk}g_{in}$; let us swap i with j and k with n in this expression to get

$$\mathbf{A} \cdot\cdot \mathbf{B} = -a^{ij}b^{kn}g_{in}g_{jk}. \tag{**}$$

Adding (*) and (**) we obtain $2(\mathbf{A} \cdot\cdot \mathbf{B}) = 0$.

Exercise 3.13

$$\mathbf{x} \cdot \left[\frac{1}{2}(\mathbf{A} + \mathbf{A}^T) \right] \cdot \mathbf{x} = \frac{1}{2} \left[\mathbf{x} \cdot \mathbf{A} \cdot \mathbf{x} + \mathbf{x} \cdot \mathbf{A}^T \cdot \mathbf{x} \right]$$
$$= \frac{1}{2} \left[\mathbf{x} \cdot \mathbf{A} \cdot \mathbf{x} + \mathbf{x} \cdot \mathbf{A} \cdot \mathbf{x} \right]$$
$$= \mathbf{x} \cdot \mathbf{A} \cdot \mathbf{x}.$$

Exercise 3.15 They are (1) $\mathbf{x} = \mathbf{a}$ with $\lambda = \mathbf{b} \cdot \mathbf{a}$, and (2) any $\mathbf{x}$ that is perpendicular to $\mathbf{b}$, with $\lambda = 0$.

Exercise 3.16 The characteristic equation $-\lambda^3 + 2\lambda^2 + \lambda = 0$ has solutions $\lambda_1 = 0$, $\lambda_2 = 1 + \sqrt{2}$, $\lambda_3 = 1 - \sqrt{2}$. The Viète formulas give

$$I_1(\mathbf{A}) = \lambda_1 + \lambda_2 + \lambda_3 = 2,$$
$$I_2(\mathbf{A}) = \lambda_1\lambda_2 + \lambda_1\lambda_3 + \lambda_2\lambda_3 = -1,$$
$$I_3(\mathbf{A}) = \lambda_1\lambda_2\lambda_3 = 0.$$

Exercise 3.17 We prove this for two eigenvectors; the reader may readily generalize to any number of eigenvectors. Let $\mathbf{A} \cdot \mathbf{x}_1 = \lambda_1\mathbf{x}_1$ and $\mathbf{A} \cdot \mathbf{x}_2 = \lambda_2\mathbf{x}_2$ where $\lambda_2 \neq \lambda_1$. Now suppose

$$\alpha_1\mathbf{x}_1 + \alpha_2\mathbf{x}_2 = \mathbf{0}. \tag{*}$$

Let us operate on both sides of (*) with the tensor $\mathbf{A} - \lambda_2\mathbf{I}$. After simplification we obtain

$$\alpha_1(\lambda_1 - \lambda_2)\mathbf{x}_1 = \mathbf{0}.$$

Because $\mathbf{x}_1 \neq \mathbf{0}$ by definition, we have $\alpha_1 = 0$. Putting this back into (*), we also have $\alpha_2 = 0$.

Exercise 3.18 Assume $\mathbf{x}_1, \mathbf{x}_2, \mathbf{x}_3$ are eigenvectors of $\mathbf{A}$ corresponding to the distinct eigenvalues $\lambda_1, \lambda_2, \lambda_3$, respectively. Let us show that any eigenvector corresponding to λ_1 must be a scalar multiple of $\mathbf{x}_1$. So suppose that

$$\mathbf{A} \cdot \mathbf{x} = \lambda_1\mathbf{x}. \tag{**}$$

By the linear independence of $\mathbf{x}_1, \mathbf{x}_2, \mathbf{x}_3$, these vectors form a basis for three-dimensional space and we can express

$$\mathbf{x} = c_1\mathbf{x}_1 + c_2\mathbf{x}_2 + c_3\mathbf{x}_3.$$

Applying $\mathbf{A}$ to both sides we have, by (**),

$$\lambda_1\mathbf{x} = \lambda_1 c_1\mathbf{x}_1 + \lambda_1 c_2\mathbf{x}_2 + \lambda_1 c_3\mathbf{x}_3 = c_1\lambda_1\mathbf{x}_1 + c_2\lambda_2\mathbf{x}_2 + c_3\lambda_3\mathbf{x}_3.$$

This simplifies to

$$(\lambda_1 - \lambda_2)c_2\mathbf{x}_2 + (\lambda_1 - \lambda_3)c_3\mathbf{x}_3 = \mathbf{0}.$$

By linear independence of $\mathbf{x}_2, \mathbf{x}_3$, we get $c_2 = c_3 = 0$. Hence $\mathbf{x} = c_1\mathbf{x}_1$.

Exercise 3.19 The characteristic equation of $\mathbf{A}$ is

$$\begin{vmatrix} 1 - \lambda & 0 & 0 \\ 1 & 1 - \lambda & 0 \\ 0 & 1 & -\lambda \end{vmatrix} = 0,$$

which is $\lambda^3 - 2\lambda^2 + \lambda = 0$. Thus $\mathbf{A}^3 = 2\mathbf{A}^2 - \mathbf{A}$.

Exercise 3.20

(a) $\mathbf{Q} \cdot \mathbf{Q}^T = \mathbf{Q}^T \cdot \mathbf{Q} = \mathbf{i}_1\mathbf{i}_1 + \mathbf{i}_2\mathbf{i}_2 + \mathbf{i}_3\mathbf{i}_3 = \mathbf{E}$. Since we can write $\mathbf{Q} = -\mathbf{i}_1\mathbf{i}^1 + \mathbf{i}_2\mathbf{i}^2 + \mathbf{i}_3\mathbf{i}^3$, the matrix of $\mathbf{Q}$ in mixed components is

$$\begin{pmatrix} -1 & 0 & 0 \\ 0 & 1 & 0 \\ 0 & 0 & 1 \end{pmatrix}.$$

Hence $\det \mathbf{Q} = -1$.

(b) The defining equation $\mathbf{Q} \cdot \mathbf{Q}^T = \mathbf{E}$ implies

$$q_{ij}\mathbf{e}^i\mathbf{e}^j \cdot q^{km}\mathbf{e}_m\mathbf{e}_k = q_{ij}q^{kj}\mathbf{e}^i\mathbf{e}_k = \delta^k_i\mathbf{e}^i\mathbf{e}_k.$$

(c) If n is a positive integer,

$$\mathbf{Q}^n \cdot (\mathbf{Q}^n)^T = \mathbf{Q}^n \cdot (\mathbf{Q}^T)^n = (\mathbf{Q} \cdot \mathbf{Q}^T)^n = \mathbf{E}^n = \mathbf{E}.$$

Exercise 3.21 Pre-dot $\mathbf{A} \cdot \mathbf{x} = \lambda\mathbf{x}$ with $\mathbf{x}$ to get

$$\lambda = \frac{\mathbf{x} \cdot (\mathbf{A} \cdot \mathbf{x})}{|\mathbf{x}|^2}.$$

Clearly $\lambda > 0$ under the stated condition.

Exercise 3.23

(a) $\boldsymbol{\mathcal{E}} \cdots \mathbf{zyx} = \epsilon_{ijk}(\mathbf{e}^k \cdot \mathbf{z})(\mathbf{e}^j \cdot \mathbf{y})(\mathbf{e}^i \cdot \mathbf{x}) = x^i(\epsilon_{ijk}y^j z^k).$
(b) $\boldsymbol{\mathcal{E}} \cdot\cdot \mathbf{xy} = \mathbf{e}^i\epsilon_{ijk}(\mathbf{e}^k \cdot \mathbf{x})(\mathbf{e}^j \cdot \mathbf{y}) = \mathbf{e}^i\epsilon_{ijk}y^j x^k.$
(c) $\boldsymbol{\mathcal{E}} \cdot \mathbf{x} = \epsilon_{ijk}\mathbf{e}^i\mathbf{e}^j x^k$ so that $(\boldsymbol{\mathcal{E}} \cdot \mathbf{x}) \cdot \mathbf{y} = \epsilon_{ijk}\mathbf{e}^i y^j x^k = \mathbf{y} \times \mathbf{x}.$

Exercise 3.28 Write $\mathbf{E} = \mathbf{e}_m\mathbf{e}^m.$

Exercise 3.30 Use the representation of the invariants through the eigenvalues of $\mathbf{A}$.

Exercise 3.31 Use Theorem 3.8, the fact that $\mathbf{B}$ in the representation is a ball tensor, and the equality $\mathbf{E} \cdot\cdot \mathbf{A}^T = \mathrm{tr}\,\mathbf{A}$.

Exercise 3.32 Introduce the nth partial sum

$$\mathbf{S}_n = \mathbf{E} + \frac{1}{1!}\mathbf{A} + \cdots + \frac{1}{n!}\mathbf{A}^n.$$

It follows that

$$\|\mathbf{S}_{n+m} - \mathbf{S}_n\| = \left\|\frac{1}{(n+1)!}\mathbf{A}^{(n+1)} + \cdots + \frac{1}{(n+m)!}\mathbf{A}^{(n+m)}\right\|$$
$$\leq \frac{1}{(n+1)!}\|\mathbf{A}\|^{n+1} + \cdots + \frac{1}{(n+m)!}\|\mathbf{A}\|^{n+m}.$$

Next, use the proof of convergence of the Taylor series for e^x from any textbook with $x = \|\mathbf{A}\|$.

Exercise 3.33 Show that

(1) $\|\mathbf{A}^n\| \leq q^n \to 0$ as $n \to \infty$;
(2) $(\mathbf{E} - \mathbf{A})(\mathbf{E} + \mathbf{A} + \mathbf{A}^2 + \mathbf{A}^3 + \cdots + \mathbf{A}^n) = \mathbf{E} - \mathbf{A}^{n+1}$;

(3) limit passage may be justified in the last equality.

Exercise 3.34 $(\mathbf{E} + \mathbf{A})^{-1} = \mathbf{E} - \mathbf{A} + \mathbf{A}^2 - \mathbf{A}^3 + \cdots + (-1)^n \mathbf{A}^n + \cdots$.

Exercise 3.35 Consider $f(\mathbf{X} + \varepsilon\mathbf{B}) = \operatorname{tr}\mathbf{X} + \varepsilon\operatorname{tr}\mathbf{B}$. Using the definition, we get

$$\frac{\partial}{\partial\varepsilon} f(\mathbf{X} + \varepsilon d\mathbf{X})\Big|_{\varepsilon=0} = \operatorname{tr}(d\mathbf{X}).$$

As $\operatorname{tr}\mathbf{B} = \mathbf{E} \cdot\cdot \mathbf{B}^T$, we get $f_{,\mathbf{X}} = \mathbf{E}$.

Exercise 3.36 Consider $f(\mathbf{X} + \varepsilon\mathbf{B}) = \operatorname{tr}\mathbf{X}^2 + \varepsilon\operatorname{tr}(\mathbf{X} \cdot \mathbf{B}) + \varepsilon\operatorname{tr}(\mathbf{B} \cdot \mathbf{X}) + \varepsilon^2\operatorname{tr}\mathbf{B}^2$. By the definition, we get

$$\frac{\partial}{\partial\varepsilon} f(\mathbf{X} + \varepsilon\mathbf{B})\Big|_{\varepsilon=0} = \operatorname{tr}(\mathbf{X} \cdot \mathbf{B}) + \operatorname{tr}(\mathbf{B} \cdot \mathbf{X}) = 2\operatorname{tr}(\mathbf{X} \cdot \mathbf{B})$$

With regard for $\operatorname{tr}(\mathbf{X} \cdot \mathbf{B}) = \mathbf{X} \cdot\cdot \mathbf{B} = \mathbf{X}^T \cdot\cdot \mathbf{B}^T$, we get $f_{,\mathbf{X}} = 2\mathbf{X}^T$.

Exercise 3.38 Use the results of Exercises 3.35 and 3.36.

Exercise 3.39 Use a consequence of the Cayley–Hamilton theorem, which is

$$I_3 = \frac{1}{3}\operatorname{tr}\left(\mathbf{X}^3 - I_1\mathbf{X}^2 + I_2\mathbf{X}\right),$$

and the results of Exercises 3.35, 3.36 and 3.37.

Exercise 3.43 Let us derive $f(\mathbf{X} + \varepsilon d\mathbf{X})$. We have

$$f(\mathbf{X} + \varepsilon d\mathbf{X}) = \frac{1}{4}\mathbf{X} \cdot\cdot \mathbf{C} \cdot\cdot \mathbf{X} + \frac{\varepsilon}{2} d\mathbf{X} \cdot\cdot \mathbf{C} \cdot\cdot \mathbf{X}$$
$$+ \frac{\varepsilon}{2}\mathbf{X} \cdot\cdot \mathbf{C} \cdot\cdot d\mathbf{X} + \frac{\varepsilon^2}{2} d\mathbf{X} \cdot\cdot \mathbf{C} \cdot\cdot d\mathbf{X}.$$

So

$$\frac{\partial}{\partial\varepsilon} f(\mathbf{X} + \varepsilon d\mathbf{X})\Big|_{\varepsilon=0} = \frac{1}{2} d\mathbf{X} \cdot\cdot \mathbf{C} \cdot\cdot \mathbf{X} + \frac{1}{2}\mathbf{X} \cdot\cdot \mathbf{C} \cdot\cdot d\mathbf{X}$$
$$= \frac{1}{2}(dx_{mn} c_{mnpt} x_{pt} + x_{mn} c_{mnpt} dx_{pt})$$
$$= \frac{1}{2}(c_{mnpt} x_{pt} + x_{ij} c_{ijmn}) dx_{mn}.$$

Equating the components of this to the components of $f_{,\mathbf{X}} \cdot\cdot\, d\mathbf{X}$, we get

$$\frac{\partial f}{\partial x_{11}} = \frac{1}{2}(c_{11pt}x_{pt} + x_{mn}c_{mn11}),$$

$$\frac{\partial f}{\partial x_{12}} = \frac{1}{4}(c_{12pt}x_{pt} + c_{21pt}x_{pt} + x_{mn}c_{mn12} + x_{mn}c_{mn21}),$$

$$\frac{\partial f}{\partial x_{22}} = \frac{1}{2}(c_{22pt}x_{pt} + x_{mn}c_{mn22}),$$

$$\frac{\partial f}{\partial x_{13}} = \frac{1}{4}(c_{13pt}x_{pt} + c_{31pt}x_{pt} + x_{mn}c_{mn13} + x_{mn}c_{mn31}),$$

$$\vdots$$

In tensor notation these formulas are

$$f_{,\mathbf{X}} = \frac{1}{4}(\mathbf{C} \cdot\cdot\, \mathbf{X} + \mathbf{C}'' \cdot\cdot\, \mathbf{X} + \mathbf{X} \cdot\cdot\, \mathbf{C} + \mathbf{X} \cdot\cdot\, \mathbf{C}'),$$

where $\mathbf{C}' = c_{mnpt}\mathbf{i}_m\mathbf{i}_n\mathbf{i}_t\mathbf{i}_p$ and $\mathbf{C}'' = c_{mnpt}\mathbf{i}_n\mathbf{i}_m\mathbf{i}_t\mathbf{i}_p$.

Exercise 3.44 Consider the tensor equality

$$\mathbf{C} \cdot\cdot\, \mathbf{X} = \frac{1}{4}(\mathbf{C} \cdot\cdot\, \mathbf{X} + \mathbf{C}'' \cdot\cdot\, \mathbf{X} + \mathbf{X} \cdot\cdot\, \mathbf{C} + \mathbf{X} \cdot\cdot\, \mathbf{C}')$$

in components.

Chapter 3 Problems

Problem 3.1 They are the components of the matrix $\begin{pmatrix} 0 & 1 & 0 \\ 0 & 0 & 0 \\ 0 & 0 & 0 \end{pmatrix}$.

Problem 3.2 $\begin{pmatrix} 0 & 1 & 0 \\ -1 & 0 & 0 \\ 0 & 0 & 2 \end{pmatrix}$

Problem 3.3 $\begin{pmatrix} -1 & 1 & -2 \\ 2 & -2 & 4 \\ -2 & 2 & -4 \end{pmatrix}$.

Problem 3.5

(a) $\frac{1}{2}(\mathbf{i}_1\mathbf{i}_2 + \mathbf{i}_2\mathbf{i}_1)$, $\frac{1}{2}(\mathbf{i}_1\mathbf{i}_2 - \mathbf{i}_2\mathbf{i}_1)$;

(b) $2\mathbf{i}_3\mathbf{i}_3$, $\mathbf{i}_1\mathbf{i}_2 - \mathbf{i}_2\mathbf{i}_1$;

(c) $\frac{1}{2}(-\mathbf{i}_1\mathbf{i}_2 - \mathbf{i}_2\mathbf{i}_1 + \mathbf{i}_1\mathbf{i}_3 + \mathbf{i}_3\mathbf{i}_1)$, $3(\mathbf{i}_1\mathbf{i}_2 - \mathbf{i}_2\mathbf{i}_1) + \frac{1}{2}(\mathbf{i}_1\mathbf{i}_3 - \mathbf{i}_3\mathbf{i}_1)$;

(d) $\frac{1}{2}(i_1i_2+i_2i_3+i_1i_3+i_2i_1+i_3i_2+i_3i_1)$, $\frac{1}{2}(i_1i_2+i_2i_3+i_1i_3-i_2i_1-i_3i_2-i_3i_1)$;

(e) $i_1i_1 + 2i_1i_2 + 2i_2i_1 + i_3i_1 + i_1i_3$, $\mathbf{0}$.

Problem 3.6

(a) $\mathbf{0}$, i_1i_2;

(b) $\mathbf{0}$, $i_1i_2 + i_2i_1$;

(c) $\frac{1}{3}\mathbf{E}$, $\frac{2}{3}i_1i_1 - \frac{1}{3}i_2i_2 - \frac{1}{3}i_3i_3$;

(d) $\frac{1}{3}\mathbf{a}\cdot\mathbf{a}\mathbf{E}$, $\mathbf{a}\mathbf{a} - \frac{1}{3}\mathbf{a}\cdot\mathbf{a}\mathbf{E}$;

(e) $\frac{1}{3}\mathbf{E}$, $2i_1i_2 + 2i_2i_1 + i_3i_1 + i_1i_3 + \frac{2}{3}i_1i_1 - \frac{1}{3}i_2i_2 - \frac{1}{3}i_3i_3$.

Problem 3.8

(a) $I_1 = \mathbf{a}\cdot\mathbf{a}$, $I_2 = 0$, $I_3 = 0$;

(b) $I_1 = 0$, $I_2 = -\frac{1}{2}$, $I_3 = 0$;

(c) $I_1 = 2$, $I_2 = 1$, $I_3 = 0$;

(d) $I_1 = 3\lambda$, $I_2 = 3\lambda^2$, $I_3 = \lambda^3$;

(e) $I_1 = 9$, $I_2 = 26$, $I_3 = 24$.

Problem 3.10

(a) $\mathbf{S} = \mathbf{S}' = |\lambda|\mathbf{E}$, $\mathbf{Q} = \mathrm{sgn}\lambda\mathbf{E}$;

(b) $\mathbf{S} = \mathbf{S}' = \mathbf{a}\mathbf{a} + \mathbf{b}\mathbf{b} + \mathbf{c}\mathbf{c}$, $\mathbf{Q} = \frac{1}{\mathbf{a}\cdot\mathbf{a}}\mathbf{a}\mathbf{a} + \frac{1}{\mathbf{b}\cdot\mathbf{b}}\mathbf{b}\mathbf{b} + \frac{1}{\mathbf{c}\cdot\mathbf{c}}\mathbf{c}\mathbf{c} \equiv \mathbf{E}$;

(c) $\mathbf{S} = \mathbf{S}' = |\lambda + a|i_1i_1 + |\lambda|i_2i_2 + |\lambda|i_3i_3$, $\mathbf{Q} = \mathrm{sgn}\,(\lambda + a)i_1i_1 + \mathrm{sgn}\,\lambda i_2i_2 + \mathrm{sgn}\,\lambda i_3i_3$;

(d) $\mathbf{S} = \mathbf{S}' = |\lambda + a|i_1i_1 + |\lambda + b|i_2i_2 + |\lambda|i_3i_3$, $\mathbf{Q} = \mathrm{sgn}\,(\lambda + a)i_1i_1 + \mathrm{sgn}\,(\lambda + b)i_2i_2 + \mathrm{sgn}\,\lambda i_3i_3$;

(e) $\mathbf{S} = \mathbf{S}' = |a|i_1i_1 + |b|i_2i_2 + |c|i_3i_3$, $\mathbf{Q} = \mathrm{sgn}\,a i_1i_1 + \mathrm{sgn}\,b i_2i_2 + \mathrm{sgn}\,c i_3i_3$.

Problem 3.12 Let $\mathbf{X}$ be a needed solution. Multiply the above equality by $\mathbf{A}$ from the left and take the trace of both sides:

$$a\,\mathrm{tr}(\mathbf{A}\cdot\mathbf{X}) + \mathrm{tr}(\mathbf{A}\cdot\mathbf{X})\,\mathrm{tr}\,\mathbf{A} = \mathrm{tr}(\mathbf{A}\cdot\mathbf{B}). \qquad (*)$$

If $a + \mathrm{tr}\,\mathbf{A} \neq 0$, then $\mathrm{tr}(\mathbf{A}\cdot\mathbf{X}) = \mathrm{tr}(\mathbf{A}\cdot\mathbf{B})/(a + \mathrm{tr}\,\mathbf{A})$. Substituting this into the above equality we get

$$\mathbf{X} = \frac{1}{a}\left(\mathbf{B} - \frac{\mathrm{tr}(\mathbf{A}\cdot\mathbf{B})}{a + \mathrm{tr}\,\mathbf{A}}\mathbf{E}\right).$$

If $a + \operatorname{tr} \mathbf{A} = 0$ then (*) reduces to $0 = \operatorname{tr}(\mathbf{A} \cdot \mathbf{B})$. So if $\operatorname{tr}(\mathbf{A} \cdot \mathbf{B}) \neq 0$ the equation has no solution $\mathbf{X}$. Let us see what happens when $a + \operatorname{tr} \mathbf{A} = 0$ and $\operatorname{tr}(\mathbf{A} \cdot \mathbf{B}) = 0$. Now $\operatorname{tr}(\mathbf{A} \cdot \mathbf{X})$ can take any value but the dev $\mathbf{X}$ is uniquely defined as follows from the initial equation: $\operatorname{dev} \mathbf{X} = \frac{1}{a} \operatorname{dev} \mathbf{B}$ as $a \neq 0$.

Answer. (1) If $a + \operatorname{tr} \mathbf{A} \neq 0$, then $\mathbf{X} = \frac{1}{a}\left(\mathbf{B} - \frac{\operatorname{tr}(\mathbf{A}\cdot\mathbf{B})}{a+\operatorname{tr}\mathbf{A}}\mathbf{E}\right)$; (2) if $a + \operatorname{tr} \mathbf{A} = 0$ and $\operatorname{tr}(\mathbf{A} \cdot \mathbf{B}) \neq 0$ then a solution does not exist; (3) if $a + \operatorname{tr} \mathbf{A} = 0$ and $\operatorname{tr}(\mathbf{A} \cdot \mathbf{B}) = 0$, then only dev $\mathbf{X}$ is uniquely defined $\mathbf{X} = \frac{1}{a} \operatorname{dev} \mathbf{B}$ and so $\mathbf{X} = \lambda \mathbf{E} + \frac{1}{a}\mathbf{B}$.

Problem 3.13 (a) $\mathbf{X} = \mathbf{C} - \frac{\operatorname{tr}(\mathbf{A}\cdot\mathbf{C})}{1+\operatorname{tr}(\mathbf{A}\cdot\mathbf{B})}\mathbf{B}$ is uniquely defined when $1 + \operatorname{tr}(\mathbf{A} \cdot \mathbf{B}) \neq 0$; (b) $\mathbf{X} = \mathbf{C}^T - \frac{\operatorname{tr}(\mathbf{A}\cdot\mathbf{C}^T)}{1+\operatorname{tr}(\mathbf{A}\cdot\mathbf{B}^T)}\mathbf{B}^T$ is uniquely defined when $1 + \operatorname{tr}(\mathbf{A} \cdot \mathbf{B}^T) \neq 0$; (c) $\mathbf{X} = \mathbf{B} - \frac{a\operatorname{tr}\mathbf{B}}{1+a\operatorname{tr}\mathbf{A}}\mathbf{A}$ is uniquely defined when $1 + a\operatorname{tr}\mathbf{A} \neq 0$.

Problem 3.14 Taking the dev-operation on the above equality, we get $a \operatorname{dev} \mathbf{X} = \mathbf{0}$. This means that dev $\mathbf{X} = \mathbf{0}$ when $a \neq 0$. If $a = 0$ then dev $\mathbf{X}$ can take any value.

Applying the trace operation to the equality, we get $(a + \operatorname{tr} \mathbf{E}) \operatorname{tr} \mathbf{X} = 0$. It is valid when (1) $a = -3$ or (2) $\operatorname{tr} \mathbf{X} = 0$.

If $a = -3$, then $\mathbf{X} = \lambda \mathbf{E}$ with any scalar λ.

If $\operatorname{tr} \mathbf{X} = 0$, then the equation reduces to $a\mathbf{X} = \mathbf{0}$ and so it can have a nonzero solution $\mathbf{X}$ only if $a = 0$.

Answer. (1) Let $a = -3$. Then $\mathbf{X} = \lambda \mathbf{E}$ with any scalar λ; (2) Let $a = 0$. Then $\mathbf{X}$ is an arbitrary tensor such that $\operatorname{tr} \mathbf{X} = 0$.

Problem 3.15 Let $\mathbf{X}$ be a solution. Calculate the trace of both sides of the equation given in the problem. It follows that $(a + \operatorname{tr} \mathbf{A}) \operatorname{tr} \mathbf{X} = 0$. There follow two possibilities: (1) $a = -\operatorname{tr} \mathbf{A}$, or (2) $\operatorname{tr} \mathbf{X} = 0$.

When $a = -\operatorname{tr} \mathbf{A}$ then $\operatorname{tr} \mathbf{X}$ can take any value; now $\mathbf{X} = \lambda \mathbf{A}$ is a solution for any scalar λ.

Let $\operatorname{tr} \mathbf{X} = 0$. Then the equation reduces to $a\mathbf{X} = \mathbf{0}$, which has a nonzero solution only if $a = 0$.

Answer. (1) $a = -\operatorname{tr} \mathbf{A}$, $\mathbf{X} = \lambda \mathbf{A}$; (2) If $a = 0$, then $\mathbf{X}$ is an arbitrary tensor such that $\operatorname{tr} \mathbf{X} = 0$.

Problem 3.19

$$(\mathbf{E} \times \boldsymbol{\omega})^2 = (\mathbf{E} \times \boldsymbol{\omega}) \cdot (\mathbf{E} \times \boldsymbol{\omega})$$
$$= \boldsymbol{\omega} \times \mathbf{E} \cdot \mathbf{E} \times \boldsymbol{\omega}$$
$$= \boldsymbol{\omega} \times \mathbf{E} \times \boldsymbol{\omega}$$
$$= \boldsymbol{\omega} \times \mathbf{i}_k \mathbf{i}_k \times \boldsymbol{\omega}$$
$$= \omega_p \mathbf{i}_p \times \mathbf{i}_k \mathbf{i}_k \times \mathbf{i}_t \omega_t$$
$$= \omega_p \omega_t \epsilon_{pkm} \epsilon_{ktn} \mathbf{i}_m \mathbf{i}_n$$
$$= \omega_p \omega_t \epsilon_{mpk} \epsilon_{tnk} \mathbf{i}_m \mathbf{i}_n$$
$$= \omega_p \omega_t (\delta_{mt} \delta_{pn} - \delta_{mn} \delta_{pt}) \mathbf{i}_m \mathbf{i}_n$$
$$= \omega_m \omega_n \mathbf{i}_m \mathbf{i}_n - \omega_p \omega_p \mathbf{i}_m \mathbf{i}_m$$
$$= \boldsymbol{\omega}\boldsymbol{\omega} - \boldsymbol{\omega} \cdot \boldsymbol{\omega} \mathbf{E}.$$

Problem 3.21 (a) $\mathbf{0}$; (b) $-2\boldsymbol{\omega}$; (c) $2\mathbf{a} \times \mathbf{b}$; (d) $\mathbf{0}$; (e) $\mathbf{0}$.

Problem 3.39 $I_1 = 3\alpha + \beta \mathbf{e} \cdot \mathbf{e}$, $I_2 = 3\alpha^2 + 2\alpha\beta \mathbf{e} \cdot \mathbf{e}$, $I_3 = \alpha^2(\alpha + \beta \mathbf{e} \cdot \mathbf{e})$.

Problem 3.40 $I_1 = 0$, $I_2 = \boldsymbol{\omega} \cdot \boldsymbol{\omega}$, $I_3 = 0$.

Problem 3.41 $I_1 = -2\boldsymbol{\omega} \cdot \boldsymbol{\omega}$, $I_2 = (\boldsymbol{\omega} \cdot \boldsymbol{\omega})^2$, $I_3 = 0$.

Problem 3.43

(a) $2\mathbf{i}_1 \mathbf{i}_1 + \mathbf{i}_2 \mathbf{i}_2 + \mathbf{i}_3 \mathbf{i}_3 - \mathbf{i}_1 \mathbf{i}_2 - \mathbf{i}_2 \mathbf{i}_1$;
(b) $c^{-1} \mathbf{i}_1 \mathbf{i}_3 + b^{-1} \mathbf{i}_2 \mathbf{i}_2 + c^{-1} \mathbf{i}_3 \mathbf{i}_1$;
(c) $a^{-1} \mathbf{i}_1 \mathbf{i}_1 + \mathbf{i}_2 \mathbf{i}_2 + \mathbf{i}_3 \mathbf{i}_3 - a^{-1} b \mathbf{i}_1 \mathbf{i}_2$
(d) $a^{-1} \mathbf{E} - a^{-2} b \mathbf{i}_1 \mathbf{i}_2$.

Problem 3.44 (a) $4\mathbf{X}^{T^3}$; (b) $\mathbf{aa}$; (c) $\mathbf{ab}$; (d) $\mathbf{B}^T$.

Problem 3.49 (a) $\mathbf{EA}$; (b) $\mathbf{EA}^T$; (c) $\mathbf{cdab}$.

Chapter 4 Exercises

Exercise 4.1

(a) Consider the derivative of the dot product. Since we are given the vectors explicitly, an easy way is to dot the vectors first:

$$\mathbf{e}_1(t) \cdot \mathbf{e}_2(t) = e^{-t}(1 - \sin^2 t) = e^{-t} \cos^2 t,$$

hence

$$[\mathbf{e}_1(t) \cdot \mathbf{e}_2(t)]' = -e^{-t} \cos t(2 \sin t + \cos t).$$

However, it is easily checked that the product-rule identity given in the text yields the same result.

(b) $[\mathbf{e}(t) \times \mathbf{e}'(t)]' = \mathbf{e}'(t) \times \mathbf{e}'(t) + \mathbf{e}(t) \times \mathbf{e}''(t) = \mathbf{e}(t) \times \mathbf{e}''(t)$.

Exercise 4.3 Use the product rule to take the indicated second derivative of $\mathbf{r} = \hat{\rho}\rho$, noting that

$$\frac{d\hat{\rho}}{dt} = \frac{d}{dt}(\hat{\mathbf{x}} \cos\phi + \hat{\mathbf{y}} \sin\phi) = (-\hat{\mathbf{x}} \sin\phi + \hat{\mathbf{y}} \cos\phi)\frac{d\phi}{dt} = \hat{\phi}\frac{d\phi}{dt}$$

and, similarly,

$$\frac{d\hat{\phi}}{dt} = -\hat{\rho}\frac{d\phi}{dt}.$$

Exercise 4.4 $v^1 = v_r$, $v^2 = v_\theta/r$, $v^3 = v_\phi/r \sin\theta$.

Exercise 4.5

(a) Fix $v = v_0$ and eliminate u from the transformation equations to get

$$y = x \tan\alpha + v_0(\sin\beta - \cos\beta \tan\alpha);$$

thus the $v = v_0$ coordinate curve is a straight line in the xy-plane making an angle α with the x-axis. Similarly, the $u = u_0$ line makes an angle β with the x-axis.

(b) $\mathbf{r} = \hat{\mathbf{x}}x + \hat{\mathbf{y}}y = \hat{\mathbf{x}}(u \cos\alpha + v \cos\beta) + \hat{\mathbf{y}}(u \sin\alpha + v \sin\beta)$ gives

$$\mathbf{r}_1 = \frac{\partial \mathbf{r}}{\partial u} = \hat{\mathbf{x}} \cos\alpha + \hat{\mathbf{y}} \sin\alpha, \qquad \mathbf{r}_2 = \frac{\partial \mathbf{r}}{\partial v} = \hat{\mathbf{x}} \cos\beta + \hat{\mathbf{y}} \sin\beta.$$

Note that these are both unit vectors. To find $\mathbf{r}^1$ we write $\mathbf{r}^1 = \hat{\mathbf{x}}a + \hat{\mathbf{y}}b$ and solve the equations

$$\mathbf{r}^1 \cdot \mathbf{r}_1 = 1, \qquad \mathbf{r}^1 \cdot \mathbf{r}_2 = 0,$$

simultaneously to get a, b. By this method we find that

$$\mathbf{r}^1 = \hat{\mathbf{x}}\frac{\sin\beta}{\sin(\beta - \alpha)} - \hat{\mathbf{y}}\frac{\cos\beta}{\sin(\beta - \alpha)},$$

$$\mathbf{r}^2 = -\hat{\mathbf{x}}\frac{\sin\alpha}{\sin(\beta - \alpha)} + \hat{\mathbf{y}}\frac{\cos\alpha}{\sin(\beta - \alpha)}.$$

These are not orthonormal vectors; moreover the coordinate system is non-orthogonal except in such trivial cases as $\beta - \alpha = \pi/2$. We have

$$g_{11} = 1 = g_{22}, \qquad g_{12} = g_{21} = \cos(\beta - \alpha),$$

and

$$g^{11} = \frac{1}{\sin^2(\beta - \alpha)} = g^{22}, \qquad g^{12} = g^{21} = -\frac{\cos(\beta - \alpha)}{\sin^2(\beta - \alpha)}.$$

(c) Write $\mathbf{z} = z^1 \mathbf{r}_1 + z^2 \mathbf{r}_2 = z_1 \mathbf{r}^1 + z_2 \mathbf{r}^2$. Substitute for the $\mathbf{r}_i$ and $\mathbf{r}^i$ in terms of $\hat{\mathbf{x}}$ and $\hat{\mathbf{y}}$ and then equate coefficients of $\hat{\mathbf{x}}, \hat{\mathbf{y}}$ to get simultaneous equations for the z_i in terms of the z^i. The answers are

$$z_1 = z^1 + z^2 \cos(\beta - \alpha), \qquad z_2 = z^1 \cos(\beta - \alpha) + z^2.$$

Exercise 4.6 $(ds)^2 = (du)^2 + 2\, du\, dv\, \cos(\beta - \alpha) + (dv)^2.$

Exercise 4.7

(a) The total differential of φ is

$$d\varphi = \frac{\partial \varphi}{\partial q^i}\, dq^i.$$

If we start at a point on the given surface and move in such a way that we stay on the surface, then $d\varphi = 0$. In the text it is shown that $dq^i = \mathbf{r}^i \cdot d\mathbf{r}$, so if we move along $d\mathbf{r}$ satisfying

$$\frac{\partial \varphi}{\partial q^i} \mathbf{r}^i \cdot d\mathbf{r} = 0$$

then we will stay on the surface. It is clear that the vector

$$\nabla \varphi = \frac{\partial \varphi}{\partial q^i} \mathbf{r}^i$$

is normal to the surface, since it is perpendicular to every tangent direction. We have

$$|\nabla \varphi| = \sqrt{\nabla \varphi \cdot \nabla \varphi}$$

$$= \sqrt{\frac{\partial \varphi}{\partial q^m} \mathbf{r}^m \cdot \frac{\partial \varphi}{\partial q^n} \mathbf{r}^n}$$

$$= \sqrt{g^{mn} \frac{\partial \varphi}{\partial q^m} \frac{\partial \varphi}{\partial q^n}},$$

hence

$$\mathbf{n} = \frac{\nabla\varphi}{|\nabla\varphi|} = \frac{\dfrac{\partial\varphi}{\partial q^i}\mathbf{r}^i}{\sqrt{g^{mn}\dfrac{\partial\varphi}{\partial q^m}\dfrac{\partial\varphi}{\partial q^n}}} = \frac{g^{ij}\dfrac{\partial\varphi}{\partial q^i}}{\sqrt{g^{mn}\dfrac{\partial\varphi}{\partial q^m}\dfrac{\partial\varphi}{\partial q^n}}}\mathbf{r}_j.$$

(b) Use the result of part (a).

(c) Use the result of part (b).

(d) Use the result of part (b).

Exercise 4.8 For cylindrical coordinates we have

$$\begin{aligned}
\nabla f &= \mathbf{r}^i \frac{\partial f}{\partial q^i} \\
&= \mathbf{r}^1 \frac{\partial f}{\partial \rho} + \mathbf{r}^2 \frac{\partial f}{\partial \phi} + \mathbf{r}^3 \frac{\partial f}{\partial z} \\
&= \mathbf{r}_1 \frac{\partial f}{\partial \rho} + \frac{\mathbf{r}_2}{\rho^2} \frac{\partial f}{\partial \phi} + \mathbf{r}_3 \frac{\partial f}{\partial z} \\
&= \hat{\rho} \frac{\partial f}{\partial \rho} + \frac{\hat{\phi}}{\rho} \frac{\partial f}{\partial \phi} + \hat{\mathbf{z}} \frac{\partial f}{\partial z}.
\end{aligned}$$

Exercise 4.9 The results are all zero by expansion in Cartesian frame.

Exercise 4.10 Use expansion in Cartesian frame.

Exercise 4.11 Use expansion in Cartesian frame.

Exercise 4.12 Equation (4.18) gives, for instance,

$$\Gamma_{221} = \frac{1}{2}\left(\frac{\partial g_{12}}{\partial q^2} + \frac{\partial g_{21}}{\partial q^2} - \frac{\partial g_{22}}{\partial q^1}\right).$$

In the cylindrical system this is

$$\Gamma_{221} = \frac{1}{2}\left(\frac{\partial 0}{\partial \phi} + \frac{\partial 0}{\partial \phi} - \frac{\partial \rho^2}{\partial \rho}\right) = -\rho.$$

Exercise 4.13 Use the results of the previous exercise and equation (4.20).

Exercise 4.14 From $\mathbf{r} = \hat{\mathbf{x}}c\cosh u \cos v + \hat{\mathbf{y}}c\sinh u \sin v$ we find

$$\mathbf{r}_u = \frac{\partial \mathbf{r}}{\partial u} = \hat{\mathbf{x}}c\sinh u \cos v + \hat{\mathbf{y}}c\cosh u \sin v,$$

$$\mathbf{r}_v = \frac{\partial \mathbf{r}}{\partial v} = -\hat{\mathbf{x}}c\cosh u \sin v + \hat{\mathbf{y}}c\sinh u \cos v,$$

and consequently

$$g_{uu} = g_{vv} = c^2(\cosh^2 u - \cos^2 v), \qquad g_{uv} = g_{vu} = 0$$

(the system is orthogonal). Then

$$\mathbf{r}^u = \frac{\mathbf{r}_u}{c^2(\cosh^2 u - \cos^2 v)}, \qquad \mathbf{r}^v = \frac{\mathbf{r}_v}{c^2(\cosh^2 u - \cos^2 v)},$$

from which we get

$$g^{uu} = g^{vv} = \frac{1}{c^2(\cosh^2 u - \cos^2 v)}, \qquad g^{uv} = g^{vu} = 0.$$

An application of (4.18) gives

$$\Gamma_{uuu} = c^2 \cosh u \sinh u,$$
$$\Gamma_{uuv} = -c^2 \cos v \sin v,$$
$$\Gamma_{uvu} = \Gamma_{vuu} = c^2 \cos v \sin v,$$
$$\Gamma_{vvu} = -c^2 \cosh u \sinh u,$$
$$\Gamma_{vuv} = \Gamma_{uvv} = c^2 \cosh u \sinh u,$$
$$\Gamma_{vvv} = c^2 \cos v \sin v.$$

Finally, from (4.20) we obtain

$$\Gamma^u_{uu} = \frac{\cosh u \sinh u}{\cosh^2 u - \cos^2 v},$$
$$\Gamma^v_{uu} = -\frac{\cos v \sin v}{\cosh^2 u - \cos^2 v},$$
$$\Gamma^u_{uv} = \Gamma^u_{vu} = \frac{\cos v \sin v}{\cosh^2 u - \cos^2 v},$$
$$\Gamma^u_{vv} = -\frac{\cosh u \sinh u}{\cosh^2 u - \cos^2 v},$$
$$\Gamma^v_{vu} = \Gamma^v_{uv} = \frac{\cosh u \sinh u}{\cosh^2 u - \cos^2 v},$$
$$\Gamma^v_{vv} = \frac{\cos v \sin v}{\cosh^2 u - \cos^2 v}.$$

Exercise 4.15 We switch to a notation in which the coordinate symbols ρ and ϕ are used for the indices. The only nonzero Christoffel symbols are

$\Gamma^\rho_{\phi\phi} = -\rho$, $\Gamma^\phi_{\rho\phi} = 1/\rho$. Then

$$\nabla_\rho f^\rho = \frac{\partial f^\rho}{\partial \rho}, \qquad\qquad \nabla_\phi f^\rho = \frac{\partial f^\rho}{\partial \phi} - \rho f^\phi,$$

$$\nabla_\rho f^\phi = \frac{\partial f^\phi}{\partial \rho} + \frac{1}{\rho} f^\phi, \qquad\qquad \nabla_\phi f^\phi = \frac{\partial f^\phi}{\partial \phi} + \frac{1}{\rho} f^\rho.$$

Exercise 4.16 Use the method shown in the text. For example,

$$\frac{\partial}{\partial q^k}(a_i^{\cdot j}\mathbf{r}^i\mathbf{r}_j) = \frac{\partial a_i^{\cdot j}}{\partial q^k}\mathbf{r}^i\mathbf{r}_j + a_i^{\cdot j}\left(\frac{\partial \mathbf{r}^i}{\partial q^k}\mathbf{r}_j + \mathbf{r}^i\frac{\partial \mathbf{r}_j}{\partial q^k}\right)$$

$$= \frac{\partial a_i^{\cdot j}}{\partial q^k}\mathbf{r}^i\mathbf{r}_j + a_i^{\cdot j}\left(-\Gamma^i_{kt}\mathbf{r}^t\mathbf{r}_j + \mathbf{r}^i\Gamma^t_{jk}\mathbf{r}_t\right)$$

$$= \frac{\partial a_i^{\cdot j}}{\partial q^k}\mathbf{r}^i\mathbf{r}_j + \left(-a_s^{\cdot j}\Gamma^s_{ki}\mathbf{r}^i\mathbf{r}_j + \mathbf{r}^i a_i^{\cdot s}\Gamma^j_{sk}\mathbf{r}_j\right)$$

$$= \left(\frac{\partial a_i^{\cdot j}}{\partial q^k} - \Gamma^s_{ki}a_s^{\cdot j} + \Gamma^j_{sk}a_i^{\cdot s}\right)\mathbf{r}^i\mathbf{r}_j.$$

Exercise 4.17 Consider, for example,

$$\nabla_k g_{ij} = \frac{\partial g_{ij}}{\partial q^k} - \Gamma^s_{ki}g_{sj} - \Gamma^s_{kj}g_{is}.$$

By (4.19), this is

$$\nabla_k g_{ij} = \frac{\partial g_{ij}}{\partial q^k} - \Gamma_{kij} - \Gamma_{kji}.$$

Elimination of Γ_{kij} and Γ_{kji} via (4.18) shows that $\nabla_k g_{ij} = 0$.

Exercise 4.18 Equate components of the first and third members of the trivial identity $\nabla(\mathbf{E} \cdot \mathbf{a}) = \nabla\mathbf{a} = \mathbf{E} \cdot \nabla\mathbf{a}$.

Exercise 4.19 In cylindrical coordinates,

$$\nabla \cdot \mathbf{f} = \frac{1}{\sqrt{g}}\frac{\partial}{\partial q^i}(\sqrt{g}f^i)$$

$$= \frac{1}{\rho}\left[\frac{\partial}{\partial \rho}(\rho f^1) + \frac{\partial}{\partial \phi}(\rho f^2) + \frac{\partial}{\partial z}(\rho f^3)\right]$$

$$= \frac{1}{\rho}\frac{\partial}{\partial \rho}(\rho f^1) + \frac{1}{\rho}\frac{\partial}{\partial \phi}(\rho f^2) + \frac{\partial}{\partial z}(f^3)$$

$$= \frac{1}{\rho}\frac{\partial}{\partial \rho}(\rho f_\rho) + \frac{1}{\rho}\frac{\partial f_\phi}{\partial \phi} + \frac{\partial f_z}{\partial z}.$$

In spherical coordinates,

$$\nabla \cdot \mathbf{f} = \frac{1}{\sqrt{g}} \frac{\partial}{\partial q^i} (\sqrt{g} f^i)$$

$$= \frac{1}{r^2 \sin \theta} \left[\frac{\partial}{\partial r} (r^2 \sin \theta f^1) + \frac{\partial}{\partial \theta} (r^2 \sin \theta f^2) + \frac{\partial}{\partial \phi} (r^2 \sin \theta f^3) \right]$$

$$= \frac{1}{r^2 \sin \theta} \left[\frac{\partial}{\partial r} (r^2 \sin \theta f_r) + \frac{\partial}{\partial \theta} (r \sin \theta f_\theta) + \frac{\partial}{\partial \phi} (r f_\phi) \right]$$

$$= \frac{1}{r^2} \frac{\partial}{\partial r} (r^2 f_r) + \frac{1}{r \sin \theta} \frac{\partial}{\partial \theta} (\sin \theta f_\theta) + \frac{1}{r \sin \theta} \frac{\partial f_\phi}{\partial \phi}.$$

Exercise 4.20

$$\nabla \times \nabla f = \epsilon^{ijk} \mathbf{r}_k \left(\frac{\partial^2 f}{\partial q^i \partial q^j} - \Gamma^n_{ij} \frac{\partial f}{\partial q^n} \right) = \mathbf{0}.$$

(Swapping two adjacent indices on ϵ^{ijk} causes a sign change; hence, whenever the symbol multiplies an expression that is symmetric in two of its subscripts, the result is zero.)

Exercise 4.22 We have

$$\left| \frac{\partial(x^1, x^2, x^3)}{\partial(\tilde{q}^1, \tilde{q}^2, \tilde{q}^3)} \right| = \begin{vmatrix} \dfrac{\partial x^1}{\partial \tilde{q}^1} & \dfrac{\partial x^1}{\partial \tilde{q}^2} & \dfrac{\partial x^1}{\partial \tilde{q}^3} \\ \dfrac{\partial x^2}{\partial \tilde{q}^1} & \dfrac{\partial x^2}{\partial \tilde{q}^2} & \dfrac{\partial x^2}{\partial \tilde{q}^3} \\ \dfrac{\partial x^3}{\partial \tilde{q}^1} & \dfrac{\partial x^3}{\partial \tilde{q}^2} & \dfrac{\partial x^3}{\partial \tilde{q}^3} \end{vmatrix} = \begin{vmatrix} \dfrac{\partial x^1}{\partial q^i} \dfrac{\partial q^i}{\partial \tilde{q}^1} & \dfrac{\partial x^1}{\partial q^i} \dfrac{\partial q^i}{\partial \tilde{q}^2} & \dfrac{\partial x^1}{\partial q^i} \dfrac{\partial q^i}{\partial \tilde{q}^3} \\ \dfrac{\partial x^2}{\partial q^i} \dfrac{\partial q^i}{\partial \tilde{q}^1} & \dfrac{\partial x^2}{\partial q^i} \dfrac{\partial q^i}{\partial \tilde{q}^2} & \dfrac{\partial x^2}{\partial q^i} \dfrac{\partial q^i}{\partial \tilde{q}^3} \\ \dfrac{\partial x^3}{\partial q^i} \dfrac{\partial q^i}{\partial \tilde{q}^1} & \dfrac{\partial x^3}{\partial q^i} \dfrac{\partial q^i}{\partial \tilde{q}^2} & \dfrac{\partial x^3}{\partial q^i} \dfrac{\partial q^i}{\partial \tilde{q}^3} \end{vmatrix}$$

$$= \left| \begin{pmatrix} \dfrac{\partial x^1}{\partial q^1} & \dfrac{\partial x^1}{\partial q^2} & \dfrac{\partial x^1}{\partial q^3} \\ \dfrac{\partial x^2}{\partial q^1} & \dfrac{\partial x^2}{\partial q^2} & \dfrac{\partial x^2}{\partial q^3} \\ \dfrac{\partial x^3}{\partial q^1} & \dfrac{\partial x^3}{\partial q^2} & \dfrac{\partial x^3}{\partial q^3} \end{pmatrix} \begin{pmatrix} \dfrac{\partial q^1}{\partial \tilde{q}^1} & \dfrac{\partial q^1}{\partial \tilde{q}^2} & \dfrac{\partial q^1}{\partial \tilde{q}^3} \\ \dfrac{\partial q^2}{\partial \tilde{q}^1} & \dfrac{\partial q^2}{\partial \tilde{q}^2} & \dfrac{\partial q^2}{\partial \tilde{q}^3} \\ \dfrac{\partial q^3}{\partial \tilde{q}^1} & \dfrac{\partial q^3}{\partial \tilde{q}^2} & \dfrac{\partial q^3}{\partial \tilde{q}^3} \end{pmatrix} \right|.$$

The result follows from the fact that the determinant of a product equals the product of the determinants.

Chapter 4 Problems

Problem 4.1 $\nabla f = f'(r)/r \, \mathbf{r}$, $\nabla^2 f = f''(r) + 2f'(r)/r$.

Problem 4.2

(a) $a\mathbf{i}_1\mathbf{i}_1 + b\mathbf{i}_2\mathbf{i}_2 + c\mathbf{i}_3\mathbf{i}_3$, $a + b + c$;

(b) $a\mathbf{i}_2\mathbf{i}_1$, 0;

(c) $a\mathbf{E}$, $3a$;

(d) $f'(r)\mathbf{e}_r\mathbf{e}_r + f(r)/r\,\mathbf{e}_\phi\mathbf{e}_\phi$, $f'(r) + f(r)/r$;

(e) $f'(r)\mathbf{e}_r\mathbf{e}_\phi - f(r)/r\,\mathbf{e}_\phi\mathbf{e}_r$, 0;

(f) $f'(r)\mathbf{e}_r\mathbf{e}_z$, 0;

(g) $f'(r)\mathbf{e}_r\mathbf{e}_r + f(r)/r\,\mathbf{e}_\phi\mathbf{e}_\phi + f(r)/r\,\mathbf{e}_\theta\mathbf{e}_\theta$, $f'(r) + 2f(r)/r$;

(h) $-\mathbf{E} \times \boldsymbol{\omega}$, 0;

(i) $f'(\phi)/r\,\mathbf{e}_\phi\mathbf{e}_z + g'(\phi)/r\,\mathbf{e}_\phi\mathbf{e}_\phi - g(\phi)/r\,\mathbf{e}_\phi\mathbf{e}_r$, $g'(\phi)/r$;

(j) $f'(z)\mathbf{e}_z\mathbf{e}_z + g'(\phi)/r\,\mathbf{e}_\phi\mathbf{e}_\phi - g(\phi)/r\,\mathbf{e}_\phi\mathbf{e}_r$, $f'(z) + g'(\phi)/r$;

(k) $\mathbf{A}^T$, $\operatorname{tr}\mathbf{A}$.

Problem 4.4 (a) $\mathbf{E}$; (b) $-6\mathbf{A}$; (c) $\operatorname{tr}\mathbf{A}$; (d) ∇f; (e) $3\mathbf{E}$; (f) $4\mathbf{r}$.

Problem 4.5

(a) $f'\mathbf{e}_r + \frac{f-g}{r}\mathbf{e}_r + h'\mathbf{e}_z$;

(b) $f'\mathbf{e}_r + 2\frac{f-g}{r}\mathbf{e}_r$;

(c) $f'\mathbf{e}_r + \frac{f-g}{r}\mathbf{e}_r$;

(d) $f'\mathbf{i}_1 + g'\mathbf{i}_2 + h'\mathbf{i}_3$;

(e) $\mathbf{0}$;

(f) $f'\mathbf{e}_\phi + \frac{f+g}{r}\,\mathbf{e}_\phi + h'\mathbf{e}_z$;

(g) $\frac{f}{r}\mathbf{e}_z + \frac{g}{r}\mathbf{e}_\phi + h'\mathbf{e}_r$.

Problem 4.6 $\mathbf{f} \times \mathbf{r}$.

Problem 4.7 Find the coordinate vectors for the new coordinates and demonstrate their orthogonality. The answers are

$$H_\sigma = a^2\frac{\sigma^2 - \tau^2}{\sigma^2 - 1}, \quad H_\tau = a^2\frac{\sigma^2 - \tau^2}{1 - \tau^2}, \quad H_z = 1.$$

Problem 4.8 $H_\sigma = H_\tau = \sigma^2 + \tau^2$, $H_\phi = \sigma^2\tau^2$.

Problem 4.9

$$H_\sigma = H_\tau = \frac{a^2}{(\cosh\tau - \cos\sigma)^2}, \quad H_z = 1.$$

Problem 4.10

$$H_\sigma = H_\tau = \frac{a^2}{(\cosh\tau - \cos\sigma)^2}, \quad H_\phi = \frac{a^2 \sin^2\sigma}{(\cosh\tau - \cos\sigma)^2}.$$

Problem 4.11

$$H_\sigma = H_\tau = \frac{a^2}{(\cosh\tau - \cos\sigma)^2}, \quad H_\phi = \frac{a^2 \sinh^2\tau}{(\cosh\tau - \cos\sigma)^2}.$$

Problem 4.18 Consider the integral

$$\int_V \nabla \cdot (\mathbf{A}\mathbf{r})\, dV.$$

Using Gauss–Ostrogradsky theorem we get

$$\int_V \nabla \cdot (\mathbf{A}\mathbf{r})\, dV = \int_S \mathbf{n} \cdot \mathbf{A}\mathbf{r}\, dS = \int_S \mathbf{g}\mathbf{r}\, dS.$$

On the other hand, with regard for the identity

$$\nabla \cdot (\mathbf{A}\mathbf{r}) = (\nabla \cdot \mathbf{A})\mathbf{r} + \mathbf{A}^T \cdot \nabla \mathbf{r} = (\nabla \cdot \mathbf{A})\mathbf{r} + \mathbf{A}^T$$

we have

$$\int_V \nabla \cdot (\mathbf{A}\mathbf{r})\, dV = \int_V (\mathbf{f}\mathbf{r} + \mathbf{A}^T)\, dV.$$

Comparing these two expressions we find

$$\int_V \mathbf{A}^T\, dV = \int_S \mathbf{g}\mathbf{r}\, dS - \int_V \mathbf{f}\mathbf{r}\, dV,$$

from which the answer follows. The answer is

$$\int_S \mathbf{r}\mathbf{g}\, dS - \int_V \mathbf{r}\mathbf{f}\, dV.$$

Chapter 5 Exercises

Exercise 5.1

(a) $\mathbf{r}'(t) = -\mathbf{i}_1 \sin t + \mathbf{i}_2 \cos t + \mathbf{i}_3$, $|\mathbf{r}'(t)| = \sqrt{2}$, so the required length is

$$\int_0^{2\pi} \sqrt{2}\, dt = 2\pi\sqrt{2}.$$

(b) The ellipse can be described as the locus of the tip of the vector

$$\mathbf{r}(t) = \mathbf{i}_1 A \cos t + \mathbf{i}_2 B \sin t \qquad (0 \le t < 2\pi).$$

Hence

$$s = \int_0^{2\pi} (A^2 \sin^2 t + B^2 \cos^2 t)^{1/2}\, dt.$$

The integral on the right is an elliptic integral and cannot be evaluated in closed form.

Exercise 5.3 $s(t) = \int_0^t \sqrt{2}\, dt = t\sqrt{2}$, so

$$\mathbf{r}(s) = \mathbf{i}_1 \cos(s/\sqrt{2}) + \mathbf{i}_2 \sin(s/\sqrt{2}) + \mathbf{i}_3 (s/\sqrt{2}).$$

The unit tangent is

$$\mathbf{r}'(s) = \frac{1}{\sqrt{2}} \left[-\mathbf{i}_1 \sin\left(\frac{s}{\sqrt{2}}\right) + \mathbf{i}_2 \cos\left(\frac{s}{\sqrt{2}}\right) + \mathbf{i}_3 \right].$$

Exercise 5.4 The curve is an exponential spiral. We have

$$\mathbf{r}'(t) = e^t[\mathbf{i}_1(\cos t - \sin t) + \mathbf{i}_2(\cos t + \sin t)],$$

hence $\mathbf{r}'(\pi/4) = \mathbf{i}_2 e^{\pi/4}\sqrt{2}$. Also

$$\mathbf{r}(\pi/4) = \frac{\sqrt{2}}{2} e^{\pi/4}(\mathbf{i}_1 + \mathbf{i}_2),$$

so the tangent line is described by

$$x = \frac{\sqrt{2}}{2} e^{\pi/4}.$$

Exercise 5.5 Assuming the curve can be expressed in the form $\rho = \rho(\theta)$, we can write the position vector as

$$\mathbf{r}(\theta) = \mathbf{i}_1 \rho(\theta) \cos\theta + \mathbf{i}_2 \rho(\theta) \sin\theta.$$

Differentiation gives

$$\mathbf{r}'(\theta) = \mathbf{i}_1[-\rho(\theta)\sin\theta + \rho'(\theta)\cos\theta] + \mathbf{i}_2[\rho(\theta)\cos\theta + \rho'(\theta)\sin\theta].$$

(a) The equation $\mathbf{r} = \mathbf{r}(\theta_0) + \lambda\mathbf{r}'(\theta_0)$ gives upon substitution and matching of coefficients

$$\rho\cos\theta = \rho(\theta_0)\cos\theta_0 + \lambda[-\rho(\theta_0)\sin\theta_0 + \rho'(\theta_0)\cos\theta_0],$$
$$\rho\sin\theta = \rho(\theta_0)\sin\theta_0 + \lambda[\rho(\theta_0)\cos\theta_0 + \rho'(\theta_0)\sin\theta_0],$$

where (ρ, θ) locates a point on the tangent line. The desired analogues of (5.2) can be obtained by (1) squaring and adding, whereby we eliminate θ from the left-hand side, and (2) dividing the two equations, whereby we eliminate ρ from the left-hand side. Elimination of λ instead gives

$$\frac{\rho\cos\theta - \rho(\theta_0)\cos\theta_0}{-\rho(\theta_0)\sin\theta_0 + \rho'(\theta_0)\cos\theta_0} = \frac{\rho\sin\theta - \rho(\theta_0)\sin\theta_0}{\rho(\theta_0)\cos\theta_0 + \rho'(\theta_0)\sin\theta_0}$$

for the analogue of (5.3).

(b) We compute $\mathbf{r}(\theta) \cdot \mathbf{r}'(\theta)$ to get

$$\mathbf{r}(\theta) \cdot \mathbf{r}'(\theta) = \rho(\theta)\rho'(\theta).$$

We also find

$$|\mathbf{r}(\theta)| = \rho(\theta), \qquad |\mathbf{r}'(\theta)| = [\rho^2(\theta) + \rho'^2(\theta)]^{1/2},$$

so that by definition of the dot product

$$\cos\phi = \frac{\mathbf{r}(\theta) \cdot \mathbf{r}'(\theta)}{|\mathbf{r}(\theta)||\mathbf{r}'(\theta)|} = \frac{\rho'(\theta)}{[\rho^2(\theta) + \rho'^2(\theta)]^{1/2}}.$$

Use of the identity

$$\tan\phi = \frac{[1 - \cos^2\phi]^{1/2}}{\cos\phi}$$

allows us to get

$$\tan\phi = \frac{\rho(\theta)}{\rho'(\theta)}.$$

Exercise 5.7 Let $\mathbf{r}'(t_0) = \mathbf{0}$. For determining a tangent vector at t_0 we can use the Taylor expansion of $\mathbf{r} = \mathbf{r}(t)$:

$$\mathbf{r}(t_0 + \Delta t) = \mathbf{r}(t_0) + \mathbf{r}'(t_0)\Delta t$$
$$+ \frac{1}{2!}\mathbf{r}''(t_0)(\Delta t)^2 + \cdots + \frac{1}{n!}\mathbf{r}^{(n)}(t_0)(\Delta t)^n + o(|\Delta t|^n).$$

If the nth derivative is the first that is not zero at $t = t_0$, then

$$\mathbf{r}(t_0 + \Delta t) = \mathbf{r}(t_0) + \frac{1}{n!}\mathbf{r}^{(n)}(t_0)(\Delta t)^n + o(|\Delta t|^n).$$

When $\Delta t \to 0$, the direction of the vector $\mathbf{r}(t_0 + \Delta t) - \mathbf{r}(t_0)$ tends to the tangential direction of the curve at $t = t_0$. Thus a tangent vector $\mathbf{t}$ to the curve at $t = t_0$ can be found as

$$\mathbf{t} = \lim_{\Delta t \to 0} \frac{\mathbf{r}(t_0 + \Delta t) - \mathbf{r}(t_0)}{(\Delta t)^n} = \frac{1}{n!}\mathbf{r}^{(n)}(t_0).$$

Exercise 5.8

$$x = x(t_0) + \lambda \frac{1}{n!}x^{(n)}(t_0),$$

$$y = y(t_0) + \lambda \frac{1}{n!}y^{(n)}(t_0),$$

$$z = z(t_0) + \lambda \frac{1}{n!}z^{(n)}(t_0).$$

Exercise 5.9 This time $s(t) = t\sqrt{\alpha^2 + \beta^2}$, hence

$$\mathbf{r}(s) = \mathbf{i}_1 \alpha \cos\left(\frac{s}{\sqrt{\alpha^2 + \beta^2}}\right) + \mathbf{i}_2 \alpha \sin\left(\frac{s}{\sqrt{\alpha^2 + \beta^2}}\right) + \mathbf{i}_3 \beta \frac{s}{\sqrt{\alpha^2 + \beta^2}}.$$

This gives

$$\mathbf{r}'(s) = \frac{1}{\sqrt{\alpha^2 + \beta^2}}\left[-\mathbf{i}_1 \alpha \sin\left(\frac{s}{\sqrt{\alpha^2 + \beta^2}}\right)\right.$$

$$\left. + \mathbf{i}_2 \alpha \cos\left(\frac{s}{\sqrt{\alpha^2 + \beta^2}}\right) + \mathbf{i}_3 \beta \right],$$

$$\mathbf{r}''(s) = \frac{\alpha}{\alpha^2 + \beta^2}\left[-\mathbf{i}_1 \cos\left(\frac{s}{\sqrt{\alpha^2 + \beta^2}}\right) - \mathbf{i}_2 \sin\left(\frac{s}{\sqrt{\alpha^2 + \beta^2}}\right)\right],$$

and hence

$$k = |\mathbf{r}''(s)| = \frac{\alpha}{\alpha^2 + \beta^2}.$$

Exercise 5.10 The principal normal is

$$\boldsymbol{\nu} = -\mathbf{i}_1 \cos\left(\frac{s}{\sqrt{\alpha^2 + \beta^2}}\right) - \mathbf{i}_2 \sin\left(\frac{s}{\sqrt{\alpha^2 + \beta^2}}\right).$$

The binormal is obtained as

$$\boldsymbol{\beta} = \begin{vmatrix} \mathbf{i}_1 & \mathbf{i}_2 & \mathbf{i}_3 \\ -\dfrac{\alpha}{\sqrt{\alpha^2+\beta^2}} \sin\left(\dfrac{s}{\sqrt{\alpha^2+\beta^2}}\right) & \dfrac{\alpha}{\sqrt{\alpha^2+\beta^2}} \cos\left(\dfrac{s}{\sqrt{\alpha^2+\beta^2}}\right) & \dfrac{\beta}{\sqrt{\alpha^2+\beta^2}} \\ -\cos\left(\dfrac{s}{\sqrt{\alpha^2+\beta^2}}\right) & -\sin\left(\dfrac{s}{\sqrt{\alpha^2+\beta^2}}\right) & 0 \end{vmatrix}$$

and is

$$\boldsymbol{\beta} = \frac{1}{\sqrt{\alpha^2+\beta^2}} \left[\mathbf{i}_1 \beta \sin\left(\frac{s}{\sqrt{\alpha^2+\beta^2}}\right) - \mathbf{i}_2 \beta \cos\left(\frac{s}{\sqrt{\alpha^2+\beta^2}}\right) + \mathbf{i}_3 \alpha \right].$$

We can determine a plane in space by specifying the normal to the plane and a single point through which the plane passes. If the plane passes through the point located by the position vector $\mathbf{r}_0$, and $\mathbf{N}$ is any normal vector, then the plane is described by the vector equation

$$\mathbf{N} \cdot (\mathbf{r} - \mathbf{r}_0) = 0$$

(that is, all points whose position vectors $\mathbf{r}$ satisfy the above equation will lie in the plane). Hence the equations for the three fundamental planes are

$$\begin{aligned} \boldsymbol{\beta} \cdot (\mathbf{r} - \mathbf{r}_0) &= 0 && \text{osculating plane,} \\ \boldsymbol{\tau} \cdot (\mathbf{r} - \mathbf{r}_0) &= 0 && \text{normal plane,} \\ \boldsymbol{\nu} \cdot (\mathbf{r} - \mathbf{r}_0) &= 0 && \text{rectifying plane.} \end{aligned}$$

The rectifying plane for the helix is found to be $x \cos t_0 + y \sin t_0 = \alpha$.

Exercise 5.11 Writing $\mathbf{r} = \mathbf{r}(t)$, we differentiate with respect to t and use the chain rule to obtain

$$\frac{d\mathbf{r}}{dt} = \boldsymbol{\tau} \frac{ds}{dt}, \qquad \frac{d^2\mathbf{r}}{dt^2} = \boldsymbol{\tau} \frac{d^2 s}{dt^2} + \boldsymbol{\nu} k \left(\frac{ds}{dt}\right)^2.$$

These give

$$\frac{d\mathbf{r}}{dt} \times \frac{d^2\mathbf{r}}{dt^2} = \boldsymbol{\beta} k \left(\frac{ds}{dt}\right)^3$$

or

$$\mathbf{r}'(t) \times \mathbf{r}''(t) = \boldsymbol{\beta} k |\mathbf{r}'(t)|^3.$$

Dotting this equation to itself we obtain

$$k^2 = \frac{[\mathbf{r}'(t) \times \mathbf{r}''(t)]^2}{|\mathbf{r}'(t)|^6}.$$

The formula in the text was presented in such a way that the absolute value function (with its inherent discontinuity at the origin) would be avoided.

Exercise 5.12

(a) $R = 5\sqrt{10}/3$. $R = \infty$.

(b) $x = -b/2a$, the vertex of the parabola.

Exercise 5.15 No; construct counterexamples.

Exercise 5.16 First show that $\boldsymbol{\beta}$ is a constant $\boldsymbol{\beta}_0$. Then use the definition of a plane as a set of points described by a radius vector $\mathbf{r}$ such that $\mathbf{r} \cdot \mathbf{n} = c$, where c is a constant.

Exercise 5.17 Straightforward using equations (5.7) and (5.10).

Exercise 5.18 The vectors $\mathbf{r}'(t)$ and $\mathbf{r}'''(t)$ reverse their directions, while $\mathbf{r}''(t)$ does not. So $\boldsymbol{\tau}$ reverses direction, $\boldsymbol{\nu}$ maintains its direction, and therefore $\boldsymbol{\beta}$ reverses direction. The moving trihedron changes its handedness. The formulas for k_1 and k_2 show that these quantities do not change sign.

Exercise 5.19 We have

$$k_2 = -\frac{\beta}{\alpha^2 + \beta^2}.$$

Note that the ratio of the curvature to the torsion is constant for the helix.

Exercise 5.20

$$\frac{d^3\mathbf{r}}{ds^3} = -k_1^2\boldsymbol{\tau} + \frac{dk_1}{ds}\boldsymbol{\nu} - k_1 k_2 \boldsymbol{\beta}.$$

Exercise 5.22 We have

$$\mathbf{r}'(t) = \mathbf{i}_1 \sin t + \mathbf{i}_2 + \mathbf{i}_3 \cos t, \qquad |\mathbf{r}'(t)| = \sqrt{2},$$

and can find the length parameter for the curve by setting

$$s(t) = \int_0^t |\mathbf{r}'(x)|\, dx = \int_0^t \sqrt{2}\, dx = \sqrt{2}\, t.$$

So $t = s/\sqrt{2}$ and we have

$$\mathbf{r}(s) = \mathbf{i}_1\left[1 - \cos\left(\frac{s}{\sqrt{2}}\right)\right] + \mathbf{i}_2\left(\frac{s}{\sqrt{2}}\right) + \mathbf{i}_3\sin\left(\frac{s}{\sqrt{2}}\right),$$

$$\mathbf{r}'(s) = \mathbf{i}_1\frac{1}{\sqrt{2}}\sin\left(\frac{s}{\sqrt{2}}\right) + \mathbf{i}_2\left(\frac{1}{\sqrt{2}}\right) + \mathbf{i}_3\frac{1}{\sqrt{2}}\cos\left(\frac{s}{\sqrt{2}}\right),$$

$$\mathbf{r}''(s) = \mathbf{i}_1\frac{1}{2}\cos\left(\frac{s}{\sqrt{2}}\right) - \mathbf{i}_3\frac{1}{2}\sin\left(\frac{s}{\sqrt{2}}\right).$$

The desired decomposition of the acceleration is

$$\mathbf{a} = s''(t)\boldsymbol{\tau} + k_1 v^2\boldsymbol{\nu}.$$

In the present case we have $s''(t) = 0$, while $k_1 = |\mathbf{r}''(s)| = 1/2$ and $v^2 = (ds/dt)^2 = 2$. Therefore $\mathbf{a} = \boldsymbol{\nu}$.

Exercise 5.23 The sphere is described by

$$\mathbf{r} = \hat{\mathbf{x}}a\sin\theta\cos\phi + \hat{\mathbf{y}}a\sin\theta\sin\phi + \hat{\mathbf{z}}a\cos\theta.$$

We have

$$\frac{\partial\mathbf{r}}{\partial\theta} = \hat{\mathbf{x}}a\cos\theta\cos\phi + \hat{\mathbf{y}}a\cos\theta\sin\phi - \hat{\mathbf{z}}a\sin\theta,$$

$$\frac{\partial\mathbf{r}}{\partial\phi} = -\hat{\mathbf{x}}a\sin\theta\sin\phi + \hat{\mathbf{y}}a\sin\theta\cos\phi,$$

hence

$$E = a^2, \qquad F = 0, \qquad G = a^2\sin^2\theta,$$

so that $(ds)^2 = a^2(d\theta)^2 + a^2\sin^2\theta(d\phi)^2$.

Exercise 5.24

(a) The differentials of coordinates for elementary curves on the coordinate lines are $(du^1, du^2 = 0)$ and $(du^1 = 0, du^2)$. The answer is given by $\cos\theta = F/\sqrt{EG}$.

(b) The desired angle is $\psi = \pi/4$.

Exercise 5.25 We find

$$\mathbf{r}_1 = \mathbf{i}_1\cos u^2 + \mathbf{i}_2\sin u^2 + \mathbf{i}_3,$$

$$\mathbf{r}_2 = -\mathbf{i}_1 u^1\sin u^2 + \mathbf{i}_2 u^1\cos u^2,$$

hence $E = 2$, $F = 0$, $G = (u^1)^2$. Then

$$S = \int_0^{2\pi}\int_0^a \sqrt{2}u^1\,du^1\,du^2 = \pi\sqrt{2}a^2.$$

Exercise 5.26 Computing $d\mathbf{r}/d\rho$ and $d\mathbf{r}/d\phi$, we find that

$$E = 1 + [f'(\rho)]^2, \qquad F = 0, \qquad G = \rho^2.$$

From these we get

$$(ds)^2 = \{1 + [f'(\rho)]^2\}(d\rho)^2 + \rho^2(d\phi)^2$$

and

$$S = \int_A \sqrt{EG - F^2}\, d\rho\, d\phi = \iint \sqrt{1 + [f'(\rho)]^2}\,\rho\, d\rho\, d\phi.$$

Exercise 5.30 From

$$\mathbf{r} = \hat{\mathbf{x}}\rho\cos\phi + \hat{\mathbf{y}}\rho\sin\phi + \hat{\mathbf{z}}f(\rho)$$

we find that

$$\mathbf{n} = \frac{\frac{\partial \mathbf{r}}{\partial \rho} \times \frac{\partial \mathbf{r}}{\partial \phi}}{\left|\frac{\partial \mathbf{r}}{\partial \rho} \times \frac{\partial \mathbf{r}}{\partial \phi}\right|} = \frac{-\hat{\mathbf{x}}f'(\rho)\cos\phi - \hat{\mathbf{y}}f'(\rho)\sin\phi + \hat{\mathbf{z}}}{\sqrt{1 + [f'(\rho)]^2}}$$

and

$$\frac{\partial^2 \mathbf{r}}{\partial \rho^2} = \hat{\mathbf{z}}f''(\rho),$$

$$\frac{\partial^2 \mathbf{r}}{\partial \rho \partial \phi} = -\hat{\mathbf{x}}\sin\phi + \hat{\mathbf{y}}\cos\phi,$$

$$\frac{\partial^2 \mathbf{r}}{\partial \phi^2} = -\hat{\mathbf{x}}\rho\cos\phi - \hat{\mathbf{y}}\rho\sin\phi.$$

Hence

$$L = \frac{f''(\rho)}{\sqrt{1 + [f'(\rho)]^2}}, \qquad M = 0, \qquad N = \frac{\rho f'(\rho)}{\sqrt{1 + [f'(\rho)]^2}}.$$

Exercise 5.31 In Exercise 5.23 we found that $E = a^2$, $F = 0$, and $G = a^2\sin^2\theta$ for the sphere. The outward normal to the sphere is

$$\mathbf{n} = \hat{\mathbf{x}}\sin\theta\cos\phi + \hat{\mathbf{y}}\sin\theta\sin\phi + \hat{\mathbf{z}}\cos\theta,$$

and this yields the values $L = -a$, $M = 0$, $N = -a\sin^2\theta$. The mean curvature is $H = -1/a$, and the Gaussian curvature is $K = 1/a^2$.

Exercise 5.35 The answers include

$$\mu_1^{\cdot 1} = \frac{-f''(\rho)}{\{1 + [f'(\rho)]^2\}^{3/2}}, \qquad \mu_1^{\cdot 2} = \mu_2^{\cdot 1} = 0, \qquad \mu_2^{\cdot 2} = \frac{-f'(\rho)}{\rho\sqrt{1 + [f'(\rho)]^2}},$$

and

$$\Gamma_{11}^1 = \frac{f'(\rho)f''(\rho)}{1 + [f'(\rho)]^2}.$$

Exercise 5.40 A logarithmic spiral

$$\mathbf{r} = ce^{a\phi}[\mathbf{i}\cos(\phi + \phi_0) + \mathbf{j}\sin(\phi + \phi_0)],$$

where the constant c is defined by the initial values of the curve.

Chapter 5 Problems

Problem 5.1 The curve is called the *astroid* (Fig. B.1(a)). Its parametric representation is $x = a\cos^3 t$, $y = a\sin^3 t$. The singular points are $(0, a)$, $(0, -a)$, $(a, 0)$, and $(-a, 0)$.

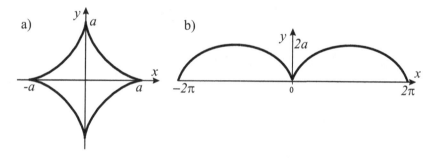

Fig. B.1 (a) astroid curve, $t \in [0, 2\pi]$; (b) cycloid curve, $t \in [-2\pi, 2\pi]$.

Problem 5.2 The curve is a *cycloid* (Fig. B.1(b)). The singular points are $(0, 2\pi k)$ for $k = 0, \pm 1, \pm 2, \ldots$.

Problem 5.3

$$\frac{x}{0} = \frac{y}{1} = \frac{z - 1}{0}.$$

Problem 5.4 $s = 6a$.

Problem 5.5 $s = 8a$.

Problem 5.6 $s = 16a$.

Problem 5.7 $s = a\sqrt{2}\sinh t$.

Problem 5.8

$$k_1 = \frac{1}{4}\sqrt{1 + \sin^2 \frac{t}{2}}.$$

Problem 5.9

$$k_1 = \frac{1}{2a \cosh^2 t}, \qquad k_2 = \frac{1}{2a \cosh^2 t}.$$

Problem 5.10 $k_2 = a \cosh t$.

Problem 5.12 Differentiating the relation $\mathbf{Q} \cdot \mathbf{Q}^T = \mathbf{E}$ with respect to s, we obtain

$$\mathbf{Q}' \cdot \mathbf{Q}^T + \mathbf{Q} \cdot \mathbf{Q}'^T = \mathbf{0}.$$

Hence $\mathbf{Q}' \cdot \mathbf{Q}^T$ is an antisymmetric tensor, so it takes the form $\mathbf{Q}' = \mathbf{d} \times \mathbf{Q}$ where $\mathbf{d}$ is the corresponding conjugate vector. Thus $\mathbf{Q}' \cdot \mathbf{Q}^T = \mathbf{d} \times \mathbf{E}$. Now refer to the result of Problem 3.21 (b).

Problem 5.13

$$\frac{-2du\, dv}{\sqrt{u^2 + 1}}.$$

Problem 5.17 $2H = \operatorname{tr} \mathbf{B} = -\operatorname{tr} \widetilde{\nabla}\mathbf{n} = -\widetilde{\nabla} \cdot \mathbf{n}$.

Problem 5.18

$$\mathbf{0} = \widetilde{\nabla} \cdot \mathbf{E} = \widetilde{\nabla} \cdot (\mathbf{A} + \mathbf{nn}) = \widetilde{\nabla} \cdot \mathbf{A} + (\widetilde{\nabla} \cdot \mathbf{n})\mathbf{n} = \widetilde{\nabla} \cdot \mathbf{A} - 2H\mathbf{n}.$$

Problem 5.20

$$K = \frac{f_{xx}g_{yy}}{\left(1 + f_x^2 + g_y^2\right)^2}.$$

Problem 5.21

$$K = -\frac{1}{AB}\left[\left(\frac{A_v}{B}\right)_v + \left(\frac{B_u}{A}\right)_u\right],$$

where indices u and v denote partial derivatives with respect to u and v respectively.

Problem 5.22 Let us consider Stokes' formula from Chapter 4 for a tensor field $\mathbf{X}$ and a vector field $\mathbf{x}$:

$$\oint_\Gamma \boldsymbol{\tau} \cdot \mathbf{X}\,ds = \int_S \mathbf{n} \cdot (\nabla \times \mathbf{X})\,dS, \quad \oint_\Gamma \boldsymbol{\tau} \cdot \mathbf{x}\,ds = \int_S \mathbf{n} \cdot (\nabla \times \mathbf{x})\,dS, \quad (B.1)$$

where we have used the relations $d\mathbf{r} = \boldsymbol{\tau}\,ds$ and $(\mathbf{n} \times \nabla) \cdot \mathbf{X} = \mathbf{n} \cdot (\nabla \times \mathbf{X})$. Using the identities

$$\mathbf{A} = -\mathbf{n} \times (\mathbf{n} \times \mathbf{A})$$

and

$$\mathbf{A} = \mathbf{E} \cdot \mathbf{X} = (\mathbf{A} + \mathbf{nn}) \cdot \mathbf{X} = -\mathbf{n} \times (\mathbf{n} \times \mathbf{A}) \cdot \mathbf{X} + \mathbf{nn} \cdot \mathbf{X},$$

we represent the tensor $\mathbf{X}$ as

$$\mathbf{X} = \mathbf{n} \times \mathbf{X}_1 + \mathbf{nx}_2 \qquad\qquad (B.2)$$

where $\mathbf{X}_1 = -\mathbf{n} \times \mathbf{X}$ and $\mathbf{x}_2 = \mathbf{n} \cdot \mathbf{X}$. Note that the surface and spatial gradient operators $\widetilde{\nabla}$ and ∇ are related by

$$\nabla = \widetilde{\nabla} + \mathbf{n}\frac{\partial}{\partial z},$$

where z is the distance coordinate along the normal to S. Then

$$\mathbf{n} \cdot (\nabla \times \mathbf{X}_1) = \mathbf{n} \cdot (\widetilde{\nabla} \times \mathbf{X}_1).$$

Formula (B.1) for $\mathbf{X}_1$ yields

$$\int_S \mathbf{n} \cdot (\widetilde{\nabla} \times \mathbf{X}_1)\,dS = \oint_\Gamma \boldsymbol{\tau} \cdot \mathbf{X}_1\,ds. \qquad (B.3)$$

In view of the identities $\boldsymbol{\tau} = \boldsymbol{\nu} \times \mathbf{n}$ and $(\boldsymbol{\nu} \times \mathbf{n}) \cdot \mathbf{X}_1 = \boldsymbol{\nu} \cdot (\mathbf{n} \times \mathbf{X}_1)$, equation (B.3) reduces to

$$\int_S \widetilde{\nabla} \cdot (\mathbf{n} \times \mathbf{X}_1)\,dS = \oint_\Gamma \boldsymbol{\nu} \cdot (\mathbf{n} \times \mathbf{X}_1)\,ds. \qquad (B.4)$$

Using the relation

$$\widetilde{\nabla} \cdot (\mathbf{nx}_2) = (\widetilde{\nabla} \cdot \mathbf{n})\mathbf{x}_2 = -2H\mathbf{x}_2,$$

from (B.2) and (B.4) we derive the final formula

$$\int_S \left(\tilde{\nabla} \cdot \mathbf{X} + 2H\mathbf{n} \cdot \mathbf{X} \right) dS = \oint_\Gamma \boldsymbol{\nu} \cdot \mathbf{X} \, ds.$$

Problem 5.23 Similar to Problem 5.22.

Problem 5.24 Applying (5.58) to the tensor $\mathbf{AX}$, we obtain the first formula

$$\int_S \left(\tilde{\nabla} \mathbf{X} + 2H\mathbf{n}\mathbf{X} \right) dS = \oint_\Gamma \boldsymbol{\nu} \mathbf{X} \, ds.$$

The second one,

$$\int_S \left(\tilde{\nabla} \times \mathbf{X} + 2H\mathbf{n} \times \mathbf{X} \right) dS = \oint_\Gamma \boldsymbol{\nu} \times \mathbf{X} \, ds,$$

follows from this. Substituting the tensor $\mathbf{nX}$ into the last formula, we obtain the third formula

$$\int_S \tilde{\nabla} \times (\mathbf{nX}) \, dS = \oint_\Gamma \boldsymbol{\tau} \mathbf{X} \, ds.$$

Chapter 6 Exercises

Exercise 6.5 Start with Hooke's law: $\boldsymbol{\sigma} = \lambda \mathbf{E} \operatorname{tr} \boldsymbol{\varepsilon} + 2\mu\boldsymbol{\varepsilon}$. Apply the trace operation to it and find $\operatorname{tr} \boldsymbol{\varepsilon}$:

$$\operatorname{tr} \boldsymbol{\varepsilon} = \frac{1}{3\lambda + 2\mu} \operatorname{tr} \boldsymbol{\sigma}.$$

Substituting this into Hooke's law, obtain $\boldsymbol{\varepsilon}$:

$$\boldsymbol{\varepsilon} = \frac{1}{2\mu} \left[\boldsymbol{\sigma} - \frac{\lambda}{3\lambda + 2\mu} \mathbf{E} \operatorname{tr} \boldsymbol{\sigma} \right] = \frac{1}{2\mu} \left[\boldsymbol{\sigma} - \frac{\nu}{1 + \nu} \mathbf{E} \operatorname{tr} \boldsymbol{\sigma} \right].$$

Now

$$W = \frac{1}{2} \boldsymbol{\sigma} \cdots \boldsymbol{\varepsilon} = \frac{1}{4\mu} \left[\boldsymbol{\sigma} \cdots \boldsymbol{\sigma} - \frac{\nu}{1 + \nu} (\operatorname{tr} \boldsymbol{\sigma})^2 \right].$$

Exercise 6.8 Use d'Alembert's principle, which states that in dynamics the system of all forces — including the inertia forces — are in equilibrium. So in the equilibrium equations, change

$$f_i \to f_i - \rho \frac{d^2 u_i}{dt^2}.$$

Exercise 6.12 Consider $\mathbf{A}(\mathbf{k})$ and $\mathbf{C}$ in a Cartesian frame. By definition,

$$
\begin{aligned}
\mathbf{A}(\mathbf{k}) &= \mathbf{k} \cdot \mathbf{C} \cdot \mathbf{k} \\
&= k_m \mathbf{i}_m \cdot c_{pqrt} \mathbf{i}_p \mathbf{i}_q \mathbf{i}_r \mathbf{i}_t \cdot k_s \mathbf{i}_s \\
&= k_m k_s c_{mqrs} \mathbf{i}_q \mathbf{i}_r \\
&= k_m k_s c_{mrqs} \mathbf{i}_q \mathbf{i}_r \\
&= k_m k_s c_{mqrs} \mathbf{i}_r \mathbf{i}_q \\
&= \mathbf{A}(\mathbf{k})^T.
\end{aligned}
$$

Exercise 6.13 It is sufficient to show that $\mathbf{a} \cdot \mathbf{A}(\mathbf{k}) \cdot \mathbf{a} > 0$ for $\mathbf{a} \neq \mathbf{0}$. The tensor $\mathbf{C}$ is such that $\boldsymbol{\varepsilon} \cdot\cdot \mathbf{C} \cdot\cdot \boldsymbol{\varepsilon} > 0$ whenever $\boldsymbol{\varepsilon} \neq \mathbf{0}$. Substitute into this $\boldsymbol{\varepsilon} = \mathbf{k}\mathbf{a}$. Because $\mathbf{C}$ is symmetric, the identity $(\mathbf{k}\mathbf{a}) \cdot\cdot \mathbf{C} = (\mathbf{a}\mathbf{k}) \cdot\cdot \mathbf{C}$ holds. Thus

$$
\boldsymbol{\varepsilon} \cdot\cdot \mathbf{C} \cdot\cdot \boldsymbol{\varepsilon} = (\mathbf{k}\mathbf{a}) \cdot\cdot \mathbf{C} \cdot\cdot (\mathbf{k}\mathbf{a}) = \mathbf{a} \cdot (\mathbf{k} \cdot \mathbf{C} \cdot \mathbf{k})\mathbf{a} = \mathbf{a} \cdot \mathbf{A}(\mathbf{k}) \cdot \mathbf{a} > 0.
$$

Exercise 6.14 From Hooke's law it follows that

$$
\operatorname{tr} \boldsymbol{\varepsilon} = \varepsilon_{11} + \varepsilon_{22} = \frac{\sigma_{11} + \sigma_{22}}{\lambda + 2\mu}.
$$

So we have

$$
\begin{aligned}
\varepsilon_{11} &= \frac{1}{2\mu}\sigma_{11} - \frac{\lambda}{2\mu(\lambda + 2\mu)}(\sigma_{11} + \sigma_{22}), \\
\varepsilon_{22} &= \frac{1}{2\mu}\sigma_{22} - \frac{\lambda}{2\mu(\lambda + 2\mu)}(\sigma_{11} + \sigma_{22}), \\
\varepsilon_{12} &= \frac{\sigma_{12}}{\mu}.
\end{aligned}
$$

Exercise 6.15 By Hooke's law we get

$$
\sigma_{11} = \lambda \operatorname{tr} \boldsymbol{\varepsilon} + 2\mu\varepsilon_{11}, \qquad \sigma_{22} = \lambda \operatorname{tr} \boldsymbol{\varepsilon} + 2\mu\varepsilon_{22}, \qquad 0 = \lambda \operatorname{tr} \boldsymbol{\varepsilon} + 2\mu\varepsilon_{33},
$$
$$
\sigma_{12} = \mu\varepsilon_{12}, \qquad\qquad\quad 0 = \mu\varepsilon_{13}, \qquad\qquad\qquad 0 = \mu\varepsilon_{23}.
$$

It follows that

$$
\varepsilon_{13} = 0, \quad \varepsilon_{23} = 0, \quad \varepsilon_{12} = \frac{\sigma_{12}}{\mu}, \quad \varepsilon_{33} = -\frac{\lambda}{\lambda + 2\mu}(\varepsilon_{11} + \varepsilon_{22}).
$$

Substituting ε_{33} into the first two equations, we obtain

$$\sigma_{11} = \frac{2\lambda\mu}{\lambda + 2\mu}(\varepsilon_{11} + \varepsilon_{22}) + 2\mu\varepsilon_{11}, \quad \sigma_{22} = \frac{2\lambda\mu}{\lambda + 2\mu}(\varepsilon_{11} + \varepsilon_{22}) + 2\mu\varepsilon_{22}.$$

These equations coincide with the equations of the plane deformation problem when we change λ to $\lambda^{\star} = \frac{2\lambda\mu}{\lambda + 2\mu}$. Then we have

$$\varepsilon_{11} + \varepsilon_{22} = \frac{\sigma_{11} + \sigma_{22}}{\lambda^{\star} + 2\mu} = \frac{\lambda + 2\mu}{4\mu(\lambda + \mu)}(\sigma_{11} + \sigma_{22}),$$

$$\varepsilon_{11} = \frac{1}{2\mu}\sigma_{11} - \frac{\lambda^{\star}}{2\mu(\lambda^{\star} + 2\mu)}(\sigma_{11} + \sigma_{22}),$$

$$\varepsilon_{22} = \frac{1}{2\mu}\sigma_{22} - \frac{\lambda^{\star}}{2\mu(\lambda^{\star} + 2\mu)}(\sigma_{11} + \sigma_{22}).$$

Chapter 6 Problems

Problem 6.1

(a) On sides AB and CD: $\sigma_{11} = \sigma$, $\sigma_{12} = 0$; on sides AD and BC: $\sigma_{22} = 2\sigma$, $\sigma_{12} = 0$.

(b) On sides AB and CD: $\sigma_{11} = 0$, $\sigma_{12} = -\tau$; on sides AD and BC: $\sigma_{22} = 0$, $\sigma_{12} = -\tau$.

(c) On sides AB and CD: $\sigma_{11} = -p$, $\sigma_{12} = 0$; on sides AD and BC: $\sigma_{22} = \sigma$, $\sigma_{12} = 0$.

(d) On sides AB and CD: $\sigma_{11} = 0$, $\sigma_{12} = \tau$; on sides AD and BC: $\sigma_{22} = -p$, $\sigma_{12} = \tau$.

Problem 6.2 On AB we have $\mathbf{n} = -\mathbf{i}_2$. The boundary condition is $\mathbf{n} \cdot \boldsymbol{\sigma} = \mathbf{0}$. Using the representation

$$\boldsymbol{\sigma} = \sigma_{11}\mathbf{i}_1\mathbf{i}_1 + \sigma_{12}(\mathbf{i}_1\mathbf{i}_2 + \mathbf{i}_2\mathbf{i}_1) + \sigma_{22}\mathbf{i}_2\mathbf{i}_2,$$

we get

$$\mathbf{n} \cdot \boldsymbol{\sigma} = -\sigma_{22}\mathbf{i}_2 - \sigma_{12}\mathbf{i}_1 = \mathbf{0}.$$

So the conditions on side AB are: $\sigma_{22} = 0$, $\sigma_{12} = 0$.
    On BC we have $\mathbf{n} = \mathbf{i}_1$. The conditions on side BC are

$$\mathbf{n} \cdot \boldsymbol{\sigma} = \sigma_{11}\mathbf{i}_1 + \sigma_{12}\mathbf{i}_2 = \mathbf{0}.$$

On side AC the task is a bit more difficult. On AC we have

$$\mathbf{n} = -\sin\alpha\mathbf{i}_1 + \cos\alpha\mathbf{i}_2.$$

The boundary condition on AC is $\mathbf{n} \cdot \boldsymbol{\sigma} = \mathbf{0}$. Using the representation

$$\boldsymbol{\sigma} = \sigma_{11}\mathbf{i}_1\mathbf{i}_1 + \boldsymbol{\sigma}_{12}(\mathbf{i}_1\mathbf{i}_2 + \mathbf{i}_2\mathbf{i}_1) + \sigma_{22}\mathbf{i}_2\mathbf{i}_2,$$

we get

$$\begin{aligned}
\mathbf{n} \cdot \boldsymbol{\sigma} &= (-\sin\alpha\mathbf{i}_1 + \cos\alpha\mathbf{i}_2) \cdot (\sigma_{11}\mathbf{i}_1\mathbf{i}_1 + \boldsymbol{\sigma}_{12}(\mathbf{i}_1\mathbf{i}_2 + \mathbf{i}_2\mathbf{i}_1) + \sigma_{22}\mathbf{i}_2\mathbf{i}_2) \\
&= \mathbf{i}_1(-\sigma_{11}\sin\alpha + \sigma_{12}\cos\alpha) + \mathbf{i}_2(-\sigma_{12}\sin\alpha + \sigma_{22}\cos\alpha) \\
&= \mathbf{0},
\end{aligned}$$

which yields the needed result.

The final answers are as follows. Side AB:

$$\sigma_{22} = 0, \qquad \sigma_{12} = 0.$$

Side BC:

$$\sigma_{11} = -\gamma y, \qquad \sigma_{12} = 0.$$

Side AC:

$$-\sigma_{11}\sin\alpha + \sigma_{12}\cos\alpha = 0, \qquad -\sigma_{12}\sin\alpha + \sigma_{22}\cos\alpha = 0.$$

Problem 6.3 On side BC, the normal is

$$\mathbf{n} = \sin\beta\mathbf{i}_1 + \cos\beta\mathbf{i}_2.$$

Then

$$\begin{aligned}
\mathbf{n} \cdot \boldsymbol{\sigma} &= (\sin\beta\mathbf{i}_1 + \cos\beta\mathbf{i}_2) \cdot (\sigma_{11}\mathbf{i}_1\mathbf{i}_1 + \boldsymbol{\sigma}_{12}(\mathbf{i}_1\mathbf{i}_2 + \mathbf{i}_2\mathbf{i}_1) + \sigma_{22}\mathbf{i}_2\mathbf{i}_2) \\
&= \mathbf{i}_1(\sigma_{11}\sin\beta + \sigma_{12}\cos\beta) + \mathbf{i}_2(\sigma_{12}\sin\beta + \sigma_{22}\cos\beta).
\end{aligned}$$

The force on BC is

$$-p_3\mathbf{n} = -p_3(\sin\beta\mathbf{i}_1 + \cos\beta\mathbf{i}_2).$$

The conditions on side BC follow from the relation $\mathbf{n} \cdot \boldsymbol{\sigma} = -p_3\mathbf{n}$.

The normal to AC is

$$\mathbf{n} = -\sin\alpha\mathbf{i}_1 + \cos\alpha\mathbf{i}_2.$$

The rest is similar to the above solution for BC.

The final answers are as follows. Side AB:

$$\sigma_{22} = -p_1, \qquad \sigma_{12} = 0.$$

Side BC:

$$\sigma_{11} \sin \beta + \sigma_{12} \cos \beta = -p_3 \sin \beta,$$
$$\sigma_{12} \sin \beta + \sigma_{22} \cos \beta = -p_3 \cos \beta.$$

Side AC:

$$-\sigma_{11} \sin \alpha + \sigma_{12} \cos \alpha = p_2 \sin \alpha,$$
$$-\sigma_{12} \sin \alpha + \sigma_{22} \cos \alpha = -p_2 \cos \alpha.$$

Problem 6.4

(a)   On side AB: $\sigma_{\phi\phi} = \sigma$, $\sigma_{\phi r} = 0$.
        On a part of circle BC: $\sigma_{rr} = -p_2$, $\sigma_{r\phi} = 0$.
        On a part of circle AC: $\sigma_{rr} = -p_1$, $\sigma_{r\phi} = 0$.
        On side CD: $\sigma_{\phi\phi} = \sigma$, $\sigma_{\phi r} = 0$.

(b)   On side AB: $\sigma_{\phi\phi} = -p$, $\sigma_{\phi r} = 0$.
        On a part of circle BC: $\sigma_{rr} = 0$, $\sigma_{r\phi} = -\tau_2$.
        On a part of circle AC: $\sigma_{rr} = 0$, $\sigma_{r\phi} = -\tau_1$.
        On side CD: $\sigma_{\phi\phi} = -p$, $\sigma_{\phi r} = 0$.

Problem 6.5

(a)   On side ABCD: $\sigma_{11} = \sigma$, $\sigma_{12} = 0$, $\sigma_{13} = \tau$.
        On a part of circle BEFC: $\sigma_{33} = 2\sigma$, $\sigma_{32} = 0$, $\sigma_{31} = \tau$.
        On a part of circle DCFG: $\sigma_{22} = \sigma$, $\sigma_{21} = 0$, $\sigma_{23} = 0$.

(b)   On side ABCD: $\sigma_{11} = -p_1$, $\sigma_{12} = \tau$, $\sigma_{13} = 0$.
        On a part of circle BEFC: $\sigma_{33} = -p_3$, $\sigma_{32} = 0$, $\sigma_{31} = 0$.
        On a part of circle DCFG: $\sigma_{22} = -p_2$, $\sigma_{21} = \tau$, $\sigma_{23} = 0$.

Problem 6.6 Denote the principal axes by $\mathbf{i}_1, \mathbf{i}_2, \mathbf{i}_3$. In this Cartesian system, the normal to triangle ABC is

$$\mathbf{n} = \frac{1}{\sqrt{3}} (\mathbf{i}_1 + \mathbf{i}_2 + \mathbf{i}_3).$$

The stress tensor is

$$\boldsymbol{\sigma} = \sigma_1 \mathbf{i}_1 \mathbf{i}_1 + \sigma_2 \mathbf{i}_2 \mathbf{i}_2 + \sigma_3 \mathbf{i}_3 \mathbf{i}_3.$$

The required stress vector is

$$\mathbf{t} = \mathbf{n} \cdot \boldsymbol{\sigma}$$
$$= \frac{1}{\sqrt{3}}(\sigma_1 \mathbf{i}_1 + \sigma_2 \mathbf{i}_2 + \sigma_3 \mathbf{i}_3)$$
$$= \frac{1}{\sqrt{3}}(50\mathbf{i}_1 - 50\mathbf{i}_2 + 75\mathbf{i}_3).$$

Problem 6.7

(a)   On the upper cylinder face: $\sigma_{zz} = \sigma$, $\sigma_{z\phi} = \tau_3$, $\sigma_{zr} = 0$.

On the internal surface part: $\sigma_{rr} = 0$, $\sigma_{r\phi} = \tau_2$, $\sigma_{rz} = 0$.

On the external later surface: $\sigma_{rr} = 0$, $\sigma_{r\phi} = \tau_1$, $\sigma_{rz} = 0$.

(b)   On side ABCD: $\sigma_{\phi\phi} = 0$, $\sigma_{\phi r} = 0$, $\sigma_{\phi z} = \tau_1$.

On a part of ring BCEH: $\sigma_{zz} = -p$, $\sigma_{z\phi} = 0$, $\sigma_{zr} = 0$.

On a part of cylinder CEFD: $\sigma_{rr} = 0$, $\sigma_{r\phi} = \tau_3$, $\sigma_{rz} = 0$.

On side FEHG: $\sigma_{\phi\phi} = 0$, $\sigma_{\phi r} = 0$, $\sigma_{\phi z} = -\tau_2$.

(c)   On side ABCD: $\sigma_{\phi\phi} = 0$, $\sigma_{\phi r} = \tau_1$, $\sigma_{\phi z} = 0$.

On a part of ring BCEH: $\sigma_{zz} = \sigma$, $\sigma_{z\phi} = 0$, $\sigma_{zr} = 0$.

On a part of cylinder CEFD: $\sigma_{rr} = 0$, $\sigma_{r\phi} = 0$, $\sigma_{rz} = -\tau_3$.

On side FEHG: $\sigma_{\phi\phi} = 0$, $\sigma_{\phi r} = -\tau_2$, $\sigma_{\phi z} = 0$.

Problem 6.8

(a) $\varepsilon = \frac{1}{2}\gamma(\mathbf{i}_1\mathbf{i}_2 + \mathbf{i}_2\mathbf{i}_1)$;

(b) $\varepsilon = \lambda_1 \mathbf{i}_1 \mathbf{i}_1$;

(c) $\varepsilon = \lambda \mathbf{E}$;

(d) $\varepsilon = u'(r)\mathbf{e}_r\mathbf{e}_r + u(r)/r\,\mathbf{e}_\phi\mathbf{e}_\phi + k\mathbf{e}_z\mathbf{e}_z$;

(e) $\varepsilon = (u'(r) - u(r)/r)(\mathbf{e}_r\mathbf{e}_\phi + \mathbf{e}_\phi\mathbf{e}_r) + k\mathbf{e}_z\mathbf{e}_z$;

(f) $\varepsilon = u'(r)(\mathbf{e}_r\mathbf{e}_z + \mathbf{e}_z\mathbf{e}_r) + k\mathbf{e}_z\mathbf{e}_z$;

(g) $\varepsilon = u'(r)\mathbf{e}_r\mathbf{e}_r + u(r)/r\,\mathbf{e}_\phi\mathbf{e}_\phi + u(r)/r\,\mathbf{e}_\theta\mathbf{e}_\theta$.

Chapter 7 Exercises

Exercise 7.11   Hooke's law is given by the formula

$$\boldsymbol{\sigma} = \lambda \mathbf{E}\,\mathrm{tr}\,\varepsilon + 2\mu\varepsilon.$$

Dot-multiply this by $\mathbf{n}$ from the left and the right. From $\sigma_{33} = 0$ it follows that

$$\lambda \operatorname{tr} \varepsilon + 2\mu\varepsilon_{33} = 0.$$

Because $\operatorname{tr} \varepsilon = \operatorname{tr} \tilde{\varepsilon} + \varepsilon_{33}$ where $\operatorname{tr} \tilde{\varepsilon} = \varepsilon_1^1 + \varepsilon_2^2$, we get

$$\varepsilon_{33} = -\frac{\lambda}{\lambda + 2\mu} \operatorname{tr} \tilde{\varepsilon}.$$

So

$$\operatorname{tr} \varepsilon = \frac{2\mu}{\lambda + 2\mu} \operatorname{tr} \tilde{\varepsilon}.$$

It follows that

$$\sigma = 2\mu \left[\varepsilon + \frac{\lambda}{\lambda + 2\mu} \mathbf{E} \operatorname{tr} \tilde{\varepsilon} \right] \equiv \frac{E}{1 + \nu} \left[\varepsilon + \frac{\nu}{1 - \nu} \mathbf{E} \operatorname{tr} \tilde{\varepsilon} \right].$$

Exercise 7.12 In Cartesian coordinates we get

$$\epsilon \cdots \epsilon = \epsilon_{11}^2 + \epsilon_{22}^2 + 2\epsilon_{12}^2, \qquad \operatorname{tr}^2 \epsilon = (\epsilon_{11} + \epsilon_{22})^2 = \epsilon_{11}^2 + \epsilon_{22}^2 + 2\epsilon_{11}\epsilon_{22}.$$

So,

$$\epsilon \cdots \epsilon + \frac{\nu}{1 - \nu} \operatorname{tr}^2 \epsilon = \left(1 + \frac{\nu}{1 - \nu} \right) \epsilon_{11}^2 + \left(1 + \frac{\nu}{1 - \nu} \right) \epsilon_{22}^2$$

$$+ \frac{2\nu}{1 - \nu} \epsilon_{11}\epsilon_{22} + 2\epsilon_{12}^2$$

$$= \frac{1}{1 - \nu} \left[\epsilon_{11}^2 + \epsilon_{22}^2 + 2\nu\epsilon_{11}\epsilon_{22} \right] + 2\epsilon_{12}^2.$$

If $|\nu| < 1$ and $\epsilon_{11} \neq 0$, $\epsilon_{22} \neq 0$ then $\epsilon_{11}^2 + \epsilon_{22}^2 + 2\nu\epsilon_{11}\epsilon_{22} > 0$. This proves the required positivity.

Exercise 7.13 Doubly dot-multiplying $\mathbf{T}$ by ϵ, we get

$$\mathbf{T} \cdots \epsilon = (\mathbf{T} \cdot \mathbf{A} + (\mathbf{T} \cdot \mathbf{n})\mathbf{n}) \cdots \epsilon = (\mathbf{T} \cdot \mathbf{A}) \cdots \epsilon = T_{\alpha\beta}\epsilon^{\alpha\beta}.$$

Substituting (7.30) into this, we get

$$\mathbf{T} \cdots \epsilon = \frac{Eh}{1 + \nu} \left[\epsilon + \frac{\nu}{1 - \nu} \mathbf{A} \operatorname{tr} \epsilon \right] \cdots \epsilon = \frac{Eh}{1 + \nu} \left[\epsilon \cdots \epsilon + \frac{\nu}{1 - \nu} \operatorname{tr}^2 \epsilon \right].$$

Using the equality

$$\mathbf{M} = -M_{\alpha\beta}\rho^\alpha\rho^\beta \times \mathbf{n},$$

we find that

$$M_{\alpha\beta} = -\boldsymbol{\rho}_\alpha \cdot \mathbf{M} \cdot (\boldsymbol{\rho}_\beta \times \mathbf{n}).$$

Then

$$
\begin{aligned}
M_{\alpha\beta}\text{æ}^{\alpha\beta} &= -\boldsymbol{\rho}_\alpha \cdot \mathbf{M} \cdot (\boldsymbol{\rho}_\beta \times \mathbf{n})\text{æ}^{\alpha\beta} = -\boldsymbol{\rho}_\alpha \cdot \mathbf{M} \cdot (\boldsymbol{\rho}_\beta \times \mathbf{n})\boldsymbol{\rho}^\alpha \cdot \text{æ} \cdot \boldsymbol{\rho}^\beta \\
&= -\mathbf{M} \cdots \left[(\boldsymbol{\rho}_\beta \times \mathbf{n})\boldsymbol{\rho}_\alpha \boldsymbol{\rho}^\alpha \cdot \text{æ} \cdot \boldsymbol{\rho}^\beta \right] \\
&= \mathbf{M} \cdots \left[(\mathbf{n} \times \boldsymbol{\rho}_\beta)\text{æ} \cdot \boldsymbol{\rho}^\beta \right] = \mathbf{M} \cdots \left[(\mathbf{n} \times \boldsymbol{\rho}_\beta)\boldsymbol{\rho}^\beta \cdot \text{æ} \right] \\
&= \mathbf{M} \cdots (\mathbf{n} \times \text{æ}) \\
&= (\mathbf{M} \times \mathbf{n}) \cdots \text{æ}.
\end{aligned}
$$

In addition, we obtain the equality

$$\mathbf{M} \cdots (\mathbf{n} \times \text{æ}) = \frac{Eh^3}{12(1+\nu)} \left[\text{æ} \cdots \text{æ} + \frac{\nu}{1-\nu} \operatorname{tr}^2 \text{æ} \right].$$

Exercise 7.15 $\operatorname{tr} \widetilde{\nabla}\tilde{\mathbf{v}} = \operatorname{tr}(\widetilde{\nabla}\tilde{\mathbf{v}})^T = \widetilde{\nabla} \cdot \tilde{\mathbf{v}}$ and $\operatorname{tr} \widetilde{\nabla}\widetilde{\nabla}w = \widetilde{\nabla} \cdot \widetilde{\nabla}w$. It follows that $\operatorname{tr} \boldsymbol{\varepsilon} = \widetilde{\nabla} \cdot \tilde{\mathbf{v}} - z\widetilde{\nabla} \cdot \widetilde{\nabla}w$.

Exercise 7.16 Let the equation of the bent plate surface be $z = w(x_1, x_2)$. In vector form, the surface is given by

$$\mathbf{r} = x_1\mathbf{i}_1 + x_2\mathbf{i}_2 + \mathbf{n}w(x_1, y_1).$$

With slightly different notation, this was analyzed in Exercise 5.33; it was shown that

$$L = \frac{w_{xx}}{\sqrt{1 + w_x^2 + w_y^2}}, \quad M = \frac{w_{xy}}{\sqrt{1 + w_x^2 + w_y^2}}, \quad N = \frac{w_{yy}}{\sqrt{1 + w_x^2 + w_y^2}}.$$

We recall that $L = b_{11}$, $M = b_{12}$, $N = b_{22}$ are the components of the curvature tensor $\mathbf{B}$. If w and its derivatives are small, we get

$$\mathbf{B} = w_{xx}\mathbf{i}_1\mathbf{i}_1 + w_{xy}(\mathbf{i}_1\mathbf{i}_2 + \mathbf{i}_2\mathbf{i}_1) + w_{yy}\mathbf{i}_2\mathbf{i}_2 = -\text{æ}.$$

In other words, the second fundamental tensor of the bent plate surface coincides with æ up to a difference in algebraic sign.

Exercise 7.18 Modify the proof in three-dimensional elasticity, changing $\boldsymbol{\sigma}$ to $\mathbf{T} \cdot \mathbf{A}$, and $\mathbf{u}$ to $\tilde{\mathbf{v}}$.

Bibliography

Ciarlet, Ph. G. (1988). *Mathematical Elasticity. Vol. I. Three-Dimensional Elasticity* (North-Holland, Amsterdam).

Ciarlet, Ph. G. (1997). *Mathematical Elasticity. Vol. II. Theory of Plates* (North-Holland, Amsterdam).

Ciarlet, Ph. G. (2000). *Mathematical Elasticity. Vol. III. Theory of Shells* (North-Holland, Amsterdam).

Ciarlet, Ph. G. (2005). *An Introduction to Differential Geometry with Application to Elasticity* (Springer, Dordrecht).

Chróścielewski, J., Makowski, J. and Pietraszkiewicz, W. (2004). *Statyka i dynamika powłok wielopłatowych. Nieliniowa teoria i metoda elementów skończonych* (Wydawnictwo IPPT PAN, Warszawa).

Cosserat, E. and Cosserat, F. (1909). *Théorie des corps déformables* (Herman et Fils, Paris). English translation: NASA TT F-11, 561 (NASA, Washington, DC).

Danielson, D. A. (1992). *Vectors and Tensors in Engineering and Physics* (Addison–Wesley, New York).

Donnell, L. H. (1976). *Beams, Plates and Shells* (McGraw-Hill, New York).

Eremeyev, V. A. and Zubov L. M. (2008). *Mechanics of Elastic Shells (in Russian)* (Nauka, Moscow).

Eringen, A. C. (1999) *Microcontinuum Field Theory. I. Foundations and Solids* (Springer-Velag, New York).

Green, A. E. and Zerna, W. (1954). *Theoretical Elasticity* (Clarendon Press, Oxford).

Goldenveizer, A. L. (1976). *Theory of Thin Elastic Shells (in Russian)* (Moscow, Nauka).

Goodbody, A. M. (1982). *Cartesian Tensors, with Applications to Mechanics, Fluid Mechanics, and Elasticity* (Halsted Press, Ellis Horwood, New York).

Heinbockel, J. H. (2001). *Introduction to Tensor Calculus and Continuum Mechanics* (Trafford Publishing, British Columbia, Canada).

Jeffreys, H. (1931). *Cartesian Tensors* (Cambridge University Press, Cambridge, UK).

Kay, D. C. (1988). *Tensor Calculus* (Schaum's Outline Series, McGraw-Hill, New

York).

Knowles, J. (1997). *Linear Vector Spaces and Cartesian Tensors* (Oxford University Press, Oxford).

Koiter, W. T. On the foundations of the linear theory of thin elastic shells. I. *Proc. Kon. Ned. Ak. Wet.* B73, 169–195.

Lebedev, L. P. and Cloud, M. J. (2003). *The Calculus of Variations and Functional Analysis with Optimal Control and Applications in Mechanics* (World Scientific, Singapore).

Libai, A. and Simmonds, J. G. (1998). *The Nonlinear Theory of Elastic Shells*, 2nd ed (Cambridge University Press, Cambridge, UK).

Lipschutz, M. M. (1969). *Differential Geometry* (Schaum's Outline Series, McGraw-Hill, New York).

Lurie, A. I. (1990). *Non-linear Theory of Elasticity*, Series in Applied Mathematics and Mechanics, 36 (North-Holland, Amsterdam).

Lurie, A. I. (2005). *Theory of Elasticity* (Springer, Berlin).

McConnell, A. (1957). *Application of Tensor Analysis* (Dover, New York).

Muskhelishvili N. I., (1966). *Some Basic Problems of the Mathematical Theory of Elasticity; Fundamental Equations, Plane Theory of Elasticity, Torsion and Bending, 5th ed. (in Russian)*, Nauka, Moscow. Transl. P. Noordhoff, Groningen, 1953 (Translation of the 3rd ed., Izd. Akad. Nauk SSSR, Moscow-Leningrad, 1949).

Naghdi, P. (1972). *The Theory of Plates and Shells.* In: Flügge, S. (Ed.), Handbuch der Physik, Vol. VIa/2, Springer-Verlag, Berlin. Pp. 425–640.

Novozhilov, V. V., Chernykh, K. Ph. and Mikhailovskiy E. M. (1991). *Linear Theory of Thin Shells (in Russian)* (Politechnika, Leningrad).

Ogden, R. W. (1997). *Non-Linear Elastic Deformations* (Dover, New York).

O'Neill, B. (1997). *Elementary Differential Geometry* (Academic Press, New York).

Papastavridis, J. (1998). *Tensor Calculus and Analytical Dynamics* (CRC Press, Boca Raton).

Pogorelov, A. V. (1957). *Differential Geometry.* Translated from the first Russian ed. by L. F. Boron (P. Noordhoff, Groningen).

Rubin, M. B. (2000). *Cosserat Theories: Shells, Rods and Points* (Kluwer, Dordrecht).

Simmonds, J. G. (1982). *A Brief on Tensor Analysis, 2nd ed* (Springer, New York).

Schouten, J. A. (1951). *Tensor Analysis for Physicists* (Clarendon Press, Oxford).

Sokolnikoff, I. S. (1994). *Tensor Analysis: Theory and Applications to Geometry and Mechanics of Continua* (Wiley, New York).

Synge, J. and Schild, A. (1978). *Tensor Calculus* (Dover, New York).

Timoshenko, S. P., Woinowsky-Krieger, S. (1985). *Theory of Plates and Shells* (McGraw Hill, New York).

Truesdell, C. and Noll, W. (2004). *The Nonlinear Field Theories of Mechanics.* 3rd ed (Springer, Berlin).

Wang, C. M., Reddy, J. N. and Lee, K. H. (2000). *Shear Deformable Beams and Shells* (Elsevier, Amsterdam).

Vlasov, V. Z. (1949). *General Theory of Shells and its Applications in Technics (in Russian)* (Moscow, Saint-Petersburg, Gostekhizdat).

Vorovich, I. I. (1999). *Nonlinear Theory of Shallow Shells* (Springer, New York).

Wang, C.-C. and Truesdell, C. (1973). *Introduction to Rational Elasticity* (Noordhoof Int. Publishing, Leyden).

Young, E. (1993). *Vector and Tensor Analysis* (Marcel Dekker, New York).

Yosida, K. (1980). *Functional Analysis* (Springer, Berlin).

Zhilin, P. A. (2006). *Applied Mechanics. Foundations of the Theory of Shells (in Russian)* (St. Petersburg State Polytechnical University, St. Petersburg).

Zubov, L. M. and Karyakin M. I. (2006). *Tensor Calculus (in Russian)* (Vuzovskaya Kniga, Moscow).

Zubov, L. M. (1982). *Methods of Nonlinear Elasticity in the Theory of Shells (in Russian)* (Rostov State University, Rostov on Don).

Index

Printed in the United States
By Bookmasters